INTEGRATED CIRCUITS
FOR
COMPUTERS

INTEGRATED CIRCUITS FOR COMPUTERS
PRINCIPLES AND APPLICATIONS

WILLIAM L. SCHWEBER

McGRAW-HILL BOOK COMPANY

New York   Atlanta   Dallas   St. Louis   San Francisco   Auckland   Bogotá   Guatemala
Hamburg   Johannesburg   Lisbon   London   Madrid   Mexico   Montreal   New Delhi   Panama
Paris   San Juan   São Paulo   Singapore   Sydney   Tokyo   Toronto

Sponsoring Editor: George Z. Kuredjian
Editing Supervisor: Mitsy Kovacs
Design and Art Supervisor: Nancy Axelrod
Production Supervisor: Priscilla Taguer
Text Designer/Cover Designer: Nancy Axelrod

Dedication
To my father, who taught me about learning, books, and understanding.

As used in this book, the term Intel is a registered trademark of Intel Corp., the term Tri-state is a registered trademark of National Semiconductor Corp., and the term PAL is a registered trademark of Monolithic Memories Inc.

Library of Congress Cataloging in Publication Data

Schweber, William L.
  Integrated circuits for computers.

  Includes index.
  1. Electronic digital computers—Circuits.   2. Integrated circuits.   I. Title.
TK7888.4.S245   1986   621.395   85-24234
ISBN 0-07-053624-4

Integrated Circuits for Computers: Principles and Applications

1 2 3 4 5 6 7 8 9 0 DOCDOC 8 9 2 1 0 9 8 7 6 5

ISBN 0-07-053624-4

CONTENTS

EDITOR'S FOREWORD

The McGraw-Hill Electronic Computer Technology series is designed to prepare students for a variety of occupations in computer-based environments. The texts that constitute this series will not only help students attain technical competency, but will also serve as reference resources throughout their entire professional development.

The Electronic Computer Technology Series has been developed according to a number of guiding principles. The Series:

- Recognizes that employers need staff members who can communicate and interact knowledgeably with their colleagues and customers concerning computer-based technologies.
- Understands the backgrounds and goals of the prospective electronic computer technology graduate, and strives to concentrate on practical material without sacrificing the theory required to keep the graduate current with state-of-the-art computer technology.
- Provides the educational foundations necessary for a variety of specializations, training programs, local employment, and institutional needs.
- Encourages the recommendations of instructors and students while analyzing and integrating into the series, as applicable, the reports of professional educational groups, technological societies, and government publications dealing with electronic computer technology curriculum and programs.

Because of the constant and rapid changes in the field of electronic computer technology, it is important that the authors in this series have relevant industrial and teaching experience. Their ongoing involvement in the field provides them with valuable insight into the needs and concerns of students, instructors, and practicing professionals.

The series components are coordinated around a core curriculum designed to present the latest technological material in a way that keeps both student and professional informed about recent advances in electronic computer technology. Concepts are explained in a clear and well-illustrated manner, and the links between theory and practical applications are explored.

The dynamic nature of technological developments in all facets of the computer industry emphasizes the importance of suggestions from instructors, students, and industry professionals to the success of a series like Electronic Computer Technology. These suggestions allow the series to evolve and develop in ways that satisfy the needs of programs that themselves are growing and changing to keep pace with the field of electronic computer technology.

Gordon Silverman
Project Editor

ELECTRONIC COMPUTER TECHNOLOGY
Gordon Silverman, Project Editor

Books in this series

Mathematics for Computers by Arthur D. Kramer
Integrated Circuits for Computers: Principles and Applications by William L. Schweber
Computers and Computer Languages by Gordon Silverman and David B. Turkiew *(in preparation)*

PREFACE

The computer, which has so revolutionized society, is made possible by integrated circuits. These ICs combine anywhere from a few circuit components and functions to many thousands onto a single tiny piece of silicon. Thirty years ago, popular magazines argued that a computer with enough capacity to do useful work, such as run a factory or do an accounting payroll, would occupy a full building, require massive amounts of electricity and air conditioning, and be unreliable because there would be so many vacuum tubes that some would always be burning out. The concept of desktop-sized personal computers and computerized functions for everyday activities was basically unknown.

The IC has made the computer a low-cost, powerful, and reliable electronic unit and has also led to an explosion in uses for computers. Computer intelligence is now in appliances, motor control, heating and cooling systems, and many other applications that involve more than just calculations with numbers. These applications do not look like what we may normally think of as computers, but they use many of the same components and concepts and provide similar capabilities. Microcomputers are becoming embedded in so many different products that they will soon outnumber the light bulbs in a typical household. Virtually every company is updating its existing products and developing new ones based on microcomputers and microprocessors because of the flexibility and performance advantages that the products can have.

This text will discuss the ICs that are building blocks of computers and computer functions. The technology of computer ICs has advanced in the years since the IC was invented, and the internal complexity of the ICs has increased tremendously. Circuit functions which had to be built from dozens of individual ICs are now available on dedicated large-scale ICs which implement the same functions, or even have computerlike capability in themselves and so can offload the main processor from many chores of the overall design. These peripheral and support chips play an important part in the design and operation of systems and have many complex and subtle features of their own. Large-scale integration has also provided a wide variety of memory chips which give the processor the large amounts of storage for both program and data. These memory chips have the speed, reliability, and characteristics needed to make a high-performance, low-cost computer or computerized function.

This text is designed to help you understand what an IC is, how ICs are used, and what they can do. It covers the wide spectrum of digital ICs from small gates (where ICs started and are still used in abundance) through larger scales of integration which replace entire circuit sections and are responsible for power of computers and computer systems. You do not have to be a circuit designer to work with these components, but you do have to understand their capabilities and characteristics in order to work with them. There are also many special terms, abbreviations, and concepts that are used in association with these components, and understanding these is important to being prepared for the unexpected or new idea.

This book is arranged in three main sections. In the first section, smaller ICs are discussed. Chapter 1 explains what ICs are and how they are made, along with the two major technologies used in ICs, TTL and MOS. Boolean algebra and the concept of binary numbering is covered in Chap. 2. This discussion leads to basic gates, which are the lowest function of IC and serve as the building blocks for all others. These gates are then combined into flip-flops in Chap. 3, and the flip-flops are used to develop counters and registers. The applications and troubleshooting of these elements are also discussed.

Chapter 4 reviews some of the many other building blocks needed in most systems: decoders, multiplexers, and drivers. Buffers, simple in concept, yet essential, and logic units are discussed in Chap. 5. Each discussion includes the role the element has, the features and benefits of the function, and the approach used when dealing with these in a circuit. Chapter 6 concludes the section by investigating the major logic families and their characteristics when actually used in a system.

The second major section of the book deals with the larger integrated circuits and functions that are provided by these larger elements. The need for memory, the roles of the various types of memory (Read-Write and ROM) and the many technologies and characteristics of these are covered in Chap. 7. How memory ICs are actually used in a system and how they are examined when there may be a problem concludes the chapter.

Chapters 8 and 9 investigate the support IC functions which are so important in any practical system. These support functions do many specialized tasks which the system microprocessor can not do efficiently, and they provide the overall system performance needed for the application. They include serial and parallel input-output, counters and counter-timers, CRT controllers, IEEE-488 controllers, and error detection and correction components. The need for these, their role, and how they work as seen from the system perspective is covered.

The third and last major section discusses special functions and ties the previous chapters together. Connecting a computer system to real-world analog signals requires analog-to-digital and digital-to-analog converters. The concept behind these converters, including number representation, operation, and interfacing is included, with applications examples, in Chap. 10.

The microprocessor itself is the subject of Chap. 11. The fundamentals of processor operation, instructions and their execution, and the concepts of program steps and system memory map are covered. How a microprocessor interfaces to other integrated circuits in the system is also discussed. Single-chip microcomputers are examined, along with applications. Some typical microprocessor types are also studied.

Chapter 12 brings the previous chapters together with an emphasis on how computer-based systems are tested. The tools used—voltmeter, logic probe, oscilloscope, logic analyzer, test software, and signature analysis—are covered, as well as the function of each instrument, its good points, and its weak points. The general strategies used when approaching a nonworking system is discussed.

There are three appendixes. The first shows and discusses the data sheet of a simple IC, and the second shows and explains the data sheet from a more complicated support IC, both from leading manufacturers. The final appendix shows the logic symbols that are commonly used in industry to represent the boolean and IC functions and compares them to newer symbols that are just beginning to be used. A glossary of selected terms is also included.

In order to make sure that the materials of this textbook are thoroughly absorbed, special features have been included to test student understanding of the material and to reinforce the materials just covered. For example, questions and/or problems appear at the end of most of the sections. These are intended to reinforce comprehension of key points within the section. A student who is unable to answer any of these questions should be instructed to go back to restudy the section or a portion of the section. At the end of each chapter comprehensive Review Questions and/or Review Problems have been included. These reviews test student understanding of the entire chapter. Answers to the odd-numbered problems may be found in the back of the textbook with answers to even-numbered problems located in the Instructor's Manual. This device allows the instructor to give even-numbered problems for homework assignments and use the odd-numbered problems for classwork.

The perspective of the entire book is that of someone who will be working with digital and computer systems, either testing them in engineering development labs or repairing them in the field. This provides a perspective from the viewpoint of a system investigator, not a system designer.

Throughout this book emphasis has been placed on understanding what these ICs do, what benefits they offer, how they are organized, and the principles of their operation. There are many examples of real applications, along with names and specific models from leading manufacturers. The digital ICs used in computers are responsible for the revolution we see around us, as well as the product revolution that is much harder to see but still as massive, and understanding these revolutions means understanding electronics of the present and future.

I would like to express my thanks to Mark Markum, Robert A. Rock, James E. Teza, Lester Knapp, and Herbert M. Hoffman for their valuable comments and suggestions in preparing this book.

William Schweber

INTEGRATED CIRCUITS
FOR
COMPUTERS

INTRODUCTION TO INTEGRATED CIRCUITS

1

The basic concept of an integrated circuit (IC) is discussed. The IC is compared to circuitry made up of individual electronic components, and the advantages of the ICs are covered. The chapter also investigates the applications of ICs, both where they replace existing functions and where they allow for functions that were not practical with individual components. There is a brief discussion of how ICs are made, which is useful in understanding the complexity of the circuitry and the problems that may be encountered in dealing with them.

1-1 WHAT IS AN IC?

The IC has completely changed the nature of electronics since its invention in 1957. The IC has changed technology and the products that we see and use, and it represents a completely different way of designing, manufacturing, and servicing electronic systems. With an IC, the complete circuit is built on a single piece of semiconductor material, usually silicon. It is called *integrated* because all the components—transistors, diodes, resistors, capacitors—are part of this one piece of silicon. It is a *circuit* because the components are also interconnected as part of the fabrication process to a final form which performs the intended function. There are generally no additional steps to form the circuit function. The IC is ready to be used as a building block in the next part of the system. An IC may have anywhere from a dozen to many thousands of individual components built onto the silicon, depending on the function that the circuit is designed to perform. The typical size of the piece of silicon is 50×50 mils (thousandths of an inch). Sizes much larger, even larger than 200×200 mils, are sometimes used to achieve certain very complicated functions that require many individual transistors, diodes, resistors, and capacitors on the IC.

The IC itself is a component in the final system. It comes packaged in one of several ways. Most common is the *dual inline package* (DIP) made of ceramic or

Fig. 1-1 The Dual Inline Package (DIP) is a very common package for the fragile silicon of the integrated circuit. *(Photo Courtesy of Motorola, Inc.)*

plastic (Fig. 1-1). Other methods of packaging include the flat pack, which can be used for smaller ICs that have up to about 16 pins, or leads. In recent years, the trend to more complicated ICs with more functions has led to the development of new packages such as the leadless chip carrier (Fig. 1-2). These new packages require new mounting, soldering, and probing techniques but provide many ad-

Fig. 1-2 More complicated ICs with more pins require packages other than the DIP. The Leadless Chip Carrier (LCC), shown as square packages next to the rectangular DIP, requires less board space than the DIP but is more costly. *(Photo Courtesy of Motorola, Inc.)*

vantages for large ICs. The DIP package is the one that will predominate for smaller to medium-size ICs for the foreseeable future. Depending on how complicated the circuit on the IC is, the IC package can have 8, 14, 16, 18, 20, 24, 28, 40, or 64 pins.

The package of an IC serves several important purposes. First, it protects the fragile piece of silicon that is the heart of the IC (Fig. 1-3). The IC has to be protected from fingers, probes, dirt, light (the silicon has some critical operational characteristics that are affected by light energy), and circuit board cleaning solutions. The package also brings out the signals from the silicon to wire leads that are large enough to be mounted on a circuit board, soldered, and probed with oscilloscope or voltmeter leads if required. The silicon chip of the IC has extremely fine wires soldered to it that are brought and soldered to the outside leads. Because the package sizes are identical, it is possible, if the production volume is high, to use an automatic machine to insert the ICs into the printed circuit board of the final product. Otherwise, the boards can be "stuffed" by hand. The package also provides a place for the manufacturer name, chip model number, and a manufacturing date code.

The IC can be attached to the printed circuit board either by being directly soldered into the board or put into a socket that is soldered into the board. The advantages of sockets are:

- They allow components to be removed for troubleshooting.
- They allow additional components to be installed to add an extra feature or option, such as more memory, that the customer wants after the initial purchase.

Fig. 1-3 The silicon die is protected from dirt, finger contact, light, and moisture by the IC package shown in DIP form with the cover removed. The signals to and from the die are brought to the IC pins by extremely fine wires. *(Photo Courtesy of Analog Devices, Inc.)*

However, sockets have some disadvantages:

- They add cost to the board.
- They prevent automatic insertion of the IC onto the board.
- They can become a source of unreliability, since every socket adds many additional physical contacts between the IC and the circuit board. These can corrode, or the IC can vibrate loose and cause intermittent contact.
- The production step of washing off the excess solder flux after the circuit board has been soldered is much more difficult because the flux and cleaner must be kept out of the socket.
- Sockets make it difficult to tell if unauthorized persons have been inserting and removing ICs in order to repair the board or try new components.

In general, the electronics industry uses sockets only when there is an overriding concern for test and repair. On a typical circuit board with a wide variety of ICs—microprocessors, memory, counters, peripheral chips—only a few will be in sockets, such as the microprocessor (since removing by unsoldering it is costly) and those ICs that the test personnel have identified as important to be able to remove easily in order to run tests and troubleshoot the circuit. In most circuits, there are a few key ICs that, if removed, make the circuit easy to test. The rest of the ICs are soldered directly into the board.

QUESTIONS FOR SECTION 1-1

1. What is a typical size for the piece of silicon which is the heart of the IC?

2. What is the most common type of package for the IC?

3. What are the functions of the package?

4. What are the advantages and disadvantages of sockets for ICs?

1-2 ICS VERSUS DISCRETE COMPONENTS

A circuit made up of discrete components has the individual transistors, diodes, resistors, and capacitors soldered into the circuit board. This is different from a product made up primarily of ICs (which have the transistors, resistors, and capacitors built onto the single piece of silicon) soldered onto the circuit board.

Because ICs have so many advantages over discrete components, the IC-based system has almost completely eliminated the discrete component design, except for some special cases. What are the advantages and disadvantages of one over the other?

Advantages of ICs

- They allow fitting many components and functions into a space that would otherwise fit only a single component. For example, the IC that forms a simple AND gate is about 0.75 inch (in) long and 0.2 in wide, including its package. To build the same function out of discrete components might take an area about 2×2 in (Fig. 1-4). A more complicated memory chip takes only slightly more

Fig. 1-4 The AND gate function built out of discrete components occupies a small circuit board approximately 3×7 inches, in comparison to an IC version which is approximately 0.4×1 inch. *(Photo by Jill Manca.)*

space than the AND gate, but if made up of discrete components a 16-bit memory (which is a very small amount) would take up a circuit board about 4×5 in. Along with space savings come weight savings too, of course, which is important in airplane, space, and many earth-bound applications.

- The power used by an IC is a small fraction of that used by the equivalent circuit made up of discrete components. This is because the power is related, among other factors, to the physical size of the component. A single IC with dozens of transistors uses about the same amount of power as a single discrete transistor. Power is an important limitation in any system for two reasons. First, there is a limit to how much can be supplied, either from the battery or other power supply. Second, much of the power is transformed into heat and must be gotten rid of so that the system does not "cook" itself. Getting rid of all this heat can be a major problem, requiring vent holes, fans, or other complicated cooling schemes.

- The reliability of an IC is much higher than a discrete design because an IC only has interconnections at the leads of the chip, whereas a discrete design has interconnections between every component. Interconnections are the primary source of failures; they break, corrode, vibrate loose, and so on. The fewer connections in any system, the more reliable that system will be.

- Costs for an IC are far lower than for the discrete equivalent. ICs are mass-produced in batches, and the cost of production is divided over the thousands of ICs produced in a single batch. This low cost can be compared to the discrete circuit, where many individual components, each of which may cost anywhere from a few pennies to several dollars, must be assembled and soldered into the circuit board. The IC cost is far less.

- The operating speed of the IC is much faster than the speed of the discrete circuit. The transistors of the IC are much smaller than discrete transistors, and therefore the electrons have less distance to move within the transistor, and certain other factors that cause slowness (undesired capacitance, for example) are less. Equally important, the transistors, diodes, resistors, and capacitors of the IC are close to each other. They are separated by millionths of an inch. In the discrete design, these distances can be several inches. The electric impulses in circuits travel at about 1 foot/nanosecond (1 ft/ns) (a nanosecond is equal to a billionth of a second). Therefore, in a discrete design a lot of time is lost by the signals traveling from one component to another. ICs make possible circuits that can operate at higher speeds and give improved performance.

- The IC is a complete functional block. This means that the designer has an easier time producing a design and can get the new product out onto the market in shorter time. He or she also does not have to "reinvent the wheel" by designing a circuit that does something that has already been designed. Instead, the designer can select an IC that does the job and then use that in the circuit. With discrete components, the designer would have to do a new design every time a similar function is needed.

In summary, ICs let you do things that would be impractical or impossible with discrete components. For example, a small computer with central processor, memory, and support chips might be made out of 50 ICs and fit on a single board. These ICs represent about 100,000 discrete components, and if the circuit were built out of these discrete components, it would occupy a whole room; use hundreds of amps of power; require an air conditioner for cooling; be unreliable; and could not be designed, built, and tested in any reasonable period of time.

What, then, are the disadvantages of ICs? Are there any cases where they still have not taken over for discrete components? The answer is that there are some specific cases where discrete components are still superior. In general, ICs cannot control large amounts of power or voltage (such as required to drive loudspeakers to full volume) without overheating and burning out. This is because the size and spacing of the IC transistors, diodes, and resistors is very small, so the heat has nowhere to go and large voltages can burn through the various layers of the IC, just as wires that carry high voltage require thicker insulation than wires carrying only a few volts. ICs also are more sensitive to physical damage from current or voltage spikes (such as electrical noise on power lines) than discrete components. Discrete components are larger and handle these larger values more easily. Therefore, products that require control of larger amounts of power or voltage or that may be used in places where there is the possibility of voltage spikes often use discrete components. However, in order to still make use of all the advantages that ICs provide, a typical design will use ICs wherever possible. The part of the circuit that faces the outside world may be built of discrete components, and these in turn will interface to the heart of the circuit, which is made up of ICs. The trend, of course, is to fewer and fewer discrete components. A typical high-fidelity stereo receiver uses ICs as the lower level preamplifiers and for overall front panel control, and these in turn drive discrete components that provide the main audio amplification power, as well as resist short circuits across the speaker lines which might blow out a regular IC.

QUESTIONS FOR SECTION 1-2

1. What are the advantages of ICs over discrete components?

2. Do discrete components have any advantages over ICs? In what cases?

3. What are two reasons why power consumption is a concern?

4. Why is the cost of an IC low?

5. Why are ICs faster than discrete components?

1-3 WHAT COMPUTER ICS ARE USED FOR

It might seem that digital ICs would be used only in computers, and that alone would be reason for their tremendous success and importance. It turns out that these same ICs are used for many other applications. These include:

- Calculators.
- Controls for mechanical devices such as elevators.
- Systems which measure real-world signals (temperature, fuel usage, and so on), perform calculations, and then provide controlling outputs (increase temperature, use less fuel).
- Digital telephone switching systems, where the routing of the calls and management of the individual phone lines are under control of a computer and are done digitally with improved performance and lower cost.
- Improved electric motor controls, which not only get greater efficiency out of the motor but also monitor certain conditions that indicate motor problems and prevent the motor from burning out.
- User-programmed thermostats which can handle a variety of temperature settings according to the time of day and day of week.
- Televisions and radios which tune stations digitally and can store preset stations in any desired order.

Of course, the list could go on for many pages and we still would not cover more than a fraction of the applications.

The applications fall into some general categories, and we will now take a look at those categories and a specific example in each. These categories are:

- Logic replacement.
- Smart controllers.
- Computers of all sizes: microcomputers, minicomputers, and larger mainframe computers.

The use of computer ICs to replace control circuitry, but not really do anything beyond what the existing circuitry was doing, is *logic replacement*. Logic replacement forms a very large but relatively invisible area of applications for computer ICs. It is also the easiest to understand and troubleshoot, since the component ICs involved are the simplest of all digital ICs. Logic replacement is sometimes called *hard-wired* because the function of the IC version is determined by the interconnection wiring between the ICs and cannot be changed except by redesigning the circuitry.

In logic replacement the computer ICs are used to provide a lower cost, lower power, and a smaller space in an application that is presently using some other method of control. This other method usually involves relays, which are "digital" elements since they have only two states—on and off. In these logic replacement applications, speed of operation is not the problem, since the function of the circuit is to control some mechanical device and mechanical devices do not move at fast speeds (as far as electrical pulses are concerned). A good example would be an elevator control.

An elevator system needs a control that recognizes when someone pushes a button asking for service, directs the elevator to the proper floor, makes sure that broken elevators are not sent anywhere, and handles some of the safety interlocks built into the elevator system. The problem of controlling elevators becomes much more complicated as the number of floors and the number of elevators in the building increases. The elevator control has to send only one elevator to answer a request for a ride, and it should be the nearest one heading in that direction, or an elevator that is idle. Since elevators use so much power, it is also important not to send too many empty or lightly loaded elevators up and down long distances if there is a partially full one that could be chosen. When you think of all the possibilities, you can see that controlling a bank of elevators in a modern office building is not a trivial job. There are lots of rules, exceptions to rules, and special cases (such as an emergency override by first aid personnel with a special key).

Traditionally, the control of elevators was handled through relay-based circuits. The relay circuits, unfortunately, got extremely complicated, as did the buildings. One result of this complexity was that the control circuits needed thousands of relays, used lots of power, generated considerable heat, and often failed because relays are electromechanical and subject to all sorts of problems (sticking contacts, burned coils). Doing the initial design was a very time-consuming process, and any changes that might have to be made in the control to accommodate a slightly different elevator configuration in the new building was a design nightmare. Finally, the action time of an individual relay was short compared to the physical motion of the elevator (a relay takes about one-fiftieth of a second to open or close); but when there was a complex system with thousands of relays interconnected, the overall action and response time could be seconds, by which time the elevator could have moved considerably and stopped between floors.

The solution is to use digital ICs as relay replacements. These ICs are usually relatively small, and each replaces anywhere from two to six relays. The wiring between the ICs is very similar to that previously done between the relays, except that there is much less because some of the connections are already made within the IC. Also, unlike the coils of a relay, the IC-based system requires no extra wiring or power. The system containing several thousand relays, which occupied a small room, is now replaced by a system of perhaps six circuit boards, each 8×12 in.

Signals from the elevator pushbuttons are converted by special interface circuitry to voltage levels that are compatible with the ICs used. The outputs of the ICs, which are used to actually control the elevators and turn lamps on or off for the use of the elevator, are generated at a low level by the IC and then boosted by special interface circuitry to the higher levels needed to control the motors and lamps. The final result is a smaller, more reliable system that essentially replaces the previous relay-based system.

Intelligent controllers are the next logical step in improving performance beyond hard-wired logic replacements. The smart controller adds a central processing unit (CPU), memory, and a program to the logic replacement so that the controller can do a better job by being more adaptable to the various situations the controller will have to handle. Consider the elevator control problem again. The number of people using the elevators, and the ways they use them, will vary with the time of day and the day of the week. In the morning from about 7:30 to 9:30, there will be large numbers of people getting on in the lobby and going to various floors. In the late afternoon, from 4:30 to 5:30, there will be the reverse situation. On Saturday traffic will be light, and on Sunday it will be almost nonexistent. The way that the elevators are used to move the most people in the shortest time, and to minimize elevator energy use, will have to change with time of day and day of week. A hard-wired control will do the same logic functions all the time and so might not be best for the various situations.

The intelligent controller (often called "smart" in the electronics field) uses a program rather than a hard-wire approach. The program consists of instruction steps that the controller must go through, and these instructions tell the controller what to do. Typically, they first tell the circuitry to report what is happening on the outside lines, which indicate which elevator buttons are pushed (Fig. 1-5). When these inputs have been brought into the controller, the instructions then tell the controller how to decide what is best to do for this particular time and day. This plan or the strategy is called the *algorithm*, and the smart controller can have more than one to use depending on the circumstances. The algorithm instructions are implemented by the controller, which decides what controlling signals to send to which elevators via the controller hardware. The controller repeats the sequence of events many times each second, so it is constantly aware of what is happening in the outside world.

There are two major advantages to the smart controller approach over the logic replacement method. First, the strategy that is used can be fairly complicated, yet it takes no more circuitry than a simple strategy, only a more complicated set of instructions. These instructions are written at the factory at the time the controller is designed, and the cost of producing a memory chip to hold the instructions does not really increase much with more or fewer instructions. Second, changes can be made quickly and cheaply. If a problem is found in the strategy, or a unique or new requirement for a different elevator and building configuration comes up, the manufacturer can develop a smart controller with the necessary changes simply by changing the instructions in the smart controller memory chip and replacing the old chip with the new one. No other changes to the circuit design or production have to be made. This is a tremendous advantage, since a manufacturer can build essentially the same circuit board and use it to control different size or style

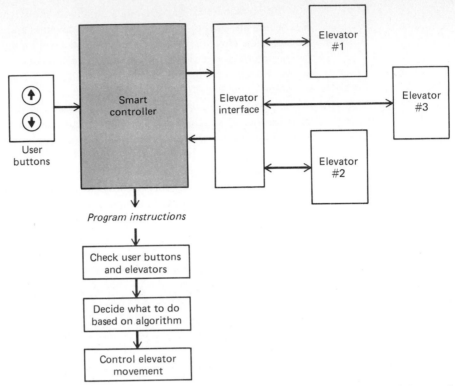

Fig. 1-5 The "smart controller" is connected between the elevator buttons and the actual elevators and motion controls. It controls the elevators using various strategies depending on the time of day and day of week.

products. This results in higher-volume and lower-cost production, and also lower support costs because the field service staff does not have to learn about so many different circuit boards.

Smart controllers are a very important application for computer ICs because they really are computers and thus let designers do things that were impractical or impossible with relay logic or hard-wired replacements. They are computers because they have a processor, memory, and input-output (I/O). They differ from ordinary computers because they are designed to do one job and do that job only in a very effective and inexpensive way. They cannot be used to do accounting or tax returns.

Look at a function that could not be done before smart controllers. The function of robotics involves motion control, where the arm of the robot is powered by an electric motor. Somehow, the robot must be able to direct the arm to a specific position with high accuracy and without going too fast and then overshooting the desired final position. This must happen even though the item may be of a different weight each time. Different amounts of motor power must be applied to position the arm with the desired accuracy in the shortest time and under varying loads. This is a complicated problem of control.

The type of motor often used is a servomotor. A servomotor is a combination of

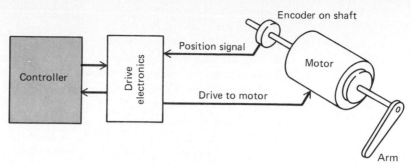

Fig. 1-6 A controller used with a servomotor. The controller not only directs the motor drive electronics to power the motor, but it receives a signal from a shaft encoder on the motor itself which indicates the position of the motor shaft and how fast it is turning.

a motor and some kind of feedback transducer which lets the circuit that is controlling the motor measure the speed of the motor or how many turns it has made, and thus how far it has moved the arm (Fig. 1-6). There is a need for some kind of feedback because the same amount of applied motor power will result in different amounts of motor movement, depending on the load the arm is carrying, and the controller has to know where the arm is versus where the controller desires the arm to be.

The smart controller is able to provide excellent performance because it can tailor the amount of motor power applied to give the best choice between motor speed, accuracy of final position, and overshoot past the position (Fig. 1-7). Say that the robot needs to move its arm to a point 6 in from where it is right now. The controller applies some value of power to the motor according to the program stored in its memory. The program then checks the feedback and sees how far the arm has actually moved. If it is below the expected amount, the load on the arm is heavy and the controller increases the power to the motor. The opposite happens if

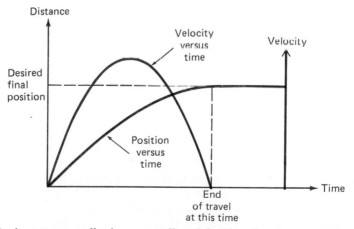

Fig. 1-7 A motor controller has a complicated function. It must get the motor to the desired position as quickly as possible, and yet not stop too soon or too late.

the arm has moved too far or too fast. The controller then continuously compares the arm position to the desired final position, and when the arm nears the final position, the controller cuts the motor drive power back to a much lower value. In this way, the arm slows down as it reaches the target and will not overshoot. If there is a small overshoot, the controller is aware of this because it knows the actual arm position. Therefore, it sends a little reverse power to the motor to bring the arm back to proper position. The result of the action of the smart controller is that the arm is positioned precisely and made to move so as to reach the target quickly. At the same time, any effects of friction, mechanical tolerances in the arm, and errors in the absolute value of drive power are of little concern, since the controller adjusts the drive power constantly in order to reach the target in the best way. This means that the motor drive need not be precisely calibrated and a lower-cost motor and power supply can be used. The smart controller has provided better performance with lower-cost parts. The robot manufacturer can use the same smart motor controller for a different-size motor simply by changing some motor-size-related values in the instructions of the controller.

Computers use the largest number and types of ICs. A computer is a general purpose electronic machine that is capable of doing many different types of jobs. The same computer can do accounting and payroll and then later be used for analyzing scientific data and controlling instruments in the laboratory. What makes the computer able to do these different jobs using exactly the same components and circuits is the computer program, which is a set of instructions. The program tells the computer what to do, in what order, what numbers to look at, and what types of decisions to make. The program is not a permanent part of the computer. Instead, it is stored separately on a removable, changeable medium such as a tape cartridge, magnetic tape, or disk. When the computer user wants to do accounting, he or she loads the accounting program into the computer memory. When the accounting activity is over, a new program to do a completely different function can be loaded into the computer.

The computer is an essential part of our modern society, and computers are the major use for digital ICs. They use large quantities of ICs in order to have enough capability for the many jobs they are expected to handle. They use many different types of ICs for the many internal functions they must perform, and most of them are covered in more detail later in this book. The internal design of computers and their use of memory, CPU, screen, and keyboard are actually very similar regardless of the type of computer (Fig. 1-8). Understanding computer architecture, as this internal structure is called, is a great help in understanding almost any system based on computer and digital technology. In some applications certain parts of the computer may not be needed, but the interaction between the various pieces is still generally the same. Some products do not look like computers because they do not have a screen and keyboard, but otherwise they perform all the same functions that a computer does—they receive numbers, do calculations, and make decisions (Fig. 1-9). In fact, many of these "noncomputer" computers allow the service technician to hook up a screen and keyboard, or disk drive, in order to run special programs that are used to troubleshoot the system when something does not seem to be working right. The parts that are not wanted are left off in order to save cost to the customer, but they can be reattached when needed.

Fig. 1-8 A typical computer contains internal elements as well as a user screen, keyboard, and provision to connect to a printer.

Fig. 1-9 A single board computer is often used as a dedicated controller. This board can receive signals, process them, and generate output signals based on its instructions. *(Photo Courtesy of Analog Devices, Inc.)*

QUESTIONS FOR SECTION 1-3

1. What are the three categories of IC use?

2. In a logic replacement application, what is replaced? By what?

3. What are the advantages of a smart controller over just logic replacement?

4. What does a smart controller do in a motion control application?

5. Describe another application that would benefit from a smart controller.

6. What are the elements of a computer?

7. What are three advantages of a smart controller to a user?

8. What are the differences between a smart controller and a general purpose computer? The similarities?

1-4 HOW ICS ARE MADE: FABRICATION FROM A SILICON WAFER

The process of making an IC is extremely complicated and requires expensive, specialized equipment. A general understanding of how ICs are made is useful because it helps make clear some of the features of the ICs that may seem unusual or different from circuits made of discrete components. The fabrication rules of an IC make some ordinary circuit design techniques very impractical. At the same time, some design rules which are difficult with discrete components become very easy with IC designs.

The building of an IC is completely different from the production of any electronic or mechanical device that is built by the assembly of small parts or components, such as a radio or television. Most digital ICs are made out of a single piece of silicon that has been carefully manipulated in several ways. There is no assembly as we would normally think of it—in fact, because the IC is a single piece, it is often referred to as *monolithic* (from the Greek words *monos*, ''single,'' and *lithos*, ''stone'').

Making an IC requires much more than the person who designs the circuit. Specialists in many areas are needed. These include:

- Designers, who actually develop the circuit diagram of the IC.
- Physicists, who understand the whole process and help plan it, since the fabrication involves the careful measurement and control of atoms, molecules, and temperatures.
- Chemists, who provide the specialized chemicals needed in very pure form.
- Optical engineers, who build and operate the precise optical magnifiers and projectors needed.
- Clean-room specialists, who are responsible for making sure that the area of IC fabrication is absolutely free of any sort of dirt, dust, or unwanted chemicals. The smallest particle of dust can ruin one or more ICs while they are in fabrication.
- Mechanical engineers, who design and operate the machines that are used in the complicated process.

The actual production of the IC begins with silicon. It is interesting that an IC, which can do so much and is so complicated, begins with one of the most common and inexpensive minerals on earth—sand! The sand is melted at very high temperatures and purified to get rid of any nonsilicon elements that may be in it. This ultrapure sand is then ''grown,'' in a special bath, into a long, solid tube that is anywhere from 4 to 6 in in diameter. The grown silicon is then sliced very thin, much like a salami. These slices, or wafers, shown in Fig. 1-10, are the starting point for the ICs. Each wafer will have several thousand ICs built on it at the same

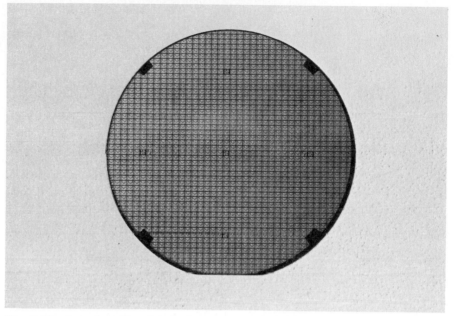

Fig. 1-10 A wafer with a 3-inch diameter being made into ICs. Each wafer contains hundreds of identical patterns which will become the ICs themselves. *(Photo by Jill Manca.)*

time. This is why ICs can be so powerful and yet so cheap—the production may be complicated, but the cost of the production is divided over the thousands that can be made in one batch.

Making the IC consists of many additional steps, some of which are repeated several times at different stages. One of the key steps is to coat the IC surface with a thin layer of silicon dioxide, better known as glass. Then, selected areas of the glass layer are etched away with powerful chemicals like hydrofluoric acid. In order to cause some areas to be etched away and others protected, a special optical and chemical process called *masking and etching* is used. A photonegative or mask of the circuit (Fig. 1-11) is exposed onto the surface of a wafer that has been coated with photoresist, a special chemical. This is done in almost the exact same way that a negative from a regular camera is used to make the positive print that you look at by exposing the negative on photographic paper (Fig. 1-12). In the case of the IC wafer, wherever light hits, the photoresist is hardened, and wherever light does not hit, the photoresist stays soft. Then the photoresist and wafer itself are washed. Where the photoresist is still soft, it rinses off, but where it is hard, it stays. The glass under the areas where the photoresist stays is protected from being etched away by the hardened photoresist. In this way, the makers of the IC can control what is left and what is removed at each step in the process.

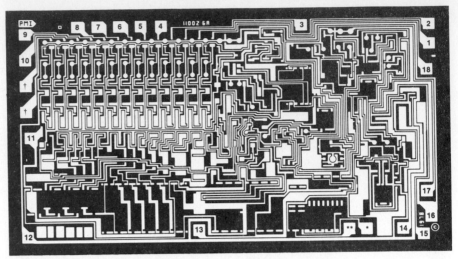

Fig. 1-11 An IC mask contains the pattern of the IC desired. Its use is similar to a photographic negative. This mask is for a die 0.163×0.090 inch. *(Photo Courtesy of Precision Monolithics, Inc.)*

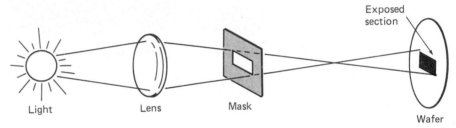

Fig. 1-12 The mask pattern is exposed onto the silicon wafer which has been coated with a special light sensitive "resist," which then hardens where the light hit it. This operation identifies where the IC wafer is to be protected from exposure to the special IC chemicals.

After each step with the glass and photoresist, the wafer is put in an oven at several thousand degrees and exposed to special chemicals such as boron, arsenic, gallium, and phosphorus. These spread, or diffuse, into the wafer wherever there is an opening in the top layer, just as a drop of paint will spread into a piece of paper as the time passes. These special chemicals form the transistors, diodes, resistors, and capacitors of the IC. The effect of the whole procedure is to build the functions into the silicon with each step.

The last step is to make the final electrical connections on the surface of the IC. The wafer is exposed to aluminum atoms, which form a top layer connecting the last remaining exposed points and the edges of the IC. It is to these edge pads that the wires will be attached in the next step. Before the wires can be attached, the silicon wafer is cut into thousands of individual ICs with a diamond saw, just as a pan of brownies is cut into the individual brownies. The ICs, called *dice*, are shown in Fig. 1-13 and are mounted on bases so they can be handled and tested by machines that automatically apply test signals and measure test voltages at the

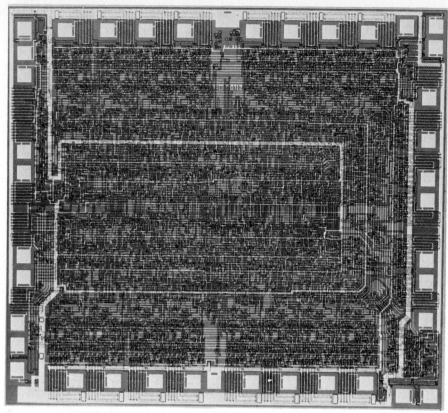

Fig. 1-13 The die of a wafer after the wafer has been cut. The wires to the IC package are connected to small square pads called "lands," on the outside edge of the die. *(Photo Courtesy of Zilog, Inc.)*

aluminum pads. If the IC is good, it is mounted in the final package, and hair-thin wires are soldered from the edge pads to the pins of the package. Finally, the package is sealed and is ready for use in a circuit board.

As you can see, the fabrication of an IC is extremely complicated. Figure 1-14 gives a summary of the steps involved. There are many things that can go wrong at each step: temperature not right, chemicals too weak or too strong, negative pattern not exactly in the right place, and so on. Despite all this, with careful work a manufacturer can usually get anywhere from 70 to 90 percent of the individual ICs to work properly. Those that do not work are sent to a lab for analysis to see if the reason can be found. Perhaps the temperature in the ovens was a few degrees too high, or a speck of dust got on the negative or lens and put false lines on the wafer. It is important that the reason be found, since the manufacturer can recover all of the costs only if most of the ICs on the wafer work.

One of the interesting things about ICs is that, within certain limits, the actual number of transistors, diodes, or resistors in the circuit is not too much of a concern, since the whole IC is made in a batch, not one transistor at a time. This means that the internal design of the IC circuit may use more transistors than if the

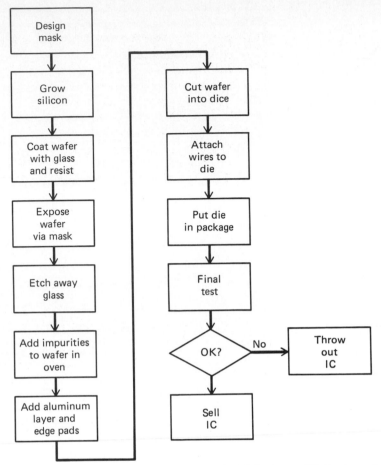

Fig. 1-14 A summary of the many steps involved in fabricating an IC. Each step involves great precision and the entire process can easily have problems which may result in many or all of the wafer dice not working properly.

product was going to be assembled out of discrete transistors. Also, there are some circuits that work very well if it is possible to get transistors with identical performance and characteristics, such as signal gain or sensitivity to temperature, into the same circuit. With discrete transistors this is almost impossible, since two discrete transistors may not be identical and their temperature in the circuit will always differ by a small amount. In an IC, the transistors are really built at the same time, under identical conditions, into the same piece of silicon, so they match almost 100 percent. This is one reason why IC designs can give such high performance.

IC designs also use transistors instead of resistors and capacitors because it is harder to make resistors and capacitors on the silicon than transistors. The values of resistors and capacitors that can be made is also very limited. Resistors can be made in values from about 50 to 50,000 ohms (Ω), and sometimes special circuits are designed so that they will not need values outside this range, although a few more transistors may be needed instead. However, the resistors that are formed in the circuit can be made to follow each other with temperature, which means that good performance is possible over a wide temperature range.

We have seen that the fabrication of an IC is a complicated process with many special steps. Despite this, the advantages of ICs are so great that manufacturers are willing to invest the money and effort to produce ICs that are in great demand. In today's technology, there are two fundamental IC classes that are most important: TTL and MOS. These will be explained next.

QUESTIONS FOR SECTION 1-4

1. How does IC manufacturing differ from assembling a circuit board?

2. What sort of specialists are needed for IC production?

3. What is the starting element of the IC?

4. At what temperatures are ICs made?

5. How are the connections of the IC connected to the pins of the package?

6. What is different about IC design as compared to discrete component design?

1-5 THE TWO MAJOR IC TECHNOLOGIES: TTL AND MOS

When computer ICs are fabricated on a wafer of silicon, it is possible to make the wafer have either regular transistors or special ones called *field-effect-transistors* (FETs). Another name for the conventional transistor is a *bipolar transistor*, since the transistor action is caused by both positive and negative charges flowing in the transistor. The FET, on the other hand, relies mostly on only one type of charge flowing.

Using bipolar transistors, digital circuits are built on the wafer. The most common IC technology type is called *transistor-transistor logic*, or TTL. It gets this name because the circuit has one transistor which serves as the next-to-final stage, or driver, for the final-output-stage transistor. There are nearly a thousand standard TTL ICs available from many manufacturers. TTL is one of the preferred types of IC because:

- It is very versatile and can be used for many types of functions. These range from chips with only one function to large, multifunction ICs.
- Circuits using the TTL design are very easy to interface to other ICs.
- They are not sensitive to damage from static electricity where there is low humidity or lots of handling.
- They are capable of operating at high speeds, which can result in very fast circuit operation.
- They use a single fixed voltage of +5 volts (V), which is easy to provide from a power supply.

One drawback to TTL ICs is that they consume a lot of power, especially as the IC function becomes complicated with many transistors. The power is a problem because it has to be supplied by the system power supply and the supplied power generates heat in the IC. This heat must be removed or the IC will overheat and burn out.

The ICs made of FETs are called *metal-oxide-semiconductor*, or MOS, types. This is because the FETs are formed on the silicon wafer using metals on oxides. The MOS type of IC is also very popular for good reasons:

- It uses very little power and therefore needs only a small power supply and little or no cooling.
- It is possible to make the entire circuit from FETs only, without resistors, diodes, and capacitors. The resulting circuit can squeeze many FETs onto a small piece of silicon and thus provide a complicated circuit function in a small IC and many ICs on a single wafer.
- MOS ICs can run on a wide range of voltages (typically 3 to 18 V). This makes them very useful in many circuits, including those where the supply voltage may vary.

Unfortunately, MOS-type ICs have some drawbacks. They tend to run slower than TTL types unless special designs are used. They can be very sensitive to static electricity and must be handled carefully, using special plastic carriers and bags. Also, MOS-type circuits can be harder to interface to other types of ICs than the TTL IC. The interface circuits of MOS ICs have very little extra power and they are sensitive to too much voltage or short circuits.

Nevertheless, the appeal of MOS-type ICs is very high because of the large number of functions that can be designed onto the IC and low power that is used. One solution to the interface problem is to provide a MOS IC with input and output pins that look like TTL, while the rest of the IC is MOS. These ICs are referred to as *TTL-compatible,* since from the user standpoint they look just like TTL. Even within the MOS family, there are distinct types. The negative MOS (NMOS) type is capable of good performance and is not too difficult to interface. Another, the complementary MOS (CMOS) type, uses the least power of all IC types but runs slower than the other types, is more difficult to interface than NMOS types, and is the most sensitive to static electricity.

Each IC type, TTL and MOS (NMOS and CMOS), has a distinct place in the world of computer ICs because of features they offer for specific applications. Systems which need the maximum speed use NMOS or TTL, while battery-powered computers or calculators use CMOS to save power and consequently may run a little slower. There is one series of TTL ICs that is most common in circuit designs, and similarly there is a standard family of CMOS ICs that is used in most applications which require CMOS.

Now that we understand what an IC is, how ICs are made, and the different types of ICs, we can begin to explore the specific functions that ICs provide. These low-level functions are the building blocks for all more advanced IC functions.

QUESTIONS FOR SECTION 1-5

1. What are the advantages of TTL? The disadvantages?

2. What are the advantages of MOS? The disadvantages?

3. What type of component is the heart of the MOS IC?

SUMMARY

This chapter has introduced the IC, which is the basis for nearly all modern electronic systems and computers. The IC has completely revolutionized electronics since its invention and has made products which were science fiction just a few years ago not only real, but affordable and reliable. The extremely complicated process used to fabricate ICs results in a very low cost product, regardless of what IC technology is used. ICs have replaced relays, vacuum tubes, and similar electronic components. They have allowed the practical implementation of products that have intelligence, that can actually use varying strategies depending on the circumstances, and can interact with the user in complex ways.

The production of ICs from sand is a very complicated process. Many special and highly precise skills are needed for a working product to result. More complex ICs use the same process as the simpler ones, but the area of the IC die is larger with more transistors, capacitors, and resistors designed into the pattern. The two major IC families, TTL and MOS, use differing technologies to produce ICs with various technical advantages, depending on the requirements of the application. The production process is essentially the same for both TTL and MOS, and often MOS ICs are designed so that the actual connections to the IC are TTL-compatible.

REVIEW QUESTIONS

1. What is an IC? Why is it called by that name?

2. Give three reasons why the silicon of the IC must be put in a package.

3. Give two reasons why sockets may be used and two reasons why they are usually avoided.

4. What is the ratio of area occupied by an AND gate built of individual components versus one made as an IC?

5. What is a major factor in system reliability? Why are ICs more reliable than discrete-component-based systems?

6. Why is the cost for an IC less than that for a discrete circuit?

7. Why is an IC faster in operation than a discrete circuit?

8. Where and why are discrete components still used?

9. Compare the role of ICs in a logic replacement function versus an intelligent controller.

10. How is an intelligent controller superior to logic replacement?

11. How is an intelligent controller like a computer? How is it different?

12. Explain why the amount of circuitry in an intelligent controller is not directly related to the complexity of the controller function. Why is this an advantage?

13. Explain how a controller can accurately position a robot arm with the aid of a feedback signal from the arm.

14. What types of skills are needed to design and fabricate ICs?

15. How is raw sand converted to pure silicon wafers? How many ICs can be on a typical wafer?

16. Why is the cost of an IC low if the design and fabrication is so complicated and difficult?

17. What is the mask? How does the IC wafer get the proper pattern of the circuit on its surface?

18. How do the desired impurities get into the silicon? How is the wafer made into usable IC components?

19. Why can an IC provide matched performance in components while discrete components cannot? Why is this feature important?

20. Discuss three characteristics of TTL which are benefits and are drawbacks in MOS. Repeat for the opposite case.

21. Why is low power use important in a product? Why is it important in the IC, even if the system can provide the power?

22. Why are TTL-compatible MOS ICs popular?

SMALL-SCALE INTEGRATION

2

Digital numbers are used in the ICs which form computer circuits. The concept of digital versus analog representation of values is discussed, and the differences are explored. The advantages of digital signals and ICs in practical systems are explored. Boolean algebra, which makes use of digital representation, and the boolean functions such as OR, AND, and NOT lead to a practical understanding of what can be accomplished with digital signals and values.

The binary system of numbers also requires understanding of the digital world. Binary numbers can take the place of the decimal numbers that are more familiar, and decimal numbers can be converted to binary easily. Some other forms of digital numbers, including binary-coded decimal (BCD), octal, and hexadecimal also are used in some systems. Finally, IC gates which actually implement the boolean logic functions are introduced.

2-1 DIGITAL NUMBERS

Computer integrated circuits are digital. All the signals that they process, and all the signals that they generate, are digital signals. Before we can understand the digital IC, we must understand what digital signals are and how they are used.

All the information that we deal with in the real world comes in two main types: analog and digital. Analog information is data which can take on any value within the overall range. It is continuous over the whole range. For example, a standard glass thermometer can be used to read a temperature anywhere from the lowest value, about $-30°$ F, to the highest value, about $130°$ F. If you look at the thermometer, you can read the temperature anywhere in between and even try to estimate the temperature to a value finer than the markings, $47\frac{1}{2}°$ or $47\frac{1}{4}°$, for example. Another analog value is the reading you make with an ordinary ruler. The ruler is marked in inches and fractions of an inch, and you read the ruler to these markings and sometimes even between the markings. In both cases of the

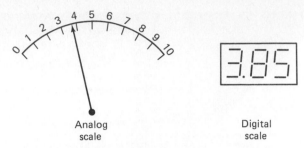

Analog scale

Digital scale

Fig. 2-1 An analog scale showing a reading and a digital readout of the same value. Note that the analog reading requires more interpretation than the digital reading.

thermometer and the ruler, any value may occur. You read the value to whatever resolution you can (Fig. 2-1). Note that while you can read to any resolution, the absolute accuracy of the reading is only as good as the system accuracy. (Accuracy is the absolute "correctness" of the reading as compared to a guaranteed standard of measurement.)

In a digital system, only certain values of the data are allowed. You probably have more experience with digital systems than you realize. For example, counting by whole numbers (integers) is a digital process. (In fact, the word *digital* comes from the word *digit*, meaning "finger," which was the original counting mechanism.) When you count out the number of people coming to a party, you are performing a digital operation because only integer values are allowed. These allowable values are called the *discrete values* of the system. The discrete values are not restricted to integers, but they must be limited to a specific group. For example, a cash register is digital and only has numbers with dollars and cents (two positions to the right of the decimal point). A cash register will not show $8.753, and you cannot read or estimate values between cents.

Anytime you are allowed only certain discrete number values, you have a digital system. This contrasts with an analog, or continuous, system where any value is allowed. A watch with hands is an analog watch because you can read the time to any value. In comparison, a watch that is digital has numbers only and you can only read the numbers that are shown. If the watch has numbers for the hours and the minutes you simply cannot try to estimate between the minutes and figure out the seconds. With an analog watch, you might estimate the seconds from the position of the minute hand between the minute marks.

There are many examples of both analog and digital things in the real world. Talking is an analog activity because there is an infinite number of possibilities of loudness, types of sounds, voice pitch, and so on. Voting is a digital activity because, on the other hand, there are only certain candidates you can vote for—you vote yes for one and no for the others.

There is a special case of digital systems that is of most interest in computer ICs. That case is called the binary case. In the binary system, the allowable values are reduced to just two values. Instead of being able to choose any whole number, everything is expressed in terms of just a 0 or a 1. Binary is really the simplest form of digital, and very often the two terms are used interchangeably in computer systems. All computer ICs use the binary system.

Other expressions for the 1 or 0 of a binary system are on and off and high and low. You already have some experience with binary systems; whenever you have to decide yes or no, or whenever you have turned a light switch on or off. (Note that a lamp that is on a dimmer is an analog lamp, but one with an on-off switch is digital.)

QUESTIONS FOR SECTION 2-1

1. Explain the difference between analog information and digital information.

2. Which of the following are analog, and which are digital: channel selector on a television, speedometer on most cars, the length of a piece of paper measured with a ruler, the number of pages in a book, a household thermostat? Explain why.

3. What is an advantage of a digital value over an analog one? What about the opposite—the advantage of an analog value over a digitally displayed one?

4. What is binary? How is it a special case of digital numbers?

5. What are the common expressions for the possible values in the binary system?

2-2 ADVANTAGES OF DIGITAL NUMBERS

Why are binary digital signals used in computer ICs when they seem to limit the choices or opportunities for counting and measuring? There are many reasons why a binary scheme is superior to a digital scheme with many levels, and why a digital scheme is in turn preferred to an analog system. Look at these reasons:

- Any electronic system has some small unwanted signals called *electrical noise* added to the signal lines, similar to the noise and static sometimes heard on the telephone or radio. This noise can come in from the outside world, it can be generated within the circuit itself, and it can be coupled from one part of the circuit to another. In an analog system, this noise adds to the signal you are trying to look at, and in fact cannot be separated from it. In other words, the noise corrupts the signal and makes the signal less accurate (Fig. 2-2). There are special ways to design circuits to make them noise-resistant, but these methods are very expensive and still cannot take care of all the noise. In an analog system there will always be some loss of the "goodness" of the signals as a result of noise, and this hurts the system performance. For example, consider a stereo music system where an analog signal of 0 through 10 V represents the loudness of the music. Suppose from one note to the next the loudness should double. However, noise might add to the second note, making it more than twice as loud. This distortion would mean you would not hear the music as it was intended to be heard. Since the noise is always changing—getting larger and smaller as well as being positive and negative—the original audio signal would be continuously corrupted and distorted.

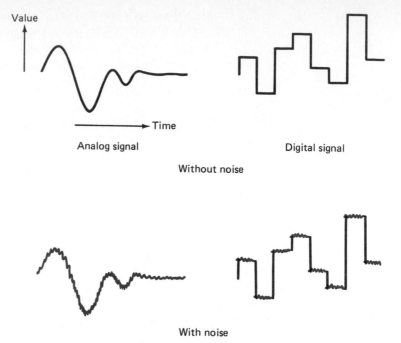

Fig. 2-2 Noise can corrupt analog and digital signals and make their exact value harder to determine accurately.

- In a binary system, in contrast, all numbers are represented by high signals and low signals. These high and low signals are separated by an area of voltage where the signal is neither considered high or low, but instead has no meaning. In typical binary ICs, a low signal is any value from 0 to about 1 V, while a high is a value from about 3 to 5 V. The range in between, from 1 to 3 V, does not affect circuit performance. As a result of this arrangement, a fraction of a volt of noise does not matter, since it is not enough to cause a low to look like a high, or a high to look like a low. In other words, the wide separation of values of low from high means that noise does not corrupt the signal. The system tolerates this noise without problem.
- In any system there will be tiny variations in the value of a component from one system to the next. These variations mean that the signals are not processed exactly the same way in each unit and thus some deviations from the original goal occur. In a digital system, the effects of component variations are much smaller than the zone between high and low, so input signals are processed identically.

  What this means is that binary systems are easier to design because they are not as sensitive to noise, and they are more exact because any number value entering into the system does not get changed unless the circuit specifically is designed to change that value. The end result is a better system.
- Binary systems also allow very elementary functions, called *gates,* to be built and then used over and over again in many different ways. These gates, which we will study in detail, are the building blocks of the IC system, much like the bricks that make up a wall. The bricks can be arranged in many ways to make

different walls with different patterns and sizes. The same is true of binary gates. It is possible to construct many digital functions using the same gates. This makes the design of complicated systems much easier, and it also makes the physical design of the IC much easier since the same photo-mask patterns can be used again. (IC design can become extremely complicated for many reasons, both theoretical and practical.)

- Circuits built of binary digital functions are easier to analyze than analog circuits. This is because the binary signals can be represented by either a 1 or 0, corresponding to the high value or the low value. It is then practical to describe the signal at every point as either a 1 or 0, and to describe the effect on the various parts of the digital circuit by what it does to the 1 or 0—it either keeps it the way it is or converts it to the other value. The rules by which the parts of the digital circuit do this are the rules of boolean algebra.

QUESTIONS FOR SECTION 2-2

1. How are binary values indicated in a circuit?

2. What are the advantages of digital signals versus analog in real systems?

3. What are the problems that noise can cause? Why is noise a limiting factor in the accuracy and confidence in the reading of the system?

4. How sensitive are analog circuits to minute variations in the values of components? How sensitive are digital circuits?

5. Why are digital circuits easier to design and analyze than analog circuits?

6. What electronic function is the basic building block of digital circuits?

2-3 BOOLEAN ALGEBRA

Boolean algebra, named after the mathematician George Boole who first invented it in 1854, is a mathematics system designed as a shorthand method to test logical statements and show if their result is true or false or yes or no. An example would be a statement such as "If it is raining and I have my raincoat, I will stay dry." Boolean logic would then be used to see if the person is wet or dry depending on whether it is raining, and if the person did have a raincoat. It can also be used with numbers which only have two values, which we can call 1 and 0.

You have probably used some of the rules of boolean algebra in routine activity without knowing it. The fundamental boolean logic operations are OR, AND, and NOT. From these we can also define some other important boolean operations, such as NOT OR, NOT AND, and EXCLUSIVE OR. The boolean operations are defined by a *truth table*. The truth table shows all the possible combinations of input values that the boolean function can have, and the resulting output. It is very useful for defining functions and seeing the effects of combined boolean functions (and, therefore, digital logic combined into a larger system). The left-hand column of truth tables shows the inputs to the boolean function, and the right-hand column shows the result of the operation.

$$Y = A + B$$

Fig. 2-3 Logic symbol for an OR function, with inputs A and B and output Y.

The OR Function

The OR function has two or more inputs and states "the output is a 1 if any one or more of the inputs is a 1. If all the inputs are 0, then the output is a 0." In schematic diagrams, a special symbol is used to represent the OR function (Fig. 2-3) and by a + in mathematical expressions. The truth table shown below represents the two-input case, where A and B are the inputs and Y is the output:

A	B	Y = A OR B
0	0	0
0	1	1
1	0	1
1	1	1

The output is a 1 for three of the four input possibilities. In your own experience, you have used a boolean logic function if you have said something like "If any one of us goes, then we will need a car." In that case, you are saying that if any one, or more, of you go, then you will need the car, but if none of you goes then you will not need the car. An OR function also exists when two switches, connected in parallel, control the flow of electrical power to a load such as a lamp (Fig. 2-4). If

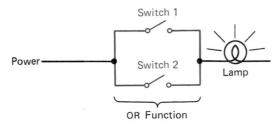

Fig. 2-4 The OR function is equivalent to controlling a light from either of two switches connected in parallel.

either or both switches are on, electrical current will flow and the lamp will be on. If both switches are open, however, no power will reach the lamp. This agrees with the truth table (if we call a closed switch a 1, an open switch a 0, a lamp which is lit a 1, and a dark lamp a 0).

The OR function applies to two or more inputs. Here is the truth table for a three input case (Fig. 2-5).

A	B	C	Y = A OR B OR C
0	0	0	0
0	0	1	1
0	1	0	1
0	1	1	1
1	0	0	1
1	0	1	1
1	1	0	1
1	1	1	1

Note that for the three-input case there are eight input possibilities, as compared to only four for a two-input situation.

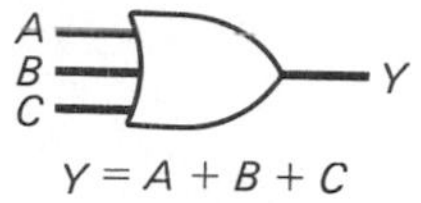

$$Y = A + B + C$$

Fig. 2-5 The OR function can have more than two inputs. The symbol for a 3-input OR function is shown.

The AND Function

The AND function says "The output is a 1 if and only if *all* the inputs are 1. Otherwise, the output is a zero." This is like saying that "We will go *only if* we all decide to go." The AND logic function is represented by a special symbol in

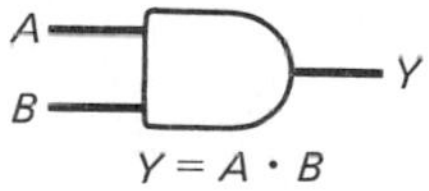

$$Y = A \cdot B$$

Fig. 2-6 Logic symbol for an AND function, with inputs *A* and *B* and output *Y*.

drawings (Fig. 2-6), and by a · or × in mathematical expressions. The truth table for a two-input AND function is:

A	B	Y = A AND B
0	0	0
0	1	0
1	0	0
1	1	1

The output is a 1 only for the case where both inputs are a 1. An example is "The car engine will start only if gasoline and a spark reach the engine. Either one alone is not enough, and if both are missing the car certainly will not start."

Electrical switches can also provide the AND function. If two switches are wired in series with the load, as shown in Fig. 2-7, then both switches have to be

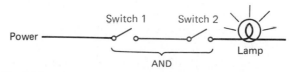

Fig. 2-7 The AND function is equivalent to controlling a light via two switches connected in series. Both must be on for the light to be on.

on for current to flow and the lamp to light. If no switches are on, or only one is on, then the lamp stays dark. Using the same on/1, off/0 notation as was used for the OR, the switch arrangement provides the functions of the AND truth table.

$$Y = A \cdot B \cdot C$$

Fig. 2-8 The AND function with three inputs. All must be 1 for the output Y to be 1.

There can be more than two inputs to an AND function (Fig. 2-8). The truth table for a three-input AND looks like this:

A	B	C	Y = A AND B AND C
0	0	0	0
0	0	1	0
0	1	0	0
0	1	1	0
1	0	0	0
1	0	1	0
1	1	0	0
1	1	1	1

Of course, there can be even more than three inputs to an AND function, and while the number of input possibilities increases, the AND rule still holds: The output is a 1 only if *all* the inputs are 1.

The NOT (Inverter) Function

The NOT function is very simple and yet very important. It states that "The output is the opposite of the input." In other words, the NOT function inverts the input state, by converting a 0 to a 1 and a 1 to a 0. The NOT function is also said to provide the "complement" of the input. The NOT function is represented either by the symbol shown in Fig. 2-9a or by drawing a small circle at the appropriate

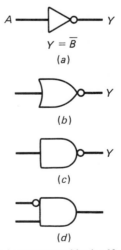

$$Y = \overline{B}$$

(a)

(b)

(c)

(d)

Fig. 2-9 The NOT function is represented by itself with a triangle and a circle, or simply a circle connected to the input or output of another logic function.

input or output of the main symbol for AND or OR (Fig. 2-9b, c, and d). In mathematical expressions, a bar (−) is put over the letter or a prime (′) is put after it ($\overline{A}$ or A′ for NOT A).

The truth table for the NOT function is very simple.

A	Y = NOT A
0	1
1	0

The NOT function can have only one input. As with OR and AND, the NOT function can be provided by a simple switch (Fig. 2-10). When the switch is open (or 0), current flows through the wires and reaches the lamp, which then turns on (logic 1). If the switch is closed (or 1), the lamp is short-circuited and all the current instead flows around the lamp, through the switch, and the lamp is off (logic 0).

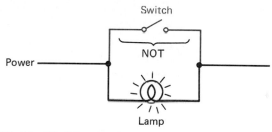

Fig. 2-10 The NOT function is equivalent to a switch across a light, so that when the switch is on the light is off and vice versa.

QUESTIONS FOR SECTION 2-3

1. What is boolean algebra? Why is it used for binary digital systems?

2. What is the rule for the OR logical function? Give a practical example of an OR situation.

3. Repeat question 2 for the AND logical function.

4. Repeat question 2 for the NOT logical function.

5. What is the value of A OR 0? A OR 1?

6. What is the value of A AND 0? A AND 1?

7. Is the AND function limited to two inputs, A and B? What about the OR function?

8. What is the rule for the NOT function? How many inputs are used?

9. What is the value of NOT 1? NOT 0?

10. What symbol is used in boolean expressions for (a) OR, (b) AND, (c) NOT?

PROBLEMS FOR SECTION 2-3

1. Show a boolean function expression, using the basic functions, that implements the statement: "If A and B are OK, or C and D are OK, only then will we go." Prove this expression using a truth table.

2. Draw the truth table for a four-input OR gate. How many input conditions are there? How many times is the output equal to 0? to 1?

3. Repeat problem 2 for a four-input AND gate.

4. Show a boolean expression for the statement: "If I go to the ballfield and it is not raining, then we will play."

5. What is the value of NOT [NOT (NOT A)] when A is 1? Why? Prove this with a truth table.

6. What is the truth table for A AND B = C, and then C AND D? How does it compare to a three-input AND function? Is the AND of two-inputs, when ANDed with a third, the same as the AND of three inputs?

7. Repeat for the OR function.

8. Draw the symbols for a two-input AND function and OR function.

9. Write the boolean expression for the logic functions shown in Fig. 2-11.

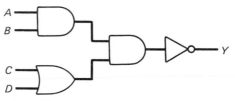

Fig. 2-11   Section 2-1, Problem 9.

10. Draw the logic functions interconnected to implement the expression: (A AND B) OR (NOT C) = Y.

2-4   Boolean Identities

Boolean identities are a set of statements of boolean algebra that are true as a result of the definitions AND, OR, and NOT, and are true for *any values* of the input. The important boolean identities for OR are:

$$A \text{ OR } B \text{ OR } C = (A \text{ OR } B) \text{ OR } C = A \text{ OR } (B \text{ OR } C)$$
$$A \text{ OR } B = B \text{ OR } A$$
$$A \text{ OR } A = A$$
$$A \text{ OR } 1 = 1$$
$$A \text{ OR } 0 = A$$

Where parentheses ( ) occur, that operation is performed first in the overall sequence. To prove one identity, fix B equal to 1 and look at the truth table for A OR B:

A	B = 1	Y = A OR B
0	1	1
1	1	1

No matter which value A takes, Y is always 1.

There are also some identities for the AND functions:

$$A \text{ AND } B \text{ AND } C = (A \text{ AND } B) \text{ AND } C = A \text{ AND } (B \text{ AND } C)$$
$$A \text{ AND } B = B \text{ AND } A$$
$$A \text{ AND } A = A$$
$$A \text{ AND } 1 = A$$
$$A \text{ AND } 0 = 0$$

We can prove the last one by constructing its truth table. Set B equal to 0 and then:

A	B	Y = A AND B
0	0	0
1	0	0

For both values of A, A AND 0 = 0. The identities A AND 1 = A and A AND 0 = 0 are important in many digital IC operations.

There are even identities for the simple NOT function:

$$A \text{ OR } (NOT \ A) = 1$$
$$A \text{ AND } (NOT \ A) = 0$$
$$NOT \ (NOT \ A) = A$$

To prove the first one with a truth table:

A	NOT A	A OR (NOT A)
0	1	1
1	0	1

The last identity shows that inverting a signal twice brings it back to its original state.

QUESTIONS FOR SECTION 2-4

1. What are boolean identities? Why are they important?

2. What are two of the boolean identities for the OR function?

3. What are two of the boolean identities for the AND function?

4. What is one of the boolean identities for the NOT function?

PROBLEMS FOR SECTION 2-4

1. Prove the identity A OR A = A.

2. Prove the identity A AND 1 = A.

3. Prove the identity A AND (NOT A) = 0.

4. Prove the identity A OR (NOT A) = 1.

2-5   BOOLEAN BUILDING BLOCKS

There are three additional boolean logic functions which are very important. They are built from the three basic functions we have discussed (AND, OR, NOT), but they are important in their own right because it turns out that they can be used to provide many additional functions. These additional logic functions are NOT OR (NOR), NOT AND (NAND), and EXCLUSIVE OR (XOR).

The NOT OR (NOR) Function

This function takes the OR function, which we have already discussed, and then adds a NOT function to the output. It is used so often that it has its own symbol (Fig. 2-12), a combination of the OR and NOT symbols. The truth table for the

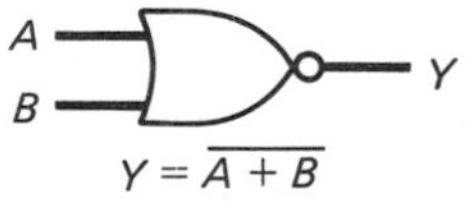

$$Y = \overline{A + B}$$

Fig. 2-12 The symbol for the NOR function is a combination of the OR and NOT symbols.

NOR is developed by first doing the table for OR, then doing the NOT operation on the result of the OR:

A	B	A OR B	NOT (A OR B)
0	0	0	1
0	1	1	0
1	0	1	0
1	1	1	0

The NOT AND (NAND) Function

The NAND function takes the AND function and adds a NOT at the output. The symbol for the NAND is a combination of the AND and NOT symbols (Fig. 2-13).

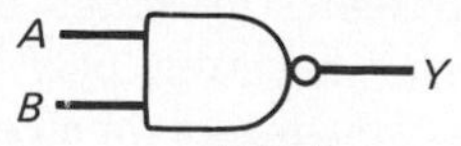

$$Y = \overline{A \cdot B}$$

Fig. 2-13 The NAND function is represented by an AND symbol with a NOT circle attached.

The truth table is made by first doing the AND operation, then performing the NOT operation on the AND result:

A	B	A AND B	NOT (A AND B)
0	0	0	1
0	1	0	1
1	0	0	1
1	1	1	0

The NOR and NAND functions are important for some not-so-obvious, but still very practical reasons. It turns out that it is easier to build a NOR function or a NAND function on an IC (depending on the IC technology used) than the OR or AND function. In order to build a simple AND, the IC designer would have to build a NAND and then follow it with a NOT, for example.

It is possible to build all other functions from either the NOT and AND functions or NOT and OR functions alone. As one example, to construct AND from NOT and OR, use this boolean expression:

A AND B = NOT [(NOT A) OR (NOT B)]

The truth table proves that the boolean expressions on both sides of the equal sign are equivalent. The truth table takes the two inputs A and B and shows the effect of successive boolean operations on them as the table goes from left to right. The results of one column are used as the input for the operation performed by the next column:

A	B	A AND B	NOT A	NOT B	[(NOT A) OR (NOT B)]	[NOT (NOT A) OR (NOT B)]
0	0	0	1	1	1	0
0	1	0	1	0	1	0
1	0	0	0	1	1	0
1	1	1	0	0	0	1

For any row (and therefore value of A and B) the result under the heading A AND B is the same as under the heading NOT [(NOT A) OR (NOT B)]. These two expressions produce identical results for all four input possibilities and are therefore equivalent. This means that the IC can be designed to use the same circuit pattern over and over, only with some different interconnections between the functions. This makes for an efficient use of the designer's time and the IC wafer layout.

The EXCLUSIVE OR (XOR) Function

Another important boolean function is the XOR function. The XOR function says: "If *either* input is a 1, but not both, then the output is a 1." If both inputs are a 1 or both are a 0, then the output is a 0. The symbol for XOR is a modified symbol

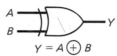

$$Y = A \oplus B$$

Fig. 2-14 The EXCLUSIVE-OR function is represented by the symbol shown, which is similar to the symbol for the OR function.

for OR (Fig. 2-14). The mathematical symbol is a + written with a circle around it: $\oplus$. The truth table is:

A	B	A XOR B
0	0	0
0	1	1
1	0	1
1	1	0

A practical example of the XOR function is a light controlled by two switches (usually one at each end of the stairs or hallway) (Fig. 2-15). If either switch is up,

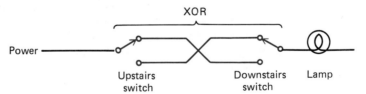

Fig. 2-15 The XOR function is equivalent to an "upstairs-downstairs" light switch, which allows the light to be turned on or off from either end of the stairs.

the lamp is on. If both are up, or both are down, then the lamp is off. This is a case of a boolean logic function being physically implemented by simple switches. You can think of the XOR function as a comparison function. If both inputs are the same, the output is a 0. If the inputs are different, the output is a 1. This makes the XOR function useful for comparing binary signals. The XOR function can be constructed in several different ways out of the AND, OR, and NOT functions we have already discussed, shown in Fig. 2-16*a*, *b*.

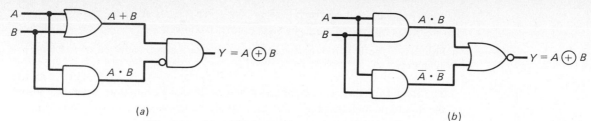

(a)   (b)

Fig. 2-16 The XOR function can be built from AND, OR, and NOT functions in several ways. Two possibilities are shown.

By developing the truth table for the function of Fig. 2-16b, for example, it is shown to agree with the definition of XOR. This truth table involves the two inputs A and B, as well as the boolean logic results of the intermediate points.

A	B	A AND B	NOT A	NOT B	[(NOT A) AND (NOT B)]	(A AND B) OR [(NOT A) AND (NOT B)]	NOT of Previous Column
0	0	0	1	1	1	1	0
0	1	0	1	0	0	0	1
1	0	0	0	1	0	0	1
1	1	1	0	0	0	1	0

The Inhibit (Select or Enable) Function

The inhibit function is not a new boolean function. It makes use of the existing boolean functions to provide a capability in boolean-based systems that is very necessary for practical operation. In a real circuit of ICs which provides boolean functions, the input signals A and B do not necessarily arrive at an IC at the same instant. As a result, the boolean output might momentarily have an incorrect value until both inputs arrive. To prevent this problem from causing misleading boolean outputs, a special input is provided, called the Inhibit input.

The Inhibit input forces the boolean output to a low state, as long as the Inhibit input is high. As soon as the Inhibit input goes low, it allows the other boolean inputs to be acted on by the boolean function, as if the Inhibit line was *not there at all*. The truth table for the Inhibit function on the AND function (Fig. 2-17) shows this effect:

A	B	Inhibit	Y	
0	0	0	0	
0	1	0	0	
1	0	0	0	
1	1	0	1	AND plus
0	0	1	0	Inhibit
0	1	1	0	
1	0	1	0	
1	1	1	0	

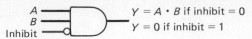

$Y = A \cdot B$ if inhibit = 0
$Y = 0$ if inhibit = 1

Fig. 2-17 The INHIBIT function shown acts as a master control on the function it is inhibiting.

The Inhibit function is sometimes known as the Select function, because it allows the input line to determine if this gate should be active or in an idle state. It is also known as the Enable line or Strobe for the same reason.

QUESTIONS FOR SECTION 2-5

1. What is the NOR function? What other boolean functions is it composed of?

2. Repeat for the NAND function.

3. Why are the NAND and NOR functions so important?

4. What is the XOR function? What symbol is used for it in expressions?

5. Give a practical example of a (a) NOR function, (b) NAND function, (c) XOR function.

6. What is the Inhibit function? Why is it important in practical circuits? What does it do?

7. What are some other names for the Inhibit function?

PROBLEMS FOR SECTION 2-5

1. Draw the truth table for the Inhibit function on an OR function. How many input conditions are there? How many output values equal 0? How many equal 1?

2. Write the expression that the functions of Fig. 2-16*a* implement. Show that this is an XOR function by using a truth table.

3. What boolean expression results when a NAND function output goes to a NOT function?

4. What function results when the inputs to a two-input NAND first go through NOTs? Show this using a diagram and a truth table.

5. What expression is implemented by the logic functions of Fig. 2-18?

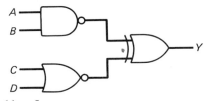

Fig. 2-18 Section 5, Problem 5.

6. Draw the logic functions which implement the expression (A NAND B) NOR (NOT C) = Y.

2-6   BINARY NUMBERS

In our study of the boolean functions, there was no relationship between the various inputs. The inputs came from different sources and were acted on by the boolean logic function. However, binary *numbers* are used to represent the numbers that the computer system must deal with. Binary format is used for bringing the numbers from one part of the computer system to another, for doing calculations, and for presenting the numbers to the final circuit that in turn presents them to the user on some sort of display. How is this done?

We are familiar with the decimal system which we use in everyday applications involving numbers. The decimal system uses the base of 10. This means that if you look at the digits of a number, starting at the right, each digit has value 10 times the preceding digit. The rightmost digit is the number of ones, the next position is the number of tens, then the number of hundreds, then thousands, and so on. The decimal number 759 really means:

$$9 + (5 \times 10) + (7 \times 100)$$

Mathematically we say that each position represents successive powers of 10: $10^0 = 1$ for the rightmost position, $10^1 = 10$ for the next position, $10^2 = 100$ follows, $10^3 = 1000$ next, and so on.

The binary system has a base of 2, since there are just the two values of 0 and 1 allowed. The value of each column is given by the powers of 2: $2^0 = 1$ for the rightmost, $2^1 = 2$, $2^2 = 4$, $2^3 = 8$, and so on. Each binary position is called a *bit*, short for *bi*nary digi*t*. A typical 4-bit binary number might be 1101, which represents $1 \times 1 + 0 \times 2 + 1 \times 4 + 1 \times 8$, or 13 (if you have to give its decimal value). The bit on the right is called the *least*-significant bit (LSB), the bit on the left is the *most*-significant bit (MSB).

As with decimal numbers, larger numbers can be represented by using more positions in binary format. A 2-bit number can have values of 00, 01, 10, up to a maximum value of 11, and so can be equal to 0, 1, 2, or 3 in decimal. Similarly, a 3-bit number can have values of 000, 001, 010, 011, 100, 101, 110, and 111, and so can show any decimal value from 0 through 7. A group of 8 bits is called a *byte*. Binary numbers of 1 byte (11111111 = 255 in decimal) and 2 bytes (1111111111111111 = 65,535 in decimal) are most common in computer systems. *Note:* To make the binary numbers easier to read, the bits are often put in groups of 4. The space has no numerical meaning. Thus, decimal 255 would be written as 1111 1111, and decimal 65,535 would be 1111 1111 1111 1111.

QUESTIONS FOR SECTION 2-6

1. What value does each position represent in the decimal number system?

2. What does the decimal number 7236 represent?

3. What value does each position represent in the binary number system?

4. What does the binary number 1011 0101 represent?

5. What is the meaning of most-significant bit? Least-significant bit?

6. What is a byte?

7. How are larger binary numbers usually written to make them more readable?

2-7 CONVERTING BINARY TO DECIMAL AND DECIMAL TO BINARY

There are many methods that can be used to convert from binary (base 2) to decimal (base 10), and to do the reverse.

To convert from binary to decimal, add up the weights of each position in the binary number where there is a 1. The easiest way is to mark the weights 1, 2, 4, 8, 16, 32, and so on, above each column of the binary number, then add up those column values with a 1 under them. Columns which have a 0 have no value in the number. Examples follow:

128	64	32	16	8	4	2	1		
1	0	1	1	0	1	0	1	= 128 + 32 + 16 + 4 + 1 =	181 decimal
0	1	0	1	1	1	0	0	= 64 + 16 + 8 + 4 =	92 decimal

To convert from decimal to binary, use the value of each successive binary column and start dividing them into the decimal number. If the decimal value is between the value of two columns, put a 1 in the binary position of the smaller one, and then subtract that value. Then go on to the next binary position value using the remainder.

EXAMPLES

Convert 47 (decimal) to binary.

Both 128 and 64 are greater than 47, so a 0 goes in the 128 column and the 64 column. However, 32 is less than 47, so a 1 goes in the 32 column, and 32 is subtracted from 47. The difference, 15, is then compared to 16, the next column value. Since 15 is less than 16, a 0 goes in the 16 column. But 15 is greater than 8, the value of the next binary column, so a 1 is placed in that column, and 8 is subtracted from 15. The difference is 7, which is greater than 4, and results in a 1 in the 4 column. Then take the difference again, $7 - 4 = 3$, and compare 3 to the next column, which has value 2. Since 3 is greater than 2, that column gets a 1, and again the difference is taken ($3 - 2 = 1$). Compare this to the last column, with column value 1. Since the remainder is 1, the last column gets a 1. Since there is no remainder, the calculation is complete. As a check, add up the values of the columns which have a 1: $32 + 8 + 4 + 2 + 1 = 47$.

Convert 105 (decimal) to binary:

	128	**64**	**32**	**16**	**8**	**4**	**2**	**1**
105	0							
− 64		1						
41			1					
− 32				0				
9					1			
− 8						0		
1							0	
− 1								1
0	**0**	**1**	**1**	**0**	**1**	**0**	**0**	**1**

QUESTIONS FOR SECTION 2-7

1. How is a binary number converted to decimal?

2. How is a decimal number converted to binary?

PROBLEMS FOR SECTION 2-7

1. Convert these decimal numbers to binary: 27, 45, 100, 247.

2. Convert these decimal numbers to binary: 87, 12, 18, 19, 75, 76, 112, 254, 256.

3. Convert these binary numbers to decimal: 0110, 1 1001, 1100, 1101.

4. Convert these binary numbers to decimal: 1 1010, 10 1100, 100 1001.

5. What is largest decimal number that can be represented by a binary number of 5 bits?

6. Repeat problem 5 for the case of 6 bits.

2-8 BINARY-CODED DECIMAL

The binary format talked about until now is "natural" or "straight" binary. It uses base 2 and has as many column positions as required to represent the number. But we live in a decimal world, and there are times when a different binary format is needed to allow easy interface between decimal numbers and binary. This is the case where users type numbers on a calculator keyboard, or expect to see numbers on a digital readout such as on an electronic scale. For these applications, a special form of binary, called binary-coded decimal (BCD), is often used.

In BCD format, each digit of the decimal number is converted to binary format, without regard for its position in the decimal number (as representing ones, tens, hundreds, etc.). Since the value of each digit can be from 0 through 9, we need 4 bits to represent each digit (0000 through 1001). The decimal number 759 would be represented in BCD as:

$$
\frac{7}{0111} \quad \frac{5}{0101} \quad \frac{9}{1001}
$$

Note that it takes more binary positions to represent a number in BCD than in straight binary. (Demonstrate this by converting 759 to natural binary.) Nevertheless, because BCD works well with people-oriented interfaces such as keyboards, digital readouts, and switches, it is used quite often. There are special ICs available which use the boolean functions to perform BCD-to-decimal and decimal-to-BCD conversions.

Other Important Number Formats

We have seen that binary numbers can be fairly long when they represent large values. Copying and discussing a 1-byte (8-bit) or 2-byte (16-bit) string of 1's and 0's can cause confusion and errors. There is a numbering system called *hexadecimal (hex) format* that is a form of shorthand used to describe these binary numbers. In hex format, the bits are formed into groups of 4. Each group of 4 bits is then converted to a hex number. Hex number means base 16, so the values of hex numbers are 0, 1, 2, 3, 4, 5, 6, 7, 8, 9, A, B, C, D, E, and F, where A through F correspond to 10, 11, 12, 13, 14, and 15, respectively. A group of 4 bits can then be referred to by its hex equivalent, as shown:

Binary	1011	0011	0101	1100
Hexadecimal	B	3	5	C

Hex format is very convenient when discussing binary numbers because conversion from one format to the other is relatively easy. For example, a 16-bit binary number is converted as follows:

Binary number	0101110100111110			
Put in groups of 4 bits	0101	1101	0011	1110
Take hex equivalent of each group	5	D	3	E

Certainly, 5D3E is easier to refer to than the original binary number. Some calculators can even handle hex numbers directly and perform mathematical and boolean operations on them.

Another common format is octal, or base 8. Like hex, octal format makes using binary numbers much more convenient. In octal, each binary number is formed into 3-bit groups. The 3-bit groups are then given equivalent values: 000 is 0, 001 is 1, and so on up to 111, which is equal to 7. As an example, convert binary 0 1110 1010 to octal:

Put in groups of 3 bits	011	101	010
Take octal equivalent of each group	3	5	2

Hex format is most commonly used in systems that use 1 or 2 bytes to represent numbers. Octal format is often used in 12-bit systems.

QUESTIONS FOR SECTION 2-8

1. What is the technical name for the type of binary numbers studied thus far?

2. What is BCD? Why is it sometimes used?

3. How does BCD represent a decimal number with 1's and 0's?

4. What is hex format? Why is it used?

5. How does hex format represent a binary number?

6. What is octal format? Why is it used?

7. What is used to represent the numbers in hex format?

PROBLEMS FOR SECTION 2-8

1. Convert these BCD numbers to their decimal equivalent: 0110 1001 0011, 0111 1001 0011, 0101 1000 0110.

2. Repeat problem 1 for 1001 1000 1000, 0001 0000 1000, 0110 1001 0011.

3. Convert these decimal numbers to BCD: 27, 45, 100, 247.

4. Repeat problem 3 for 38, 67, 200, 234.

5. Convert these hex numbers to binary: 3, 7, 9, A, D, E, F.

6. Repeat problem 5 for 6, 8, B, C.

7. Convert these binary numbers to hex: 0110, 1111, 0011, 1001.

8. Repeat problem 7 for 0000, 0010, 0101, 1011, 1110.

9. Convert these hex numbers to binary: 123, A32F.

10. Repeat problem 9 for B7E0, C3F0.

11. Convert these binary numbers to hex: 0111 0011, 1101 0010.

12. Repeat problem 11 for 1001 0001, 1010 1100.

2-9 AND, OR, NOT GATES

It is possible to implement the boolean functions physically using electromechanical devices such as relays. However, as the circuit becomes complicated, relays begin to take up a lot of space and power. They are also less reliable because of moving parts and contacts. The alternative is the *gate*.

Gates are ICs which actually perform the boolean functions that we have studied. They are available for the fundamental boolean logic functions of AND, OR, NOT, NAND, NOR, and XOR, as well as for many variations and combinations of these functions.

ICs with only a few gates, less than about 12, are called *small-scale integration,* or SSI, while those with between 12 and 100 gates are called *medium-scale integration,* or MSI. Gates implement the boolean logic functions, using one voltage value to represent a 1 and another voltage value to represent a 0.

Generally, SSI and MSI ICs come packaged in DIPs with either 14 or 16 pins. Two of the pins are required for power and ground. This leaves 12 or 14 pins for the boolean logic inputs and outputs, so the IC can contain more than a single boolean function, or a boolean function with many inputs. Some typical SSI ICs include:

Four 2-input NAND gates (Fig. 2-19)
Four 2-input NOR gates (Fig. 2-20)
Six NOT gates (Fig. 2-21)
Three 3-input NAND gates (Fig. 2-22)

These will be discussed in more detail in the chapter on common IC families.

Gates can be connected together so that the output of one becomes the input of one or more other gates. This allows the user to create more complicated boolean logic functions. When gates are connected together this way, it is called *combinational logic*.

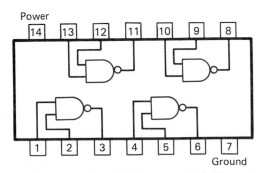

Fig. 2-19 This 14-pin IC contains four NAND gates, each with two inputs. The entire IC shares the same power and ground pins.

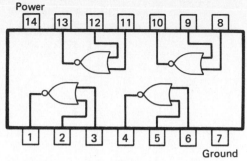

Fig. 2-20 A 14-pin IC with four 2-input NOR gates.

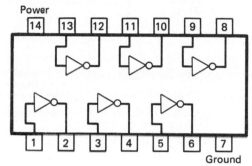

Fig. 2-21 Six NOT gates can fit into a single 14-pin IC.

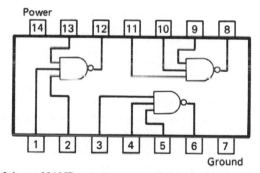

Fig. 2-22 Three 3-input NAND gates can use all the pins of the IC.

Gates which perform boolean logic do not act instantaneously. It takes some time for the input signal to flow through the gate circuit and the output to appear. This propagation delay time is very short, about 10 billionths of a second (1 billionth of a second = 1 ns). However, when many gates are connected to each other in a long chain, these delays can add up. Also, the same signal may reach different parts of a circuit at different times (Fig. 2-23), and so may cause false results in parts of the system. This is another situation where the enable function is used. The enable is used to synchronize the various parts of the system built up of a large number of ICs. In this way, the circuit only looks at the inputs after they

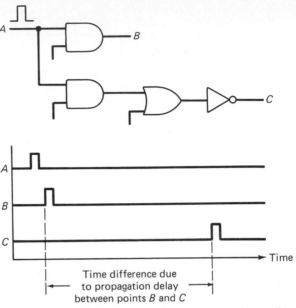

Fig. 2-23 Propagation delays cause signals to arrive at slightly different times in different parts of the system circuitry. An inhibit function can act as an enable to synchronize all the gates to the same timing.

have propagated and settled down. Synchronized ICs are very important as the internal design of the IC gets more complicated, and many synchronous IC types are available.

Troubleshooting Gate ICs

When there are problems with circuits that use digital logic gates, it is usually not hard to isolate the problem. This is because a gate takes the signals at its input, performs the boolean operation, and provides a corresponding output. In many cases the signals at the inputs of the gate are static (not changing) and this makes troubleshooting even easier.

The first thing to do is check the power supply values at the IC itself, using a voltmeter between the IC power and ground pins. This supply voltage should be within the specified value for the IC type and should not be varying. (If it is changing, the IC will work erratically and produce results that make sense sometimes and make no sense at other times.) If the voltage is OK, check the inputs to the gate and then the output. The output should be the correct value based on the boolean operation that the gate performs. As with the power supply, *correct value* means that the voltages which represent the 1 or 0 logic level should be within the specification for that type of IC.

If the inputs are OK, but the outputs are not, there are several possible reasons:

- The input circuit within the IC could be bad. It could be receiving signals properly but not passing them on further into the IC.
- The internal logic circuit within the IC is bad.
- The output circuit of the IC, which drives the next gate, may also be bad.

For any of these cases, the IC must be replaced. The problem could also be:

- The circuit wire or track (in the case of a printed circuit board) coming from the output of the IC has a short circuit of some kind, which is either forcing the output to a steady high or low (1 or 0) or causing the output to follow another signal line instead of allowing the correct output of the gate to occur.
- The next gate in the circuit has a defective input which is holding the output to some value.

For these two cases, the cause of the problem is not in the IC itself. It is further along in the circuit and the problem must be found and corrected for the IC to work properly.

Of course, most real circuits are not this simple. In many circuits, there are several things that will make troubleshooting more difficult. These are:

- Most circuits have "feedback" where the outputs of some ICs go back to the beginning of the signal flow and become the inputs of previous ICs. This means that figuring out the logical flow of signals (where to start) is difficult.
- Most circuits have a combination of the simple gates and the more complicated ICs we will discuss later. These complicated ICs are difficult to troubleshoot.
- Many circuits operate dynamically, which means that all the signal lines are constantly changing state. The circuit may have no provision for stopping the signals so that the gates can be checked statically. In this case, an oscilloscope or other special instruments may be needed. These instruments allow the service person to display many signal lines at the same time and even "freeze" the picture of the state of these lines at any instant.

Always remember that a gate "knows" only what logic values are on *its* inputs, and therefore should *always* produce the right boolean output. The gate does not know or care about the rest of the circuit. Troubleshooting gates in more complicated circuits is a problem because the system is operating at high speeds and special instruments such as logic probes, logic analyzers, and multiple-channel oscilloscopes may be needed to see the operation. If you can see all the inputs and outputs at the same time, using whatever tool is correct for the type of system, then you can troubleshoot gates.

Most digital systems today use ICs that are more complicated than these simple gates. Nevertheless, the simple gates are still very useful because they tie together the signals from these larger ICs. A typical circuit board will have a mix of larger ICs and some smaller ones. Many times the SSI ICs are referred to as "glue chips" because they bring together important signals and combine them in some way that is needed for the next IC or storage in the system.

QUESTIONS FOR SECTION 2-9

1. What are gates?

2. Why are they superior to electromechanical relays?

3. What is an SSI IC? What is an MSI IC?

4. How do gates represent the two values of boolean logic?

5. What is propagation delay? How can it cause circuit and system problems?

6. What is done to minimize the problems caused by propagation delay?

7. What is the first thing to check when trying to find out why a circuit built of gates does not work properly?

8. What are three of the causes of problems, after the first checks have been done?

9. What are the complications in finding problems with most practical circuits?

2-10 TIMING DIAGRAMS

Timing diagrams are often used to show the changes of logic level on a signal line, and the changes on one signal line with respect to another. In a timing diagram, the horizontal axis shows time, running from left to right. The vertical axis shows the logic state of one or more signals versus time. The timing diagram is needed when studying the activity of various signal points in the system and is often used in troubleshooting to understand which changes in signals will cause changes in the state of other signals.

A timing diagram for a single AND gate is shown in Fig. 2-24a. In this case, the

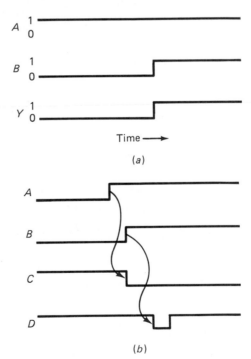

(a)

(b)

Fig. 2-24 (a) Timing diagram for an AND gate with A = 1 and B = 0; then with B changing to 1.

(b) A more complicated timing diagram often uses arrows to indicate which signals can directly cause the logic states of other signals to change.

timing diagram shows three signals: input A, input B, and output Y, and also shows that A = 1, B = 0, so that Y = 0. When B changes to 1, the output Y becomes 1.

Most timing diagrams show the change in a signal from one logic level to the other as a perfectly vertical line, which implies that the signal changed state instantaneously. Of course, this is not the case. It takes a very small but definite amount of time for the signal to change state. For most ICs, this time is anywhere from 5 to 100 ns. The diagram does not show this rise time (time to go from logic 0 to logic 1) to this fall time (time to go from 1 to 0) because that information is not the main use of the timing diagram. However, some timing diagrams do show it if the rise and fall times are critical.

Many timing diagrams have small arrows pointing from a change of logic state on one signal line to a change in another line. This is used to indicate cause and effect in a circuit between one point and another. Figure 2-24b shows this for the case of a complicated IC which generates an output = 1 a long time after a key input line goes to 0. Just looking at the timing diagram, which shows many lines in the system, might not immediately make this clear. However, the arrow shows the relationship clearly.

In many real systems, some of the signal lines are irrelevant at some points in the operation. They may be either high or low, and either it is unpredictable which state the line is in, or of no interest. This is often shown on the timing diagram by a cross-hatched area, called a "don't care" condition.

QUESTIONS FOR SECTION 2-10

1. What is a timing diagram? Where is it needed?

2. What is rise time? Fall time?

3. Why are arrows sometimes used on timing diagrams?

SUMMARY

The concept of representing signals by a group of binary digital signals uses the theories of boolean algebra developed over 100 years ago to provide some simple functions. These boolean functions are the building blocks for very complicated and powerful systems. They are very suitable for use with ICs because the same design can be used over and over to build ICs with many gates.

The common decimal number is just one way of representing the value of a signal or quantity. Natural binary, BCD, octal, and hex are some other formats used. Regardless of the format, the boolean logic operations are still used to provide a system with the required functions. Troubleshooting these systems requires an understanding of the function of a single gate as well as the function of groups of gates that are connected together. Some traditional analog techniques are still used, but many of them are not applicable and new approaches are required for digital logic.

REVIEW QUESTIONS

1. Explain why a digital system contains only a specific number of members, while an analog group contains an infinite number of members.

2. Why are digital systems often favored over analog ones?

3. Explain why the price values of items in a store is a digital value, why the reading of a mercury thermometer is analog.

4. How are binary numbers represented?

5. What is noise in a system? Why is it a problem?

6. Explain three advantages of digital circuits as compared to analog circuits.

7. What special algebra is used to analyze binary systems? What is special about it?

8. What are the three fundamental functions of binary logic? What are their rules?

9. What is a truth table? Why is it needed for binary functions?

10. What are the symbols for the three fundamental functions in an expression? In a figure?

11. What boolean functions can be used to develop all other boolean functions?

12. What boolean function does a one-input NAND provide? A one-input NOR?

13. What are boolean identities? Why are they important?

14. Give a boolean identity for the AND, OR, and NOT function.

15. Define the XOR function and give its truth table.

16. Why is the inhibit function needed? What are some other names for it?

17. What numbers are used to represent the values in a given column position in the decimal system? In the binary system?

18. What is the value of each column for the four rightmost columns in the decimal system? In the binary system?

19. For a binary representation of decimal numbers, what is the value of the LSB when the decimal number is odd?

20. Explain the term *byte*.

21. Explain the idea behind decimal to BCD conversion.

22. Repeat question 21 for the reverse conversion.

23. What is BCD format? How is a decimal number converted to BCD? How is a BCD number converted to decimal?

24. What is hex format? How is a binary number converted to hex format? How is a hex number converted to binary format?

25. What symbols are used to represent numbers in hex format? Why are they needed?

26. Which takes more 1's and 0's to represent a decimal number, binary or BCD? Why?

27. What does the term *gate* mean?

28. What term is used to describe the time it takes for a gate to perform the boolean operation on the inputs and provide an output?

29. What must always be checked first in a system that is apparently having problems?

30. What is the usefulness of a timing diagram?

31. What is the difference between propagation delay and the rise and fall times?

32. What are don't care signals? How are they shown?

REVIEW PROBLEMS

1. Show a boolean function that implements the statement: "If I go and he goes, or she goes, then we will have enough people."

2. Draw logic symbols for the boolean expression of the previous problem.

3. Develop the truth table for the expression of problem 1.

4. Implement the following with logic gates: (A AND B) OR [C AND (NOT A)].

5. Implement the following with logic gates: (A XOR B) AND (NOT A).

6. Prove by using a truth table that A OR B is the same as NOT [(NOT A) AND (NOT B)].

7. Draw the expression of problem 6 with logic gates.

8. Prove by using a truth table that (NOT A) AND (NOT B) AND (NOT C) is the same as NOT (A OR B OR C).

9. Draw the expressions of the previous problem with logic gates.

10. Prove that (NOT A) OR (NOT B) OR (NOT C) is the same as NOT (A AND B AND C).

11. Use gates to implement the expression of problem 10.

12. Convert these decimal numbers to binary: 22, 44, 63, 64, 65.

13. Convert the same numbers to BCD.

14. Convert these binary bytes to decimal: 1111 0000, 0000 1111, 1010 1010, 0101 0101.

15. Convert these hex numbers to binary: FA, 83, A3C, 3456.

16. Convert the binary values of problem 14 to hex.

17. For an OR gate, draw the timing diagram for the condition where both inputs are 0, then one of the inputs suddenly becomes a 1.

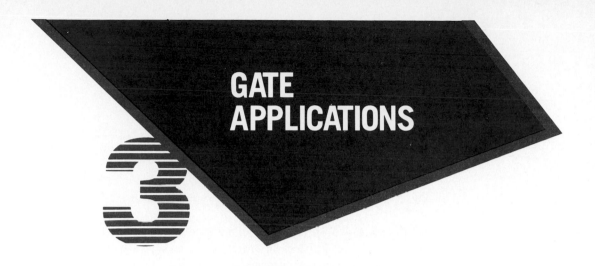

GATE APPLICATIONS

INTRODUCTION

The role of basic gates and logic functions is expanded in this chapter to build circuit elements which can perform additional important functions. These elements, called *flip-flops,* can be used to retain information after a signal has passed or changed, to delay a signal, and to change a logic level each time a new input is received. These capabilities are required when a digital system and computer need to store binary information or interconnect to other devices.

The versatility and flexibility of flip-flops is further enhanced as they are connected together in groups. Depending on how they are combined, flip-flops can form counters, which keep count of the total number of pulses received; and registers, which store multiple bits of information. The bits in registers can be retrieved all at one time either in the order received or in reverse order.

3-1   FLIP-FLOPS

All the gates and combinational logic studied to this point develop a boolean logic output based on the various inputs at each instant of time. There is no past history or memory in the circuit formed by these ICs. However, in many systems it is necessary to store the state of external lines or of gate outputs. For example, it may be necessary to know in which order several external switches were pushed, and therefore there is a need to store the switch push inputs so that they can be processed by the gates and circuit. The idea of *memory* is an important part of nearly all practical systems that use digital ICs. An IC that has memory and is built out of basic gates would be a very powerful building block for a more useful system.

A circuit called the flip-flop is a simple combination of gates that acts like a 1-bit memory element. It is sometimes called a 1-bit latch, because the state of a bit can be locked, or latched, into it. The flip-flop is a cross connection of two

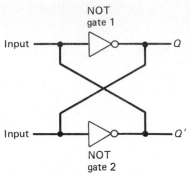

Fig. 3-1 A simple flip-flop is built from two NOT gates that are cross-connected.

NOT gates (Fig. 3-1). The output of one gate is the input of the other. The outputs are called Q and Q', since in operation these outputs will be in opposite logic states. They will be complements to each other.

If the input of NOT gate 1 is a 0, then the output of gate 1 is a 1, and this output is also the input to NOT gate 2. If gate 2 has an input of 1, its output is a 0, and this is the input of gate 1. This agrees with the original assumption that the input of gate 1 is a 0. This simple cross-connected pair of gates has really two states: one in which $Q = 0$ and $Q' = 1$, and the other in which $Q = 1$ and $Q' = 0$. [In practical flip-flops the gate is usually a one-input NAND (Fig. 3-2), instead of a NOT; they both have the same boolean truth table.]

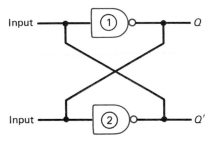

Fig. 3-2 A flip-flop can also be made from NAND gates.

Most flip-flops have some added circuitry which makes the flip-flop more useful and practical. Suppose we make the cross-coupled pair of gates out of two-input NAND gates and feed their inputs from the output of two NOT gates (made from two 1-input NANDs), with inputs S and R (Fig. 3-3). To store a 1 as output Q, make input $S = 1$ and input $R = 0$. The S input is called the *Set*, or *Preset* input, because by setting it to 1 it can set the Q output to 1. To store a 0 as output Q, make input $S = 0$ and input $R = 1$. The R input is called the *Reset* or *Clear* input because it can clear output Q to 0 by placing a 1 at the R input.

This type of very simple flip-flop serves a useful purpose, but for real circuits it does not have all the features that are needed for smooth system operation. As ICs and their systems get larger, there is need to synchronize all the various parts of the IC and system, that is, to have everything occur simultaneously, as it does in a marching band. *Synchronize* means that all gates look at their input signals lines at

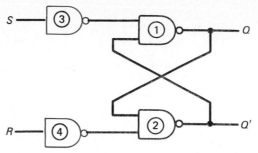

Fig. 3-3 The *S-R* flip-flop has an additional pair of NAND gates and allows the initial state of the outputs to be set.

the same time, no matter where the gates are in the circuit. Otherwise, signals may be in the process of changing from a 1 to 0 or 0 to 1 at slightly different times and there may be false or incorrect outputs in the circuit for a brief moment. This could cause malfunctions which would spread throughout the system. In large systems, the same signal from a gate output may arrive at other gates at different times because of the physical distance that the signal must travel and the time it takes to get through various other gates (propagation delay).

Circuits and systems which are synchronized are called *sequential* and are much more common than the combinational circuits studied. In a sequential system a clock oscillator puts out a continuous string of pulses that go to all parts of the circuit (Fig. 3-4). The pulses are like drumbeats and keep all ICs of the circuit in step. Between pulses the ICs are not active and no new activity or logic changes occur. Instead, the time between pulses is used to let the signals in the circuit reach their next point and settle down. It will next be shown how these clocked signals are used with flip-flops.

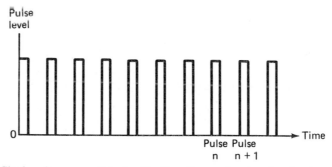

Fig. 3-4 Clock pulses are sent to the flip-flops to synchronize the circuits.

The simple flip-flop with the *S* and *R* input (Fig. 3-4) serves as the starting point. Taking the two input gates 3 and 4, make them into two-input NAND gates. The new inputs are tied together and connected to the clock (Fig. 3-5). When the Clock input is 0, the outputs of gates 3 and 4 are 1 and the outputs of gates 1 and 2 do not change. But, when the Clock input is 1, gates 3 and 4 are enabled, and the input signals on *S* and *R* travel right through gates 3 and 4 to gates 1 and 2, then get locked in when the Clock goes back to 0. Therefore, when Clock = 0, the flip-

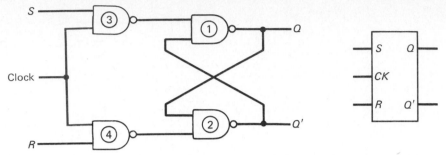

Fig. 3-5 A clocked *S-R* flip-flop reacts only when the clock pulse is present. It has a special logic symbol.

flop is just like the simple *S-R* seen previously. The Clock line provides a means to synchronize this flip-flop with others in the system.

It turns out there is one problem with this simple *S-R* flip-flop with Clock input. What happens when $S = R = 1$ and Clock = 1? Then, both the input that is passed onto gate 1 and gate 2 is 0, and the output Q of gate 1 and the output Q' of gate 2 are 0. But as seen before, the cross-coupling of the output of gate 1 and the input of gate 2 (and the same for gate 2 output and gate 1 input) means that Q is always the opposite of Q'. In other words, this flip-flop has an ambiguous output condition when $S = R = 1$ (sometimes called an undefined or indeterminate condition) and there must be a way to prevent the condition $S = R = 1$ from happening to the flip-flop. This leads to the first of the really practical flip-flops, the *J-K* type.

QUESTIONS FOR SECTION 3-1

1. What important feature is not provided by gates as studied so far?

2. What is a flip-flop? What feature does it provide?

3. How is the simplest flip-flop constructed? What is the relationship between Q and Q'?

4. Explain the operation of the *S-R* flip-flop.

5. What is the Preset input? The Clear input?

6. What is synchronization? Why is it needed?

7. What is the system Clock function?

8. How is the *S-R* flip-flop made into a synchronized flip-flop?

9. What is the problem with the *S-R* flip-flop?

PROBLEMS FOR SECTION 3-1

1. Draw an *S-R* flip-flop made of two-input NOR gates.

2. Show the truth table for the clocked *S-R* flip-flop.

In the *J-K* flip-flop, the *S-R* flip-flop has some additional input AND gates added (Fig. 3-6). The external inputs are called *J* and *K*, and the outputs of the flip-flop are fed back to the added inputs. Output *Q* is run to the AND gate that also has input *K*, while output *Q'* goes to the AND gate that has input *J*. The truth table next to the figure of the *J-K* flip-flop shows the state of the output Q_{n+1} for input J_n and K_n. The *n* refers to Clock pulse *n*, while *n* + 1 means the time *after* Clock pulse *n* has passed. The truth table shows that the previously ambiguous case for $S = R = 1$ is now defined for the equivalent case of $J = K = 1$. The output after the Clock pulse Q_{n+1}, is equal to Q_n', the complement of the output Q_n. To prove this, construct a truth table for each point of the *J-K* flip-flop. Since there are two inputs there are four input possibilities, and the Clock is a third input, so there are a total of 4 × 2 or 8 input combinations. (Three inputs means $2^3 = 8$ combinations.) The *J-K* flip-flop has exactly the same output as the *S-R* flip-flop for the first three *S-R* (or *J-K*) input cases, but the last input case is now clearly and fully defined.

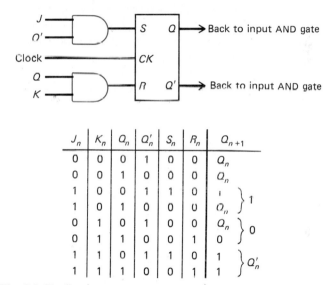

J_n	K_n	Q_n	Q_n'	S_n	R_n	Q_{n+1}	
0	0	0	1	0	0	Q_n	
0	0	1	0	0	0	Q_n	
1	0	0	1	1	0	1	} 1
1	0	1	0	0	0	Q_n	
0	1	0	1	0	0	Q_n	} 0
0	1	1	0	0	1	0	
1	1	0	1	1	0	1	} Q_n'
1	1	1	0	0	1	1	

Fig. 3-6   The *S-R* flip-flop is converted to a *J-K* flip-flop with additional AND gates. The truth table shows the effect of this.

In a practical IC that performs the *J-K* function, the new input AND gates are combined with the input NAND gate, using a three-input NAND gate. This is easier to build onto the IC and saves space on the IC silicon (Fig. 3-7). The three-input NAND gate version is shown in the figure. The figure also shows two other inputs on the NAND gates that drive the *Q* and *Q'* outputs, called Preset and Clear. These inputs are used to directly set the output state of the *J-K* flip-flop, so that the gate can be "initialized" to either $Q = 1$ or $Q = 0$. In most practical systems, it is important to have some way of setting everything up to a known state when power comes on or when a major reset of the system is required, such as

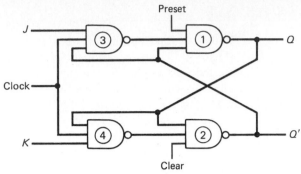

Fig. 3-7 The Preset and Clear allow the output state to be set externally and asynchronously.

when the system is being serviced. As seen in the *J-K* flip-flop truth table in Fig. 3-6, the output at one time depends on the value of the output at the previous time. This means that on power-up it is impossible to say which way the outputs will come up. But the Preset and Clear inputs give the circuit designer a way of forcing the outputs to the desired state. For Q to be 1 (called *presetting*), all that is needed is Preset = 0, Clear = 1, and Clock = 0. For the Q output to be 0 (called *clearing*), it is required that Clear = 0, Preset = 1, and Clock = 0. It is important to note that the Preset and Clear inputs do not require any synchronization with the clock. They go directly to the flip-flop, bypassing the input stage, which has the clock. This direct type of input is also called *asynchronous* because it is not related to the synchronization provided by the clock and can be performed at any time. Once the flip-flop has been initialized by the Present and Clear inputs, they are both set equal to 1 to enable the flip-flop for normal operation.

3-3 APPLICATIONS OF *J-K* FLIP-FLOPS

The *J-K* flip-flop is a very versatile circuit element. Many ICs are available which contain one or more *J-K* flip-flops. It may seem that there is very little use for a 1-bit memory element, since most real systems use thousands of bits as part of their overall memory. There are cases, however, where 1 bit is all that is needed at the interface between two circuits or parts of a system.

Consider a case where a computer system needs to know that a pushbutton switch has been used. The person pushing the switch may hold it down anywhere from a very short time of 20 ms to several seconds. The computer system needs to know that the switch has been pushed, but cannot afford to spend all of its time watching the input line from the switch. The timing of the activities of the computer side of the system is not related to the timing of when the person pushes the switch, so the computer cannot just look at the instants when the switch *might* be pushed. The solution to this problem is the *J-K* flip-flop.

The flip-flop acts as the interface between the pushbutton part of the system and the computer part of the system (Fig. 3-8). The *J* input is wired to a logic 1, and the *K* input is wired to a logic 0. The flip-flop output Q is connected to an input of the computer. The computer also has one of its outputs connected to the Clear

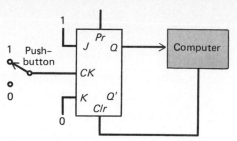

Fig. 3-8 The *J-K* flip-flop can be used to perform handshaking between a computer and a user pushbutton.

input on the flip-flop. Most important, the pushbutton is connected to the Clock input, since the human interface side of the system has no real clock except the action of the pushbutton. This simple connection solves the problem, as follows.

When the system powers up, the computer sends out a pulse to the Clear input, and thus sets the flip-flop to a known state with $Q = 0$. Periodically, the computer checks the Q output to see if the output has changed from a 0 to a 1. What would cause this to happen? When the pushbutton connected to the Clock input is pressed and goes from 0 to 1, a Clock pulse is generated and the logic level of the J input is clocked through the flip-flop to the Q output. Even after the pushbutton is released, the Q output remains at 1. Meanwhile, the computer is scanning the Q output. Some time after the pushbutton has been pushed and released, the computer sees that $Q = 1$ and takes the action it is programmed to follow when the pushbutton is pressed. After the computer completes the action, it generates a Clear pulse to set the Q output back to 0. In effect, the flip-flop acts as the door between the unpredictable world of the person pushing the outside switch, and the fast-moving, programmed computer.

The computer knows that new data has arrived and is able to clear the flip-flop when it has taken care of the new data. This type of interaction is called *handshaking* and is an important function of digital ICs that act as interfaces between different circuits within a system or different systems that must work with each other. Many times systems that do not work, or once in a while do not work properly, have problems because the handshake sequence is not done properly. All the ICs may be working, but there is a mistake in the thinking that went into the design. These types of problems can be hard to find unless the person trying to fix the system has a clear idea of what should be happening versus what really is happening. We will see more interface handshaking when we discuss serial- and parallel-interface ICs.

QUESTIONS FOR SECTIONS 3-2 AND 3-3

1. What is the difference in gates between and *S-R* and *J-K* flip-flop?

2. What problem does the *J-K* flip-flop overcome compared to the *S-R* flip-flop?

3. How does the *J-K* flip-flop use three-input NAND gates?

4. What do the Preset and Clear signals do? Are they synchronized with the flip-flop Clock signal?

5. How would a flip-flop be used to send data to a computer?

6. Give two other applications of a *J-K* flip-flop.

7. What is handshaking? Why is it needed? Give an example that is not computer- or flip-flop-related.

PROBLEMS FOR SECTIONS 3-2 AND 3-3

1. Sketch the sequence of events for a *J-K* flip-flop used as a computer interface versus time.

2. For the *J-K* flip-flop, draw the *Q* and *Q'* outputs versus time for the *J, K,* and Clock inputs shown in Fig. 3-9. (Assume output *Q* is 0 at the time the first Clock pulse is applied, and Preset and Clear are 1.)

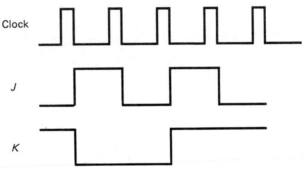

Fig. 3-9 Drawing for Sections 3-2 and 3-3, Problem 2.

3. A *J-K* flip-flop is connected as shown in Fig. 3-10. What boolean logic function is thus performed between the Clock line as input and the *Q'* signal as output?

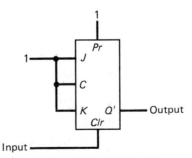

Fig. 3-10 Drawing for Sections 3-2 and 3-3, Problem 3.

The *D*-type flip-flop is built from a *J-K*-type flip-flop. The *J* input is connected to the *K* input by an inverter (NOT gate) (Fig. 3-11). Look at the truth table of the *J-K* flip-flop in Fig. 3-6 for the case where *J* and *K* have opposite values. The output on *Q* *after* the Clock pulse is the value of the input on *J* *before* the Clock pulse. Therefore, the flip-flop acts as a *delay* unit, which is why it is called the *D*-type.

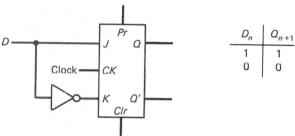

D_n	Q_{n+1}
1	1
0	0

Fig. 3-11   The *D*-type flip-flop is made from a *J-K* type and provides a delay of one clock period.

The *D*-type flip-flop is used in applications where a time delay of one Clock pulse period is needed in the system. An example might be where a signal from one source is known to arrive to the system before another part of the circuit is ready. In that case, the *D*-type flip-flop can provide a needed delay. Because the amount of delay time is equal to the period between Clock pulses, the delay is as precise and accurate as the system clock, which is usually very accurate. We can also chain together several *D*-type flip-flops and get longer amounts of delay, as well as develop other very useful ICs, as we will see in the next few sections.

Several different versions and styles of *D*-type flip-flops are available as complete ICs, so that the circuit does not require another IC package for the NOT gate inverter that goes between the *J* and *K* inputs.

In many applications, several logic lines to a computer may come from different sources or from various parts of the same circuit. They may not arrive at the computer at the same time. This can cause problems because the processor in the computer may look at these lines as a group and expect that all of them have valid data at the same time. One solution is to use a group of *D*-type flip-flops, with one for each of the logic lines (Fig. 3-12). The Clock inputs of all the flip-flops are connected to the system clock. What happens is this: The data arrives at each flip-flop at a slightly different time but is not passed through to the *Q* outputs and on to the computer. At the next Clock pulse the data on the four inputs is clocked through the flip-flops and appears on the *Q* outputs. The processor can read all of these at the same time and know that all the data is valid.

The *T*-Type Flip-Flop

The final variation of the basic flip-flop is the *T*, or *toggle*, type. This type is built very simply by connecting the *J* or *K* inputs of the *J-K* flip-flop together (Fig. 3-13). By looking at the truth table for the *J-K* flip-flop in Fig. 3-6, we can see that when *J* = *K* = 1, the output *Q* *after* the Clock pulse is the complement (opposite) of the output *before* the Clock pulse. In this way the *T*-type flip-flop toggles, or flips, between the 0 and 1 output values with each Clock pulse.

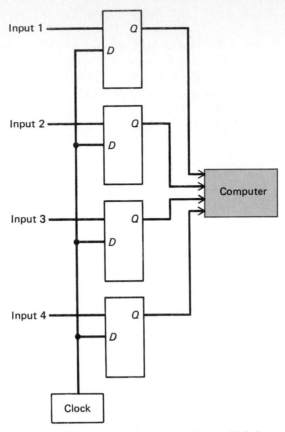

Fig. 3-12 Multiple D-type flip-flops can be used to bring multiple inputs into a computer.

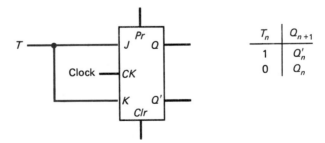

T_n	Q_{n+1}
1	Q'_n
0	Q_n

Fig. 3-13 The T-type flip-flop toggles with each clock pulse.

A simple application for this type of flip-flop is a system where a pushbutton is connected to a system, along with a light. Every time the button is pushed, the system alternately turns the light on and off. This could be built by wiring the J and K inputs to logic 1, and then connecting the pushbutton to the Clock input. If the lamp is connected to the Q output, the lamp will change state from on to off to on every time the button is pushed. This would be another way to get pushbutton information into a computer. The computer would regularly check the Q output. If it had changed state, then the button was pushed since the last time. Of course, the computer would have to remember the last state of Q in its own memory.

Because the T-type flip-flop is built by wiring the inputs of a J-K flip-flop together, there are no available ICs that perform only the T function. Instead, circuits that need the T function use a J-K flip-flop with the appropriate IC pins wired together.

QUESTIONS FOR SECTION 3-4

1. What is the function of the D-type flip-flop?

2. How is a D-type flip-flop made?

3. Why is the D-type flip-flop needed? Give two examples.

4. Repeat questions 1 through 3 for a T-type flip-flop.

PROBLEMS FOR SECTION 3-4

1. Use a truth table to show that an S-R flip-flop is a T-type if S is connected to Q' and R to Q.

2. For a D-type flip-flop, sketch the Q output versus a series of Clock pulses if the input = 1.

3. Repeat for a T-type flip-flop.

4. Sketch the output of the second T-type flip-flop versus time where the output of the first one is the input for the second one.

5. Use a truth table to show that a D-type flip-flop becomes a T-type if D is connected to Q'.

3-5 TROUBLESHOOTING FLIP-FLOP CIRCUITS

Circuits which use flip-flop ICs are usually a little harder to troubleshoot than circuits which use combinational logic only. This is because flip-flop circuits have memory and usually have a Clock function. Both of these factors make it more difficult to tell what the signals in the circuit should be doing if everything is working properly. There are some general guidelines to follow when you start to troubleshoot flip-flop-based circuits.

1. Always check the power and ground at the IC itself. Make sure the voltage level is at the proper value for the IC. An oscilloscope may also be needed to see if the power is free from noise and spikes, which would affect IC operation. A voltmeter shows only the voltage value but not the ''quality'' of the dc power.
2. Identify from the IC model number and schematic the type of flip-flops used: S-R, J-K, D, or T. Then you will know the correct truth table to use and operation to expect at the IC, as well as which pins of the IC package are the inputs, outputs, Clocks, Presets, and Clears.

3. Check the signal lines to see that they are all within the specification for a good logic 0 and a logic 1, regardless of what the flip-flops are actually doing. A line that is not in the proper range means a definite problem, just as for the simple gates.

4. If the IC has a Clock input, see if it can be disconnected from the system clock. Then the flip-flop can be checked statically or at slow speed, under direct control of the test person. (Many circuits are designed with a jumper that disconnects the clock to make it easier for the test). If the clock can be disconnected, check the inputs and outputs, advance the clock manually, and check the inputs and outputs. Compare the results against the truth table for the flip-flop.

5. Some flip-flop circuits have "feedback" loops. This means that instead of the input to the system flowing from one flip-flop to another like a chain, some outputs go to flip-flops that are back earlier in the signal flow (Fig. 3-14). This is

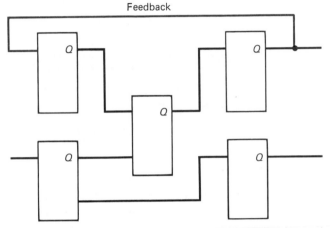

Fig. 3-14 The output of one flip-flop can be fed back to the input of a previous flip-flop. This complicates troubleshooting.

done by the designer of the circuit for two reasons: It may be the only way to get the desired function, and it may reduce the number of ICs needed and save money. But it can make troubleshooting harder, because it is more difficult to figure out what the signals should be at each point in the system. See if the feedback loops can be opened. Just like with clocks, many circuits have a removeable jumper to do this easily. If the feedback can be opened, it is much easier to trace the flow of signals and see if they are doing the right thing at each point and at each flip-flop input and output.

6. Verify the operation of flip-flops in the part of the system that does not seem to be working right. Use the truth table for the flip-flop and see if the input and output states follow the truth table. An oscilloscope or logic probe is usually used, but a voltmeter can also be used if the circuit is running without a clock or running very slowly.

We have studied how a practical flip-flop called the *J-K* type is built up out of simple logic gates. This flip-flop has two inputs, along with a Clock input and Preset and Clear inputs. The *J-K* flip-flop can also be used as a delay, or *D*-type, flip-flop with a simple inverter between the *J* and *K* inputs. The *J-K* flip-flop can also be converted to toggle, or *T*-type, flip-flop by connecting the *J* and *K* inputs together. Many versions of the *J-K* and *D* flip-flops are available as complete ICs from many manufacturers.

The flip-flops we have studied are sensitive to the actual level (1 or 0) of the input. It is possible to also build flip-flops that are not level-sensitive but instead react to the *change* in input state. These are called *edge-sensitive* types and are used in many circuits. Like level-sensitive flip-flops, edge-sensitive flip-flop ICs are also available commercially.

Just as the boolean gates are the building blocks of the important flip-flops, the flip-flop is the starting point for many important IC functions that will be studied in the next sections. They are also important interface elements by themselves.

QUESTIONS FOR SECTION 3-5

1. What two factors cause problems in troubleshooting?

2. What can be done about them?

3. What are four of the six basic steps to follow when troubleshooting flip-flop-based circuits?

4. Are flip-flops sensitive to voltage level? What other choice is there?

3-6 COUNTERS

In order to do any arithmetic, you have to be able to count. The same is true of electronic systems that are used for computer applications—counting is a fundamental operation. As input pulses enter the system, the system ICs must be able to count the number of pulses that have arrived. They must also be able to present the total counted to another circuit or part of the same circuit.

What is meant by counting? How do we count in the decimal system we are used to using? Assume we have to count up to 500. Start with 000, 001, 002, 003, and so on up to 009. Then start a second column, with the carry from the first column: 010, 011, 012, 013, and when the second column reaches 9 carry a 1 into the third column, 100, 101, 102, up to 500. The first column is the number of ones, the second is the number of tens, and the third is the number of hundreds.

Binary counting uses the same principle, but as seen in our review of binary numbers, the column values are different. The column values, starting with the rightmost (least-significant) bit, are 1, 2, 4, 8, 16, 32, 64, and so on. Note that the value of each column is half the one to the left. Counting in binary is simple: The maximum value in a column is 1 instead of the 9 in decimal. When a column and those to the right of it are 1, then you have to carry a 1 to the next column to the left. Binary counting goes like this:

       0001
       0010   Carry from ones column to twos column
       0011
       0100   Carry from twos column to fours column
       0101

```
0110
0111
1000        Carry from fours column to eights column
1001
1010
1011
1100
1101
1110
1111
```

The object is to build a circuit of gates and flip-flops that performs this counting function. Using a single flip-flop of the T-type, if the flip-flop receives a string of input pulses, the output changes state for every second input pulse (Fig. 3-15).

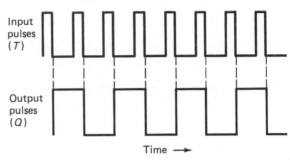

Fig. 3-15 Input versus output for a T-type flip-flop. There is an output for every second input pulse.

From the timing diagram it is seen that the number of output pulses (Q) is one-half the number of input pulses, just the same as the relation between the column values in a binary number.

Next connect several flip-flops together, with the Q output of one connected to the input of the next (Fig. 3-16). Feed the same string of pulses into the first flip-flop and the number of output pulses at each stage is one-half the number of input pulses. The timing diagram shows this clearly. Convert the timing diagram into a truth table by looking at what the timing diagram shows the Q output to be for each flip-flop after each input pulse. The result is

```
0001
0010
0011
  ⋮
1111
```

This is exactly what a binary counter should do. Therefore, a string of flip-flops connected in series acts as a counter for binary numbers.

The counter developed is called a *ripple counter*, because the input pulses go from one flip-flop stage to the next like wave ripples in the water. It is also an

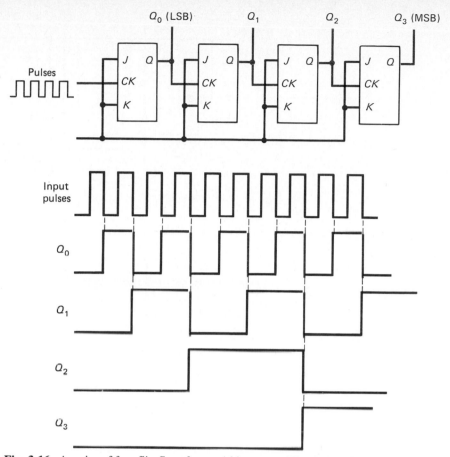

Fig. 3-16 A series of four flip-flops form a 4 bit counter. The timing diagram shows how each flip-flop counts a different bit of the four bits.

asynchronous counter because the input pulses can come at any time. There is no clock for this counter. This is useful if the counter is used for an application where the inputs may occur at any time, like the opening of a door, where the goal is to count the number of people coming into the store. In many applications within a system, though, the pulses to be counted can come only with the system clock, so synchronous counters are used. The inputs are counted only when they occur along with the system clock. This is done by using the input pulse and the Clock signal and clocked flip-flops (Fig. 3-17).

It is also important to make sure that when the circuit powers up, the counter starts at 0. (Most gates and flip-flops power up in random states, so some initialization is required.) Therefore, many counters have provisions for an external Clear signal, which sets all the internal flip-flops to $Q = 0$ (Fig. 3-18). Another feature that counters need in some applications is the ability to count down, or decrement. (When a counter counts up, it is said to increment.) Counters that count down usually are started at "full count" of all stages set to $Q = 1$, and so an external Preset signal is included.

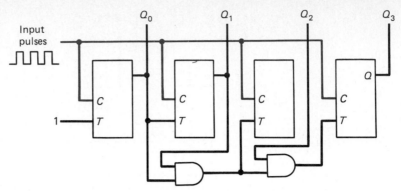

Fig. 3-17 A synchronous counter only counts pulses that occur during the system clock pulse.

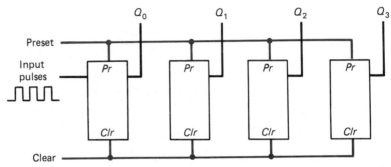

Fig. 3-18 An external Clear signal sets the count to 0000; Preset sets it to 1111.

Counter ICs are available that are synchronous, asynchronous, up, up-down, and with Preset and Clear, in a wide variety of flip-flop stages. Counters can be connected together to provide as many stages of counting as are required for the application by bringing the output from the last stage of one IC to the input of the first stage of the next IC.

QUESTIONS FOR SECTION 3-6

1. What is the rule for binary counting? How is it similar to regular decimal counting? How is it different?

2. How is a counter formed from flip-flops? What kind of flip-flop is used?

3. What is the key characteristic of the ripple counter?

4. Why is a synchronous counter used?

5. Why is Preset needed for a counter? Why is Clear needed?

6. What does the term *increment* mean? What about *decrement?*

7. In a counter, which flip-flop has the MSB, the one closest to the input pulses or farthest from the input pulses? Which one has the LSB?

PROBLEMS FOR SECTION 3-6

1. Count in binary from 1100 1111 to 1110 0000.

2. How many stages would be required to do the counting of problem 1?

3. Sketch the output of each stage of a 5-bit counter for a period of 18 input pulses.

3-7 APPLICATIONS OF COUNTERS

Counters are very versatile. They can be used to count the total number of pulses, to count a specified number of pulses, to measure time, or to measure the frequency of an input pulse stream.

Consider an example where the need is to count and indicate to a user every ninth input pulse. Start with the four-stage counter, and then add some gates to it. Since decimal 9 is 1001 in binary, take the output of the first (LSB) stage and the fourth (MSB) stage and feed them into a NAND gate (Fig. 3-19). Then, when the counter reaches 1001, the output of the NAND gate will go to 0. There is a need to reset the counter to 0000 to start counting again, which is achieved by connecting the NAND output to the Clear inputs of all the flip-flops of the counter. The counter will count 0000, 0001, 0010, . . . up to 1001, then the NAND output will go to 0, the counter will reset and begin counting again at 0000. What we have done is generate an output pulse for every nine inputs, and therefore divided the input string of pulses by 9. For this reason, a counter set up this way is called a *divide-by-N* counter (an output pulse is generated every N inputs, thereby dividing the input stream by N).

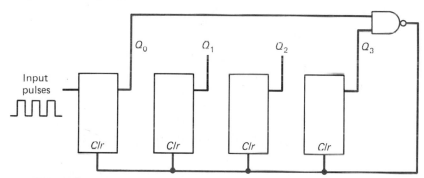

Fig. 3-19 A counter acts as a divider when some of the outputs are NANDed together to generate a Clear at a specific value.

The divide-by-N scheme discussed is "hard-wired" to divide by only one number, in this case, 9. For a flexible circuit that can divide by any number N, a NAND gate with an input from each output stage of the counter is used. The inputs go to the NAND gate through switches and the user can set the switches to represent any binary number (Fig. 3-20). In this way, the same counter can act as a divide-by-N up to the counting limit of the counter itself.

Counters are also used to build digital clocks (also known as timers). If the counter input pulses come from an accurate and stable oscillator with known frequency, then the total count represents, in binary, the amount of time that has

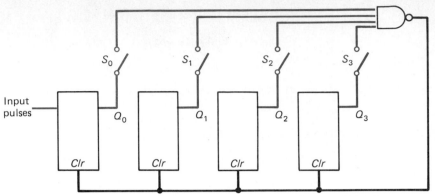

Fig. 3-20 By using switches, the user can make a counter that can divide by any number N.

passed. To make a clock that can be used for measuring to the nearest second, the oscillator would put out 1 pulse per second. A clock to measure to a millionth of a second would need an oscillator running at 1 million pulses per second [1 megahertz (MHz)]. These clocks could be read by another circuit or a microprocessor, but the binary output would be hard for a person to read. The clock, to be useful to a person, would also need circuitry to convert the binary output to a format that could be read by the user, usually decimal. The conversion circuitry to do this will be studied in a later section of this chapter.

In many microprocessor-based systems, there is a high-frequency clock that is used to pace the ICs at frequencies ranging from 1 to 10 MHz, depending on the IC types. At the same time, the microprocessor needs a slower, periodic "tick" (usually 10 or 100 times per second) in order to run some scheduled software and keep track of the actual time of day. A counter-timer is used to do this by counting the system Clock pulses and dividing them by a large value of N to provide the slower time pulses needed. A 1-MHz system clock would be divided by 10,000 to generate 100 ticks per second, and a counter with 14 stages would be needed.

Counters are also used as the heart of instruments and test equipment that measure frequency (the number of cycles or pulses per second). Before digital ICs, measuring frequency was very difficult. The instruments that were used were expensive, hard to use, and did not work very well. With flip-flops and ICs the whole situation changed quickly and completely; now frequency measurement is easy and accurate and the instruments used are compact and inexpensive. How are counters used to do this?

The simple counter used needs some circuitry added at its input. An AND gate is used to allow the pulses to be counted into the counter, or block them from the counter (Fig. 3-21). The other input of the AND gate is a precise Clock signal that lasts for a known time period (1 s for this exercise). This Clock signal is used to gate the input pulses. Only when the Clock signal is present are the input pulses counted. What the counter is doing is counting the number of pulses that occur during the Clock time period. If six pulses arrive during the gate time, we can say that 6 pulses per second is the frequency of the input signal. By using different

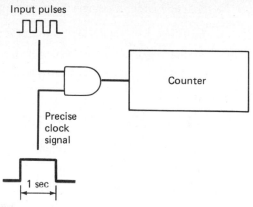

Fig. 3-21 A counter can also measure frequency—the rate at which pulses occur—by using a precise gate signal and ANDing it with the input pulses.

gate times and more stages, it is possible to construct a counter that can measure frequencies from very slow (about 1 Hz) to extremely fast (up to several million hertz). Special, more expensive ICs can be used to go beyond even this frequency.

Troubleshooting Counter Circuits

The first step is to check the power, ground, and digital signal levels at the IC, as with gates and flip-flops. Assuming these are OK, these things should be checked:

- Input pulses to the counter should be present. They can be from outside events (a mechanical switch, for example) or a clock oscillator.
- Make sure that the Preset and Clear are not causing the counter to go to all 0's or all 1's at the wrong time, or are holding the counter at all 0's or all 1's. Most counter ICs do not bring out the Preset and Clear of each stage. Instead, all the Presets go to a single IC pin and all the Clears are brought to another pin of the IC. This makes checking much easier. However, some circuits use individual flip-flops or ICs that have the Preset and Clear for each stage available separately, in order to allow the counter to be set to any value, not only all 0's or all 1's. For these types of counters, each Preset and Clear must be checked.
- Since the output frequency or count of each stage is one-half the previous stage, check these outputs as the counter gets a string of input pulses to be counted. If the clock can be controlled by you (once again, some circuits have a jumper to allow a bypass and override of the system clock), then step through the counting very slowly.

QUESTIONS FOR SECTION 3-7

1. What are two applications of counters?

2. How is a counter used to make a divide-by-N circuit?

3. How is a counter used to measure frequency?

4. What is the relation between counting and measuring the passage of time?

5. How could a counter be used as part of a system Clock function?

6. What does the term to *gate* pulses mean?

7. Once the power, ground, and digital levels are checked, what three things are checked with a counter circuit that seems to be malfunctioning?

PROBLEMS FOR SECTION 3-7

1. Sketch a divide-by-7 counter.

2. Sketch a counter used to measure frequencies of up to 100 Hz, using four stages. What gate time is needed?

3. Repeat for measuring up to 100 Hz with a 1-Hz gate clock. How many stages are needed in this case?

4. Sketch a divide-by-3 counter and a divide-by-5 counter. If the output of one is the input of the other, what is the divide by number of the combination? Why?

5. For the counter shown in Fig. 3-20, will it be able to count up to at least 18 (decimal) with S0, S1, and S2 open and S3 and S4 closed? With S0, S1, S2 closed and S3, S4 open? *Note:* Add another stage and switch S4 before answering.

6. Convert the counter of Fig. 3-21 into a frequency meter by adding a gate circuit with a gate time of 0.2 s and closing all switches. Can a frequency of up to 1000 Hz be measured? If not, what two changes could be made to make it handle 1000 Hz?

3-8 REGISTERS

The flip-flop is basically a 1-bit storage element. In many practical situations, there is a need to store more than 1 bit of data at the same time. A typical small system may have from several thousand to several hundred thousand bits of memory. There is also a need to have a small number of bits stored together so they can be used as a group. The function that does this is called a *register*.

A register is a circuit made up of flip-flops that can store anywhere from 2 to several dozen bits (or more). A typical size for a register is 4, 8, or 16 bits. The uses of registers include these:

- As a temporary storage area for data from several external switches that will then be passed on to the microprocessor of the system as a group.
- For storage of a binary number at some midway step in a calculation.
- To act as a physical interface, or buffer, between different systems or between circuits of a single system (Fig. 3-22) to allow data to flow from one to the other.
- To store bits that signal, or flag, information which can be checked at any later time by another operation or sequence of the system. For example, the bits of a 4-bit register may each correspond to a different door being monitored by a

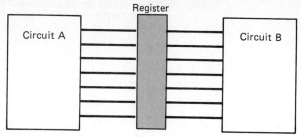

Fig. 3-22 A register is often used as a physical interface between two circuits of a single system.

computer-based alarm system. When any door is opened, a flag bit in the register gets set and stays set even after the door is closed. Periodically, the microcomputer checks each flag to see what the door status is. By checking these flags, the system is informed of which door was opened.

- To allow a user to latch, or set, a desired set of conditions and choices. The computer-based coffee vending machine may have a bit of its register assigned to each of the following: type of coffee (regular or decaffeinated), sugar or no sugar, cream or no cream, strong or weak (Fig. 3-23). Then, when the last button is pushed, the machine starts making the coffee and checks what choices the coffee buyer wanted. The register lets the person buying the coffee make choices in any order, since there is a special bit reserved for each item. The machine does not care about the order of selection because it checks the register in one operation, when it is making the coffee.

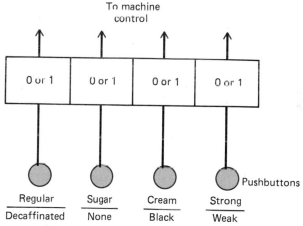

Fig. 3-23 How a register might be used to store customer choices in a coffee machine.

How a Register Is Constructed

A register is built by taking a group of flip-flops and connecting them in a chain, just as is done for a counter. However, there is one important difference. For the counter, the output of each flip-flop stage was available to the user. For the register, both the output *and* the input are available (Fig. 3-24). This means that each flip-flop can be addressed separately with data, rather than the data having to flow or ripple through the chain of flip-flops. The ability to address the flip-flops

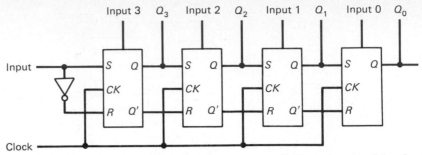

Fig. 3-24 A 4-bit register. Outputs of each stage are available to the rest of the circuit. Each stage can be written to by its Preset. (Note: Clear line not shown for clarity.)

at the same time (called a *parallel operation*) in order to store bits in the register and read them back as a group is what gives the register its versatility. As a practical point, nearly all registers used in systems are made up of clocked flip-flops. This is to ensure that all the stages really do act at the same time when they are latching data bits or presenting outputs.

3-9 APPLICATIONS OF REGISTERS

The typical applications we have discussed so far are the parallel-in–parallel-out type. This means that the bits are latched into the register as a group and read out as a group. Sometimes this is also called a *scratchpad* memory, because the register acts as temporary storage area for data bits. *Temporary* means that the bits will not be needed later, since some other part of the system will look at the bits, perform some action, and then there will be no further use for the bits. The coffee machine is one such application; once the coffee has been made, the machine does not care what the customer ordered and it has to get ready for the next order.

Registers are very flexible because of all the input and output connections they have. This flexibility allows registers to be used as serial-to-parallel converters. The input to the counters previously studied was a serial-bit stream, meaning that a string of 1's and 0's came into the counter, on a single line, as time advanced. The output of the counter was a parallel block of bits, since all the output could be checked at once. A register performs a very similar function. It allows a stream of 1's and 0's to be captured as they come in, and then read by a computer as a group. Examine a register acting as a serial-to-parallel converter in more detail, using an eight-stage register (Fig. 3-25). First, the flip-flops of the register are cleared by a pulse on the Clear input, and then the Clear and Preset inputs are both set to their inactive state. The register is now ready for normal operation. A serial data stream of 8 bits comes to the input of the register, along with Clock pulses. For this example, make the bit stream be 1101 0011 (LSB). The LSB is clocked into the first flip-flop of the register. At the next Clock pulse, the LSB is transferred to the next flip-flop, while the second bit enters the register. This process of the previous bits shifting over and a new bit entering continues until all 8 bits have been clocked in. The timing diagram and truth table show this operation for each Clock pulse. At the end of the eighth Clock pulse, all the bits in the serial stream are captured,

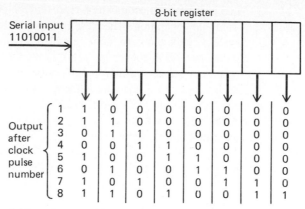

Fig. 3-25 An 8-bit register used for converting a serial stream of bits into a parallel group. This is called serial-to-parallel conversion.

in order, in the register flip-flops and can be read in parallel at the Q outputs. This is a very common application because messages between computer-based systems are often transmitted in serial order (to save on the actual number of interconnecting wires needed), but computers themselves are not efficient if all the power of the processor is devoted to picking up a single bit at a time. The register stores the bits as they come in, and then the processor can look at the whole group. In many situations, the characters being transmitted from one system to another are encoded with 8-bit binary patterns representing each letter or number. (There is a standard code commonly used, called ASCII [short for American Standard Code for Information Interchange] where each letter, number, and special symbol has been assigned a unique code number from 0 through 255. These code numbers are then sent as 8-bit binary patterns, from 0000 0000 to 1111 1111.)

Registers also perform the reverse operation: converting a parallel group of bits to a serial data steam (parallel-to-serial conversion). This would be used where a system wanted to send characters (Fig. 3-26). The processor would present the 8

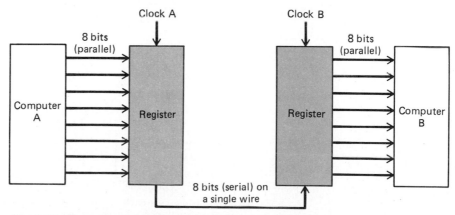

Fig. 3-26 Computer-to-computer interconnection using parallel-to-serial and serial-to-parallel registers. Only one wire is needed between computers.

bits of the ASCII code, which represent the character to be sent, to the registers on a parallel operation. Then the bits would be clocked out of the register by the system clock and onto the wire which connects the two systems. (Of course, this parallel-to-serial conversion is not limited to ASCII characters.) The details of this operation are very similar to the operation of the serial-to-parallel conversion. At each Clock pulse, the data at the Q output of a flip-flop is transferred to the next flip-flop. At the MSB flip-flop, the data is transferred out of the register entirely. After eight Clock pulses, all the data in the register has been transferred out and the flip-flop is empty.

Registers can also serve as buffers. (*Note:* the word *buffer* has very different meanings in different parts of a digital circuit. Another kind of buffer will be studied in Chap. 5.) A register, acting as a buffer, is a physical interface between two parts of a system, two systems, or the outside world and a system. The buffer is used when the clock of one system is running at a different rate than the clock of another system, or when one part of a system may not be ready for digital data from another part. In these situations, the buffer acts as a serial-in–serial-out register. Data is clocked into the register, the register fills, and then the data is clocked out. The in and out activities are independent and clocks with different frequencies can be switched in to act as the register clock (Fig. 3-27). The switching is done by each system, in order to bring in the clock it needs for its own operation. If the register is acting as a buffer for parts of the same system with only one clock frequency, then each part of the system controls the enabling or disabling of the clock. The data is first shifted into the register (serial in) and then it is shifted out at a later time (serial out). Since the data shifts out in the same order it came in, this register is sometimes called a *first-in first-out* (FIFO) register.

In many applications, several registers are used in parallel to form a "wider" register which can act as a buffer for more than one signal line. Typically, eight

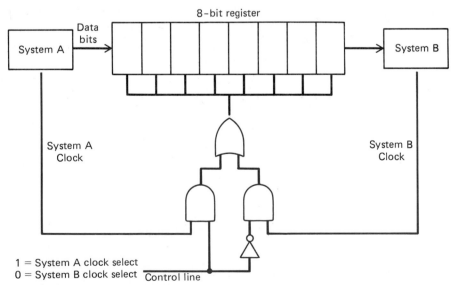

Fig. 3-27 A register acting as a buffer between two systems with different clocks. The desired clock rate can be selected by a control line.

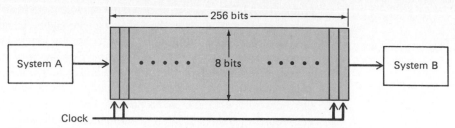

Fig. 3-28 Wide registers used as systems buffers. A 256-bit deep by 8-bit wide register is shown.

registers might be used to form a buffer that can handle a byte (8 bits) at once, so that entire bytes can be transferred. Registers are designated by their length (depth) and width (Fig. 3-28). A large register used to buffer many bytes between systems might be 256 × 8, meaning it has 256 positions, each 8 bits wide. A bit entering one end of the register will be clocked out after 256 clock cycles.

A very common situation is the interconnection between a computer and printer, so the computer can provide hard-copy output for the user of the system. This application really involves registers in three configurations: as a parallel-to-serial converter within the computer to put the computer characters to be printed onto the wire as serial bits between the computer and the printer; as a serial-to-parallel converter within the printer to reconstruct the original bits into parallel format so that the printer can decide which character to print; and as an 8-bit-wide FIFO buffer that allows the characters to be sent by the computer at a rate faster than the printer can actually print them (Fig. 3-29). This is a critical point since most computers send characters much faster than they can be printed. (Also, some printers print one line left to right and the next right to left to save return time. They must have all the characters in a buffer before printing a line.) It is possible for the printer to signal (or handshake) to the computer after each character is printed that it is ready for the next one, but this is very inefficient in terms of the computer use and this signaling also uses time and lowers efficiency. Instead, the

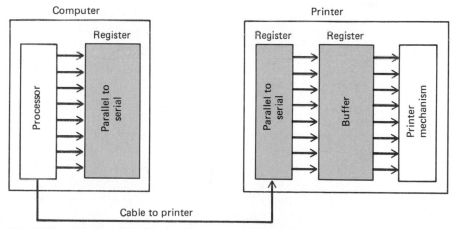

Fig. 3-29 Computer interface to printer using three buffers: parallel-to-serial, serial-to-parallel, and wide.

printer has a buffer that can store a burst of characters from the computer and, only after the buffer is full, does the printer have to signal the computer to stop sending new characters. The FIFO in the printer thereby increases system efficiency and throughput (total amount of information flowing in a given amount of time).

Sometimes, in testing digital circuits and systems, it is necessary to generate a sequence of bits over and over. For example, there may be a problem in the data link between a computer and a printer. The test technician does not know where the problem is (it could be in the computer, cable, or printer) and would like to generate characters and see if they are printed properly. The technician needs a test box that generates a known character continuously. He or she would disconnect the computer and replace it with the *pattern* or *sequence* generator and see if the printer worked (Fig. 3-30). If it did, the fault is with the computer. If not, the fault

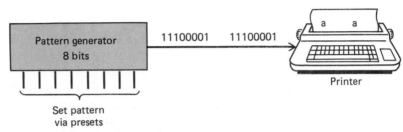

Fig. 3-30 A pattern generator used to test a printer and computer. The technician sets the binary pattern desired into the generator. In this case, pattern 11100001 is the ASCII code for the letter ''a''.

is with the cable or printer. How is this pattern of characters generated repetitively? A shift register is used, with the output of the last stage connected to the input of the first stage (Fig. 3-31). The individual stages are cleared and then

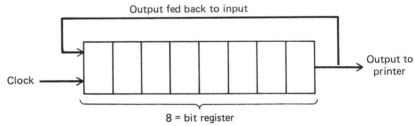

Fig. 3-31 A pattern generator. The register output is connected back to the input and the register is used in serial mode. (Note: Presets not shown.)

loaded with the binary pattern corresponding to the character to be sent by presetting the flip-flops. Then the register is clocked, and the binary pattern is clocked out of the register, where it goes to both the cable and printer, and back into the input of the shift register where it is clocked back in. These pattern generators are extremely useful in testing any sort of communications link, whether wire, telephone, or radio.

Register ICs are available in many configurations. These include 8×2 bits, 8×1, 4×4, and 256×1, among others.

Troubleshooting Registers

When checking a register, the first thing to determine is its type and function. Is it 8×2 or 4×4 (both are a total of 16 bits)? Is the function as a buffer or as temporary number storage? Be careful not to judge by the IC type alone, because two separate 8×1 ICs may be interconnected to act as a single 8×2, for example. Once the register type and function are known, check for proper operation of the register Clock, Preset, and Clear (if one exists). Finally, check that the data going into the register and the data coming out are the same. (*Note:* Since the register may make a parallel group of bits into a serial data stream you will have to find the data at various pins on the IC). Remember that the data into and out of this register must agree, because a register does not change the bits themselves. It does not look at the bits and make a decision, based on boolean logic truth tables, that changes some of the bits. Registers are said to be "transparent" in this respect.

QUESTIONS FOR SECTIONS 3-8 AND 3-9

1. What is a register? How does it differ from a flip-flop or counter?

2. Give four uses of registers.

3. What is a serial-to-parallel register? Where is it used?

4. What is a parallel-to-serial register? What are the applications for it?

5. What is the ASCII code?

6. What is a buffer? What is a wide buffer? How is the size described?

7. Why must a printer use a buffer in most cases? Why must a printer that prints left to right and right to left have a buffer?

8. What is a pattern generator? Where is it needed? How is it made from registers?

9. What are two things to check when troubleshooting registers?

10. Does a register change bit values or store and move them? Explain.

PROBLEMS FOR SECTIONS 3-8 AND 3-9

1. Sketch a block diagram of a buffer with two separate data inputs that allows a single control line to select which data line is to be used.

2. Draw a block diagram of a buffer that handles a 4-bit wide input and can buffer up to 128 such inputs.

3. Draw a pattern generator that continuously generates the pattern 1011 1011 1011. . .

4. A computer sends 10 characters to a printer at 10 characters per second and then waits 9 s. The printer can print at 1 character per second. What is the minimum FIFO size, in characters, needed to make sure that the computer can send characters and not have to wait for the printer to finish?

5. If each character in problem 4 is represented by an 8-bit ASCII code, what size FIFO is needed?

6. Sketch the different buffer configurations that can be built from four separate buffer ICs, each of 4 bits.

7. Repeat problem 6 for buffer ICs of 2×4.

8. For a 4-bit register with a serial input of 0110 0010, sketch the parallel output versus Clock pulses for 10 Clock pulses. Assume the register is cleared initially.

9. Repeat problem 8 for 12 Clock pulses and sketch the serial output.

SUMMARY

Simple AND, OR, and NOT functions are the starting points for building digital systems. By themselves they are limited in what system functions they can provide. The proper interconnection of these boolean functions leads to an extremely useful and important capability needed in any system: The ability to store a binary value and provide it as needed at a later time. Flip-flops provide this function. They also allow a system clock to pace the overall operation of the circuit so that practical problems such as propagation delay can be overcome. Flip-flops are used in different configurations to store information, provide handshaking between different parts of a system or circuit, and synchronize various circuits.

These flip-flops can also be used to develop the next level of important digital function. This includes counters, which accumulate numbers of input pulses and store the total number, as well as registers. Registers store groups of bits that can be received one at a time (serial), several at the same time (parallel), and retrieved either serially or in parallel. ICs are available which provide complete flip-flop, counter, or register combinations in a single building block.

REVIEW QUESTIONS

1. What is combinational logic? How does it differ from sequential logic?

2. What is the role of the clock in sequential logic?

3. What is the key feature that a flip-flop provides to a system?

4. What are the four types of flip-flops? What are the key characteristics of each?

5. How is a D-type flip-flop made?

6. How is a T-type flip-flop made?

7. What is an application of (a) a *J-K*, (b) *D*-type, (c) *T*-type flip-flop?

8. *Q* and *Q'* outputs of a *J-K* flip-flop are connected to the inputs of a NAND gate. What does the NAND output look like as the *J-K* flip-flop is activated?

9. Why are flip-flop circuits harder to troubleshoot than simple combinational logic circuits?

10. What tools are needed for the first checks on a circuit? What are the first things to check?

11. What is handshaking? Why is it needed? How is it done?

12. Why is binary counting used with digital circuits?

13. What logic element forms the basis of the counter?

14. What is the significance of the word *ripple* in ripple counter?

15. How is binary counting done with flip-flops?

16. How can a counter be set (initialized) to all 0's? All 1's? Why is it necessary to do this?

17. Why are synchronous counters usually used? How are they implemented?

18. How is a counter converted to a divide-by-*N* circuit? What is the relation between input and output pulses for such a circuit?

19. How can a counter be used for frequency measurement?

20. How is a counter used for measuring time intervals?

21. Why is the gate function important?

22. What is a (a) parallel-input–parallel-output register? (b) serial-input–serial-output register? (c) serial-input–parallel-output register? (d) parallel-input–serial-output register?

23. Give an application example for each of the four register types.

24. Do the bits in a register have to be related to each other? Give an example of where they are and where they are not.

25. What is the relation between a buffer and a register? How are they similar and how are they different?

26. Explain how several 16 × 2 buffers can be used for a 32 × 2 buffer; for a 16 × 4 buffer; for a 32 × 4 buffer.

REVIEW PROBLEMS

1. The input to a *T*-type flip-flop is a Clock signal that is ANDed with a signal as shown in Fig. 3-32. Sketch the flip-flop output versus inputs.

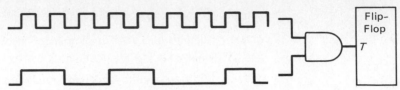

Fig. 3-32 Figure for Summary Problem 1.

2. For the *J-K* flip-flop, sketch the output Q versus Clock for the two conditions given. Assume the initial $Q = 0$. (a) For $J = K = 1$; (b) for $J = 1$, $K = 0$.

3. Count in binary from 0 1100 to 1 0000.

4. Show that a counter of N stages can count in decimal up to $2^N - 1$.

5. How many stages of flip-flops does a counter need to count from 0 through 500 (decimal)? To 525 (decimal)?

6. A counter has 6 bits. The gate time is 0.1 s. What is the maximum frequency it can measure?

7. A 6-bit counter needs to measure up to 200 Hz. What gate time would be used?

8. For Fig. 3-20, what switch settings of S0 through S3 would provide a divide-by-10 counter?

9. Data is coming into a system as an 8-bit serial byte. It needs to be converted into two groups of 4 bits, with 4 bits available in parallel, and then the next 4 bits available in parallel. Sketch a diagram of how registers could be used to do this.

10. A system computer generates a 2-byte word as a group of parallel bits. The device connected to this computer requires (a) 16 serial bits, (b) 2 serial streams of 8 bits each, at the same time. Sketch how this could be done with registers.

11. An 8-bit register receives a byte in parallel, and outputs it in serial, LSB first. For a parallel byte of 1100 0110, what is the serial output stream for the first six clock pulses?

DECODERS, MULTIPLEXERS, AND DEMULTIPLEXERS

4

INTRODUCTION

The basic logic gates can be combined to provide many necessary circuit functions. Decoders recognize specific bit patterns and provide an output when the pattern occurs. In cases where several input signals need to be combined onto a single line, a multiplexer is used. The multiplexer can be controlled to allow the user or circuit to specify how the signals should be combined. The opposite of the multiplexer is the demultiplexer, which takes signals from a single line and allows them to be directed to one of several output lines.

There is also a need for decoders which convert one type of numerical format to another. These formats include binary, decimal, hex, BCD, and the special format used to control digital displays.

4-1 DECODERS

In a typical computer or microprocessor system, the central processor has to send data to and receive data from a wide variety of other ICs in the system (Fig. 4-1).

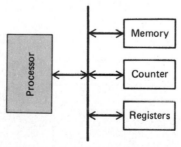

Fig. 4-1 Many ICs can be connected to a system processor. A decoder is used to select the IC to be addressed.

These include memory ICs, where the processor data is stored; counters, which are used for timing; and registers, which allow the processor to communicate with the outside world. All these ICs are on a common bus that is shared by the processor and virtually all other ICs in the system. Somehow, the processor has to be able to direct its data to the desired IC, even though all the ICs are on the bus and receiving the data. Only the desired IC should respond to the data; the other ICs should ignore data on the bus. The special function that does this is a *decoder*.

A decoder has a specific set of signals from the system coming into it. When the signals have a certain desired pattern, the decoder generates an output signal which is used to activate, or enable, the IC that the decoder is connected to. Consider a specific example: a microprocessor that has a register used for receiving input from a keyboard, located at address 1000 (binary). It also has another register used for sending output to a printer at address 1010 (binary), and it has a scratchpad register which stores four values of temporary data at locations 1100 through 1111 (binary). Since the address bus of the system is common to all these registers, how does the processor and system make sure that data being written to the scratchpad register is not also sent to the printer output register, or the data from the keyboard input register is not confused with data coming from the scratchpad register?

The decoder provides the solution. The decoder is an AND gate with many inputs, and these inputs are wired to specific address lines (Fig. 4-2). Only when the correct address appears on the address bus does the decoder provide a logical

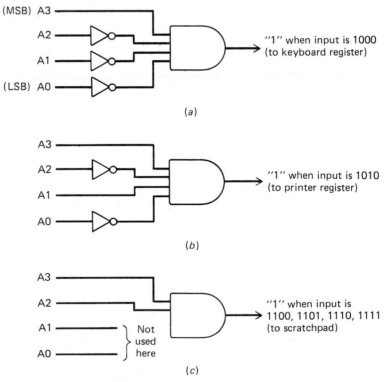

Fig. 4-2 Decoders for specific addresses. (*a*) 1000; (*b*) 1010; (*c*) 1100 through 1111. (Note: The address lines come from the processor.)

output, which then activates the register. If we call the four address bits A3, A2, A1, and A0 (LSB), then the decoder for the keyboard register would provide a logical output for A3 AND A2′ AND A1′ AND A0′. The decoder for the printer register would provide a logical 1 output for the A3 AND A2′ AND A1 AND A0′ condition. For the scratchpad, which occupies addresses 1100, 1101, 1110, and 1111, the decoder would be wired to provide the 1 output level for the input condition of A3 AND A2 only. Note that A1 and A0 are left out, since they can be either 0 or 1, and for either case there is a need to enable the scratchpad IC. The IC chip which is enabled then "listens" to the processor and either receives data from it or provides data; all the other ICs are not enabled and they ignore the processor and do not interfere with its operation.

In most practical circuits, the NAND gate is used instead of the AND. This is because most ICs require a logic 0 level on their Enable line to be activated rather than a logic 1. Most implementations of decoders also have qualifiers, or other signal lines, brought in along with the address lines. These qualifiers are signals that also must be present in order to provide the decoded output (Fig. 4-3). They

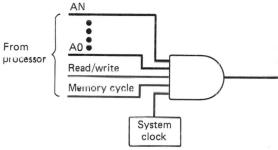

Fig. 4-3 Decoder with address lines and other qualifiers.

include a system Clock line, to make sure that the AND (or NAND) only looks at the address lines when the lines have a valid address and are not in transition from one address to another. The qualifiers also may include the processor Read-Write signal, which the processor generates to indicate to the system if it is reading or writing data. This allows the system to have a Read register and a Write register at the same address, such as would be used for communicating with a keyboard and video screen connected to the system. The keyboard is read by the processor, while the processor writes to the screen.

When designers are checking the operation of a new system, or test engineers are troubleshooting a system in the field, they often have a need to synchronize an oscilloscope or other equipment to a particular operation of the system processor. For example, if the system generally works fine but sends out incorrect characters to a printer, the test person would like to examine the bits being sent to the printer register. The equipment used to troubleshoot often has a decoder, where clip leads are attached to the processor address lines, Read-Write line, Clock, and perhaps a few other key signals (Fig. 4-4). The operator then "dials in" via manual switches the address of the output register and desired state of the qualifiers. The test equipment provides an output when that address is generated by the processor; this output is used to synchronize the scope and other test equipment.

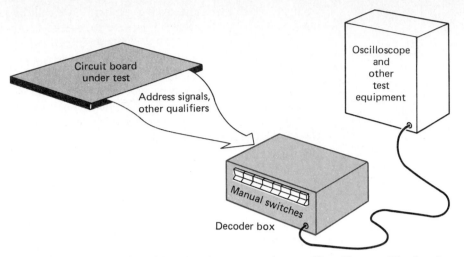

Fig. 4-4 Testing circuits with a decoder to recognize specific addresses. The decoder output is used to synchronize the oscilloscope and other test equipment.

The decoder is a very powerful and common feature of electronic systems. It is usually built out of multiple input AND or NAND gates. The output of a decoder is almost always used to enable an *entire* IC rather than a part of an IC or a group of ICs. Keep this in mind when troubleshooting a system—if the malfunction seems related to the feature or features provided by a single IC, the likely causes of the problem are either in the IC itself or the decoder circuit ICs which enable it.

QUESTIONS FOR SECTION 4-1

1. What is a decoder in a multiple IC system?

2. Why is a decoder needed? Where is it used?

3. What is the Enable signal developed by the decoder?

4. What inputs does the decoder use?

5. What logic level is most commonly used as an Enable?

6. What is a qualifier? Give two different places where it is needed.

7. What is one symptom that often points to a decoder failure? Why?

PROBLEMS FOR SECTION 4-1

1. Draw a decoder which provides a binary 0 Enable for the single address 0110.

2. Draw a decoder for reading address 1001, using the address itself, and a Read line from the processor which is high when reading an address. The decoded output should be low.

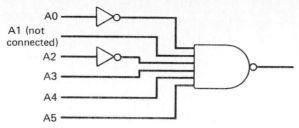

Fig. 4-5 Section 4-1, Problem 3.

3. Will the circuit shown in Fig. 4-5 act as a decoder for a block of addresses 1000 through 1010?

4. Develop a decoder for address 1000 through 1011. The Enable should be low, and there should be a single low qualifier that generated by the processor when the processor is doing a memory cycle.

5. A decoder is needed to enable one IC at address 1011 when a Read cycle is occurring (processor Read-Write line is high) and an Enable at another address 1110 when a Write cycle is occurring (processor Read-Write line is low). Draw the decoder for this with a low output.

4-2 MULTIPLEXERS

Consider a case where a microprocessor-based alarm system has to monitor the open or closed status of dozens of doors and windows. One way to do this would be to run an individual signal from each door or window to the processor, using flip-flops if necessary to latch the information and perform handshaking (Fig. 4-6). However, most processors and systems really do not have more than 8 or 16 input lines. ICs themselves usually have only 40 leads and many of these are needed for address, data, power, ground, and other signals. In general, processor systems do not have enough input pins to handle all the individual "outside world" inputs that they may have to process.

What can be done? The solution is to use a circuit called a *multiplexer* (*mux*), or data selector. The multiplexer receives a large cluster or input lines, and only lets the logic state of *one* of the lines through to the processor. Which signal gets through is determined by a multiple-bit code that is given to the multiplexer, either from the processor directly or from a counter. The multiplexer is made up of the basic gates studied: AND, OR, and NOT. These gates are arranged so that they recognize the code (called a select or address code) that determines which input bit should be passed through. Only the bit on the selected input line is allowed to pass through the circuit of the multiplexer; all others are blocked.

Looking at Fig. 4-7, there is four-line to one-line multiplexer. Two select lines, A and B, are used. Since two lines have four possibilities (00, 01, 10, and 11), this circuit can uniquely select from up to four inputs. As a selected code appears on lines A and B, only one of the AND gates will have either one or both select inputs equal to 0, which means that the output of the AND gate is 0 regardless of any

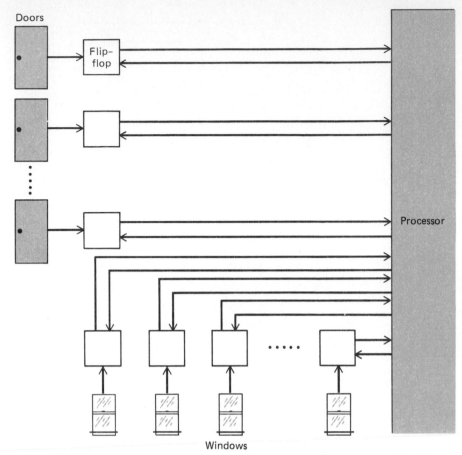

Fig. 4-6 A large number of individual wires to the processor is needed to monitor many doors and windows in an alarm system.

other inputs to the gate. In effect, one AND gate is enabled while the other three are disabled. This disabled gates have a 0 output, regardless of the data input. (Recall that A AND 0 = 0 from the study of boolean truth tables). The selected gate is enabled and the input data passes through, since A AND 1 = A. The outputs of the four AND gates are ORed together and passed on to the rest of the system or processor. Since three outputs are 0, the OR function passes the data without change (A OR 0 = A).

The result is to condense four input data lines onto a single line by providing signals to act as selects. But, since N select lines can identify 2^N inputs, there is considerable saving on the use of lines, especially as N gets large. For example, an 8-bit select code can control a 256-to-1 multiplexer (why?). Most practical multiplexers also have an additional input called a *strobe*, which serves the same function as the clock in a flip-flop-based circuit—it allows the select lines to settle down to their final value before the circuit and output are activated.

Multiplexers are often used with systems that have keyboards on which the user types in messages. A typical computer keyboard has anywhere from 80 to 100 or more keys (add up all the letters, numbers, punctuation, and special symbols). The

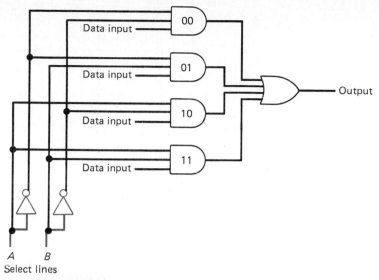

Fig. 4-7 A 4-line to 1-line multiplexer. The select code on lines *A* and *B* identifies which of the four inputs is to be passed to the output.

keys are electrically arranged in a matrix of rows and columns, and each key is connected at the junction of only one specific row and column. Assume a 100-key keyboard with 5 rows of 20 columns (Fig. 4-8). The processor, which wants to see which key is pressed by the system user, turns on a row of keys by putting a logic 1 onto that row. The processor then scans the 20 columns to see if any key or keys are depressed. It uses a multiplexer for this, rather than devote 20 wires to the columns. The processor does this by controlling a 20-line to 1-line multiplexer, which requires only 5 select bits (2^5 is 32, which is greater than the 20 required) and checking the multiplexer output after each new select code from 0000 to 1 0011, thus covering the range of 0 to 19, or 20 columns. After each select code, the processor notes if the output of the multiplexer was a 1, which indicates a depressed key, since the key is making the connection between the activated row line and the column (Fig. 4-9). The processor is preprogrammed with a table showing which key is located at each row and column intersection, so it can identify the key based on its knowledge of the activated row number and the select code (column) number. The entire row select and column select is done much more quickly than a person could push and release a key, so no keystrokes are missed.

This use of multiplexers has reduced the number of wires or input and output lines required for a 100-key system to 11—5 for the rows, 5 for the column select code, and 1 for the actual keystroke data bit. This can be reduced further by the use of demultiplexers.

There are standard ICs available which have all the circuitry necessary to provide the multiplexer function for 1 of 4, 1 of 8, or 1 of 16 inputs, using two, three, or four select lines. Larger multiplexers can be made by connecting the smaller ICs together and adding some additional logic outside the standard IC package.

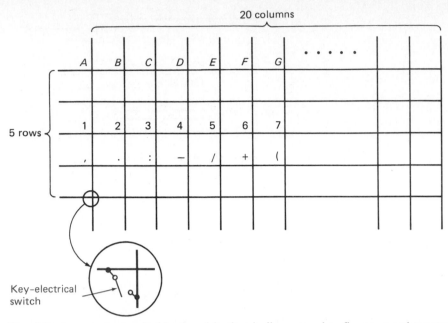

Fig. 4-8 A computer terminal keyboard is electrically arranged as five rows and twenty columns. Each row-column intersection has a switch corresponding to a character on the keyboard. Depressing the keyboard key closes the switch and connects the row to the column.

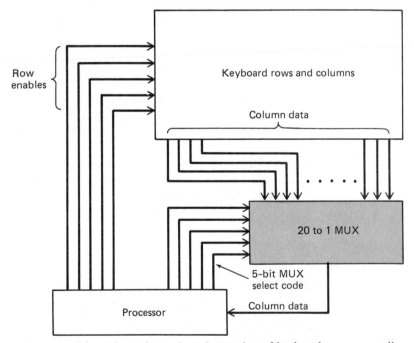

Fig. 4-9 A multiplexer is used to reduce the number of keyboard-to-processor lines required. The processor activates one row, then goes through multiplexer column-select codes 0 through 19 to allow the column information (corresponding to which keys are pushed) to reach the processor.

QUESTIONS FOR SECTION 4-2

1. What is the general problem that a multiplexer solves?

2. What does a multiplexer do that solves the problem of the previous question?

3. How does a multiplexer function? What is it built of?

4. What is the role of the select lines? The select code?

5. How many select lines are needed to have a 64-to-1 multiplexer?

6. Give an example of a multiplexer application. Explain how the multiplexer works in this application.

PROBLEMS FOR SECTION 4-2

1. There are four data inputs into a multiplexer. The data on input 0 is a 0, on 1 it is a 0, on 2 it is a 1, and on 3 it is a 1. If the select code is 01, what is the multiplexer output? Repeat for a select code of 10.

2. The data input to be selected on an eight-line to one-line multiplexer is from the third input. What should the select code be?

3. The data on the inputs of a 16-to-1 multiplexer is, starting from the 0 code, as follows: 0001 1000 1010 1100. What will be the output when the select code is 0110? When it is 1000?

4. Verify by truth table the operation of the multiplexer of Fig. 4-7.

5. Verify by truth table that the circuit shown in Fig. 4-10 of two identical 4-to-1 multiplexers acts as a single 8-to-1 multiplexer.

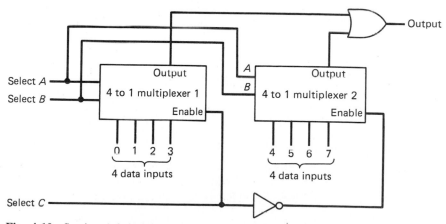

Fig. 4-10 Section 4-6, Problem 4-2.

The *demultiplexer* (*demux*) performs the opposite operation of the multiplexer. The demultiplexer receives data on a single line, and, based on a select code, routes it to one of several outputs. As with the multiplexer, the demultiplexer can control 2^N possibilities, where N is the number of select bits.

The previous section had the example of a keyboard with five rows, and it took five output lines from the processor to activate the desired row. In many cases the processor does not have this number of lines available, or the keyboard may have more rows than five and require even more processor outputs. The demultiplexer reduces the number of lines considerably.

Look at the demultiplexer in more detail by studying a two-line to four-line demultiplexer (Fig. 4-11). The single data line goes to all the AND gates of the demultiplexer. However, because of the select codes, only one of the gates will be enabled and the input data is blocked from the other ones. In effect, the demultiplexer has steered the single input to one of four possible outputs, based on the received 2-bit select code. The demultiplexer "decodes" this select code and uses it to direct the input to the proper output line.

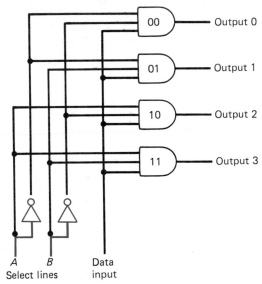

Fig. 4-11 The demultiplexer "steers" the single data line to the appropriate output. This is a two-line to four-line demultiplexer.

How does this help in the keyboard example? Assume the input data to be demultiplexed is fixed at 1. Then, let the processor that is trying to control and read the keyboard use three output lines (Fig. 4-12). As the processor steps through the select codes 000, 001, 010, 011, and 100 (decimal 0 to 4), the logic 1 on the input will be directed to the five lines from the processor, but with only three lines. By using a multiplexer and a demultiplexer, the system can read a

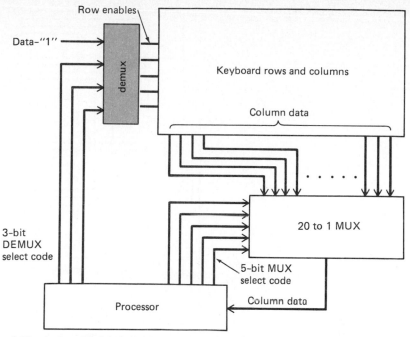

Fig. 4-12 A demultiplexer can be used to reduce the number of processor output lines needed to activate keyboard rows.

100-key keyboard with only three output lines for the demultiplexer, five for the multiplexer, and one for the key column data, for a total of only nine. This is much less than are required to read 100 unrelated points with individual wires.

The demultiplexer studied serves to take a single data line and steer it onto an output line, based on the select code. Ignore the input data line for a moment, and it really is a decoder—something that takes a binary number (the select code) and decodes it to a decimal number (Fig. 4-13). For the case studied with a 2-bit select code, the two are bits decoded to one of four outputs corresponding to the decimal equivalent of the two bits. The point to remember is that a demultiplexer and a decoder are very similar. In a demultiplexer, the input data is steered by select codes. In a decoder, the input is fixed at a 0 or a 1, and what the decoder function does is convert the select code into a decimal number by turning on the gate output corresponding to that decimal number. This demultiplexer-decoder is similar to the IC-selecting decoder studied at the beginning of this section. The same term is used for both functions.

There are ICs available that act as 2-line to 4-line, 3-line to 8-line, and 4-line to 16-line decoders-demultiplexers in a single package. Because the decoder and demultiplexer are so similar and differ only by having an input data line, the same IC is often used for either purpose. The input data line is either used for data (demultiplexer) or wired to a logic 1 or 0 (decoder).

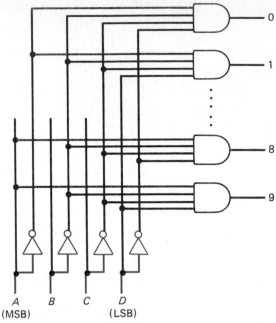

Fig. 4-13 A BCD-to-decimal decoder takes four input bits, decodes them to decimals 0 through 9, and sets the corresponding output high.

Troubleshooting Decoders, Multiplexers, and Demultiplexers

Decoders, multiplexers, and demultiplexers are generally not difficult to troubleshoot. They are constructed from boolean gates and do not have memory (as flip-flops do). They can be checked on the basis of their function and truth table alone.

As always, first check exactly the type of IC and identify the function and purpose of the IC in the circuit. The power supply and signal lines are next checked for voltage levels that are correct and proper for the type of IC. Finally, the function of the IC can be checked by examining the inputs (data, select lines, qualifiers) and the outputs, and seeing if the outputs correspond with what the truth table indicates is correct for the input state. If the system can be stopped and controlled by the test engineer, then the circuitry can be checked with a voltmeter or logic probe. Some systems have provisions for "single-stepping" the entire system so that it can be manually stepped through all operations. If the single-step operation is possible, checking out these ICs is straightforward. If the system can only be run at its normal speed, a multiple-channel oscilloscope or logic analyzer is needed to allow observation of many signal lines at the same time.

QUESTIONS FOR SECTION 4-3

1. What is a demultiplexer? Why is it needed?

2. What is the functional relation between a multiplexer and a demultiplexer?

3. What does the select code do for a demultiplexer?

4. Give an example of where a demultiplexer would be used, and why.

5. What is the functional difference between a decoder and a demultiplexer?

6. What is the meaning of the term three-line to eight-line demultiplexer?

PROBLEMS FOR SECTION 4-3

1. Verify the operation, via truth table, of the two-line to four-line demultiplexer of Fig. 4-11.

2. How many select lines are needed to steer an input to one of 32 outputs?

3. An input bit is to go to output 40 of a 64-output demultiplexer. What is the select code?

4. Show that the circuit of Fig. 4-13 properly decodes these inputs: decimal 8, decimal 9.

4-4 DECODERS FOR NUMERIC FORMATS

In a previous section decoders were used to take a combination of digital inputs and convert them to another combination of digital signals, such as from binary format to decimal. Decoders are used for many types of conversions, and they are also used in combination with other specialized circuits to provide outputs to the system user. Decoders convert a binary form that is useful in one situation (such as used within a circuit) to a form that is needed in another part of the system or product (such as a display for the operator).

First, a word about the terms used. Some circuits are encoders, and some are decoders. But whether a circuit is one or the other is sometimes confusing, and often an encoding function looked at one way is a decoder when looked at another way.

Many decoder (encoder) formats are used in actual systems, and there are ICs available that provide these functions. We will look at:

- Decimal to BCD (often considered an encoder)
- BCD to 7-segment display
- Decimal to 7-segment display
- Binary to hexadecimal 7-segment display

Decimal to BCD

The purpose of a decimal-to-BCD encoder (as it is usually referred to) is to take any 1 of 10 input lines, 0 through 9, and convert the line number to its BCD equivalent on 4 output lines. This type of circuit might be required, for example, where the operator of the control panel of some equipment has to push one of several switches (Fig. 4-14). The switch number must go into the equipment circuitry and processor and also be displayed on a small digital readout so the operator knows that he or she has pushed the right switch and the computer has "seen" it. In this case, the processor circuitry would prefer to read a 4-line BCD

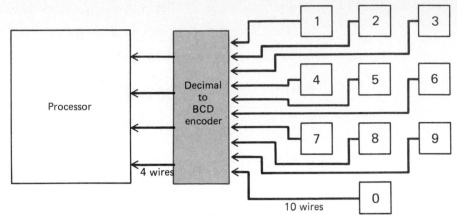

Fig. 4-14. The use of an encoder to convert 10 lines of switch information to BCD format.

format rather than a 10-line decimal format. The four lines can also go to a digital numeric display for the operator.

The decimal-to-BCD circuitry is built using a combination of boolean gates. By studying the pattern of decimal numbers and the corresponding BCD numbers, there is a clear relation between the two types of numbers observed:

	BCD
Decimal	**MSB LSB**
0	0000
1	0001
2	0010
3	0011
4	0100
5	0101
6	0110
7	0111
8	1000
9	1001

A circuit is used which takes the binary number pattern as its input, recognizes this pattern, and activates one and only one of ten output lines (corresponding to the binary input value). It usually includes a Strobe line (or Enable) which means that the outputs are all 0 until the Strobe is in the active state of 1. The purpose of the Strobe is to prevent false outputs while the decimal input lines are changing or not yet settled.

The BCD to 7-Segment Decoder

The digital readout used on most instruments is a 7-segment display. Look at a digital watch or calculator—each digit is formed by a rectangle with a bar across the middle, made up of seven individual segments (Fig. 4-15). Using these seven

Fig. 4-15. A 7-segment digit and the segments that correspond to numbers 0 through 9.

segments, it is possible to display all the digits 0 through 9, as well as some letters or special symbols. In order to convert the BCD format within the system to the format needed to activate the proper segments, a BCD 7-segment decoder is used.

It is standard industry practice to label the segments with the small letters a, b, c, d, e, f, g, as shown in the figure. Required is a circuit that does the following for each number:

Decimal Number	BCD Equivalent	7-Segment Equivalent
0	0000	a,b,c,d,e,f on; g off
1	0001	b,c on; a,d,e,f,g off
2	0010	a,b,g,e,d on; c,f off
3	0011	a,b,c,d,g on; e,f off
4	0100	b,c,f,g on; a,d,c off
5	0101	a,c,d,f,g on; b,e off
6	0110	a,c,d,e,f,g on; b off
7	0111	a,b,c on; d,e,f,g off
8	1000	all on; none off
9	1001	a,b,c,d,f,g on; e off

As each BCD number appears on the four input lines of the decoder, the output of the circuit turns on the appropriate segments for that digit (Fig. 4-16). A Strobe or Enable line is usually included so that the display does not momentarily flicker with false numbers. The 7-segment format is a binary code that is used only for externally presenting data to a user through a digit; it is never used within a circuit for computation. This is because it requires seven signal lines to represent each digit, compared to BCD which requires only four, or regular binary which is even more efficient in the use of lines than BCD.

Decimal to 7-Segment Decoder

In some cases the data for the display does not come from a source that is in BCD format. Instead, the source is already in decimal format, and so the display requires a decimal to 7-segment decoder. When any one of the 10 input lines goes

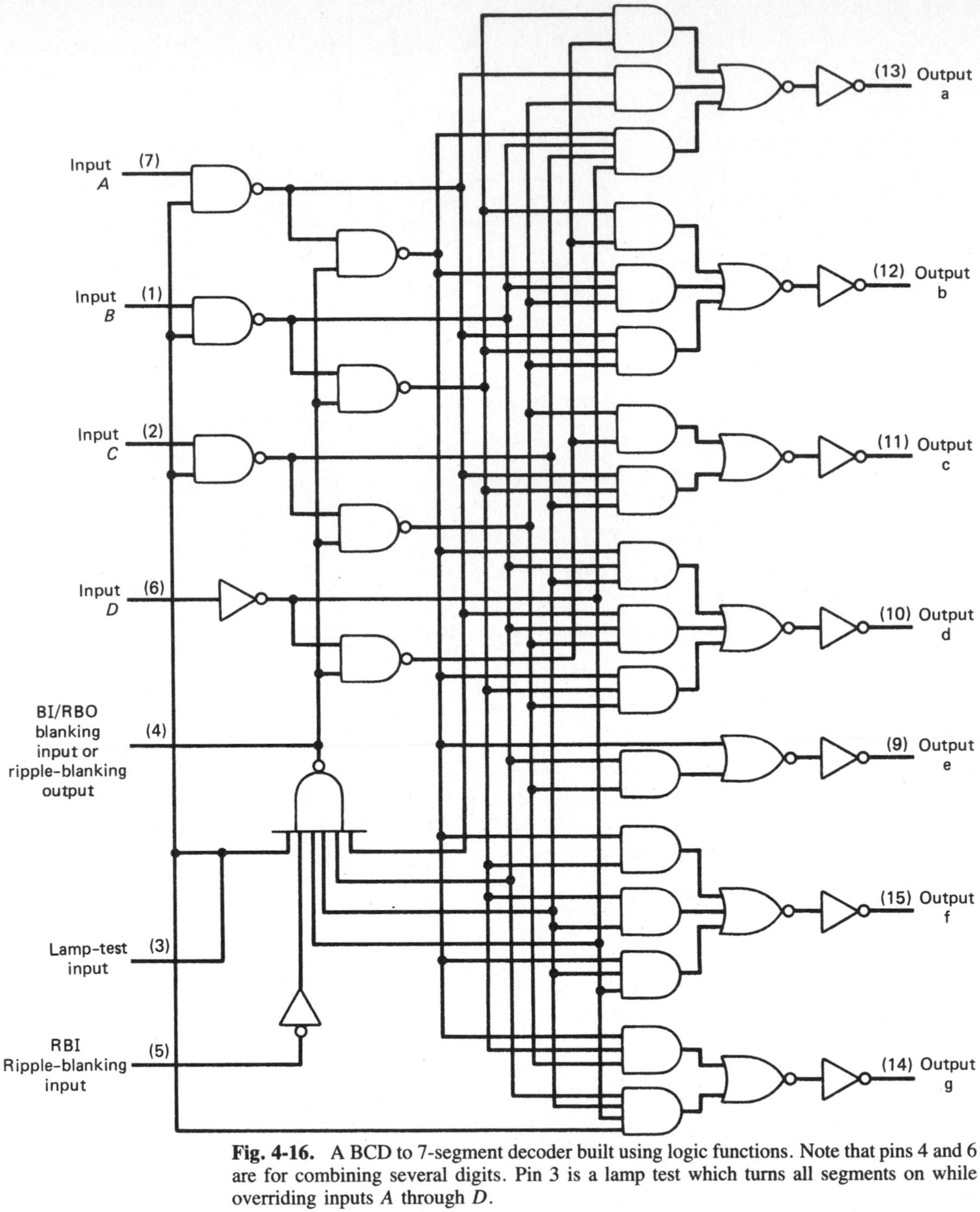
Input A (7)
Input B (1)
Input C (2)
Input D (6)
BI/RBO blanking input or ripple-blanking output (4)
Lamp-test input (3)
RBI Ripple-blanking input (5)
(13) Output a
(12) Output b
(11) Output c
(10) Output d
(9) Output e
(15) Output f
(14) Output g

Fig. 4-16. A BCD to 7-segment decoder built using logic functions. Note that pins 4 and 6 are for combining several digits. Pin 3 is a lamp test which turns all segments on while overriding inputs *A* through *D*.

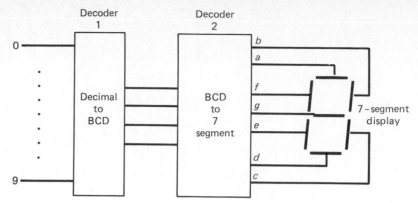

Fig. 4-17. A combination of decoders can be used to produce decimal to 7-segment decoding.

high, the appropriate segments, a through g, should be turned on by having the decoder outputs go to 1. There are two ways to do this:

- Have a decimal-to-BCD decoder and have the output of this decoder go into a decimal decoder (Fig. 4-17). This method will work but it uses a lot of circuitry and a large number of gates, since there are many similarities between both functions.
- Look at overall function desired, and build circuitry out of gates that goes directly from decimal to 7-segment format.

In actual practice, the second choice is almost always made because it uses the least amount of circuitry and the fewer ICs. The result is lower cost and higher reliability in the system.

Hexadecimal to 7-Segment Decoder

The BCD format uses a 4-bit binary number to represent 0 through 9. This leaves 10 through 15 (binary 1010 through 1111) unused. The hex format uses the full range of the four bits and is often used as a convenience when talking or writing about binary or BCD numbers. The bits are simply grouped together in 4-bit units and then the number can be referred to by the hex value of each group. Many types of test equipment used with computer systems and circuits have hex format, and some programming of microprocessors is done in hex.

Consider a test instrument that is designed to let the user watch 16 digital lines at once. It could use 16 individual indicators (either lamps or 1's and 0's on a screen), but it would be very hard to read all 16 and compare them to the correct pattern or describe the pattern to someone else. Instead, the 16 lines are shown in BCD format, which means only 4 characters instead of 16.

Each character represents 4 bits. Obviously, binary 0000 through 1001 are shown as 0 through 9 on the display, but what about binary 1010 through 1111? These are decimal 10 through 15, but only one displayed digit is allowed. Usually the letters A, B, C, D, E, and F are used, and they are shown on the 7-segment display as indicated in Figure 4-18. This set of characters on a 7-segment display

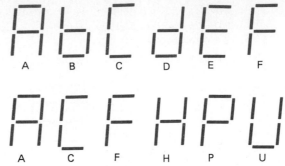

Fig. 4-18 Different sets of 7-segment letters are used in some test equipment.

can be confused with some of the numbers, so sometimes the letters A, C, F, H, P, and U are used, which are clear and unambiguous. For either situation, a decoder is needed which (like the BCD-to-decimal decoder) takes the four signal lines and decodes them to turn on the appropriate segments of the digital.

QUESTIONS FOR SECTION 4-4

1. What is the function of a decoder for number formats? For data in one format that needs to be in another format?

2. What is decimal-to-BCD decoder?

3. Where might decimal-to-BCD decoders be used?

4. Why do decoders such as decimal to BCD also have a Strobe line?

5. What is 7-segment format? How are the segments identified?

6. What is the function of a BCD to 7-segment decoder?

7. What is a decimal to 7-segment decoder? What are the two ways it can be built?

8. Where is hex format used? What characters are used to represent decimal 10 through 15 on some 7-segment displays? Why?

9. Explain the difference between the term decoder as used in this section versus how it was used in the section on decoders for enabling ICs.

PROBLEMS FOR SECTION 4-4

1. Sketch a decoder that recognizes binary 1100 only and puts it on a 7-segment display. A logic 0 turns on a digit segment.

4-5   DECODERS AND DISPLAY DRIVERS

The circuits studied provide the functions needed to decode various formats to a 7-segment display, but they do not provide the actual driving circuitry. This is because the physical display itself has unique requirements. Look at two very common types of displays: the light-emitting diode (LED) types and the liquid crystal display (LCD) type.

Both the LED and the LCD have advantages over the other in a particular application. Generally, the LED is bright, usable in the dark, and requires little special drive circuitry. On the other hand, it uses a lot of power, which is a major drawback for battery-powered equipment. The LCD uses very little power but requires special drive circuitry and this circuitry becomes extremely complex as the number of digits increases.

The gates of the decoder ICs cannot drive either LED or LCD displays directly. LEDs usually require more current than a standard gate can provide, while maintaining the logic 0 or 1 voltage. Instead, the gate is used to drive a transistor (Fig. 4-19), which can control and provide the current [around 10 milliamperes (mA) for a typical LED]. When the gate output is a 1, the transistor turns on and allows current to flow to the segment of the LED; the transistor is off when the gate output is a 0 and no current can flow to the LED segment. The transistors which drive the segments can be discrete (or individual) transistors wired into the circuit. Special ICs are also available, called *drivers*, which consist only of the seven transistors needed to provide current to or drive the seven segments.

There is a second problem in driving these LEDs. Since a digit requires seven transistors (or the equivalent in a single IC), a display with several digits would require many transistors. A display with four digits would require $4 \times 7 = 28$ transistors. This number of components would take up considerable space on the

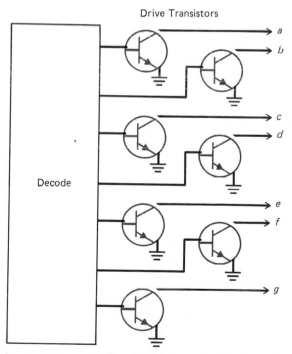

Fig. 4-19 Transistors usually are used to control the current flowing to the segments. The transistors are controlled by the decoder.

circuit board and add to the system cost. The solution is, once again, to save lines by multiplexing. Here is how it is done.

All the a segments of each digit are wired in parallel, connected to the same driving transistor, and likewise for the b segments, etc. So far, this has used seven transistors, regardless of the number of digits. However, when the transistor turns on, all the a segments of all the digits will also turn on—certainly not a useful scheme! We also need a separate way of controlling the digits individually. A separate transistor is used as the control for the power to each digit (Fig. 4-20). This means that a segment will only turn on if two things happen: the transistor that controls the segment is on, and the transistor that controls the digit is also on. This has reduced the number of transistors needed to just seven (one for each segment) and one for each digit. For a four-digit display, this is $7 + 4 = 11$, compared to the 28 we needed previously. The savings are even greater as the number of digits increases.

The transistors that control the overall power to each digit are in turn controlled by separate demultiplexer circuits that turn each digit on for a short period. During this time, the four BCD input lines have the code corresponding to the value of that

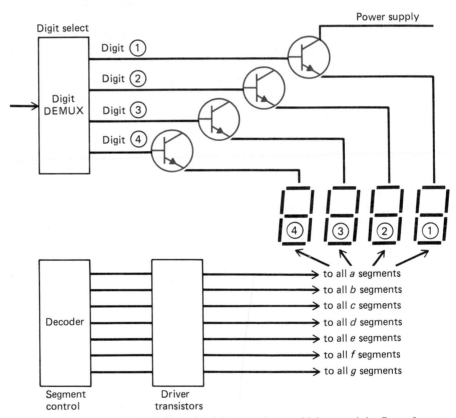

Fig 4-20 A fully multiplexed drive circuit has transistors which control the flow of current to the entire digit as well as to the segments. This is a circuit for four 7-segment display digits.

digit. This is decoded into the 7-segment information and the appropriate segments of that digit only light up. Then, the information for the next digit is presented, and the demultiplexer circuitry advances by one and turns on the next digit. In effect, the display digits are lit up one at a time—each digit is turned on by the demultiplexer and the proper segments of that digit only are turned on by the BCD to 7-segment decoder. The scan through the digits happens so fast (usually 30 to 60 times per second) that the eye of the observer thinks the digits are on all the time.

This overall scheme is called a *multiplexed decoder-driver*. It is very commonly used in multiple-digit displays (such as on a calculator or frequency meter). There are ICs available which combine all these functions onto a single chip. The entire circuit is a simple, easy-to-use building block in IC form.

There is a drawback to this multiplexed decoder-driver; it can be difficult to troubleshoot. First, because the digits all *appear* to be on at once, it may be hard to relate what you see on the display with what the various test instruments you are using indicate. This is because your eye sees over a longer time period, while the instruments show very short time intervals. Second, a slightly different approach to thinking about what is observed is needed. For example, if all the a segments in all the digits are off, then most likely the decoder or a driver transistor is bad. On the other hand, if only one digit is dark but all the others seem to be working properly, then probably the demultiplexer output or driver transistor for that digit is bad. The real problem in checking 7-segment displays is in knowing what the display should be showing compared to what it is actually showing. If the displayed number should be a 9, but instead the b segment is bad, you will see another number, 5. Unless you know that 9 should be showing, you will not realize the problem. In some systems, there is no way for you to directly control the display and get it to show all the digits, and so you have no control over what is shown. This is different from a simple calculator where you push a number on the keyboard and that number should show on the display. For this reason, most systems have an "all 8's" test, which is initiated by the service technician. This test turns all digits to 8, so the display shows 8888, and all segments of all digits should be showing. If they are not, you can then begin to diagnose the problem logically.

The LCD type of digital display requires a special drive signal which alternatively goes positive and negative. This signal can only be provided by a special driver IC, which has some fairly complicated circuitry. Generally, LCD displays are not multiplexed as LEDs are, since the LCDs cannot be turned on and off at a fast enough rate to make them appear always on to the eye. LCDs usually have a wire to each segment of each digit (Fig. 4-21). This is not as bad as it may seem, since the LCD requires only a tiny amount of drive current and does not need an external transistor to provide it. LCD drivers are large ICs with signal lines for the digit input codes and a line for every segment of the display. Some LCD drivers allow the user circuitry to multiplex the input digit values so that not as many BCD lines are needed from the system; others require four input bits on separate wires for each digit. Obviously, a LCD with eight digits would usually use the multiplexed BCD values. Otherwise the IC would need 4×8 pins for the digit values and 7×8 for the segments, plus power and ground, for a total of 90 pins.

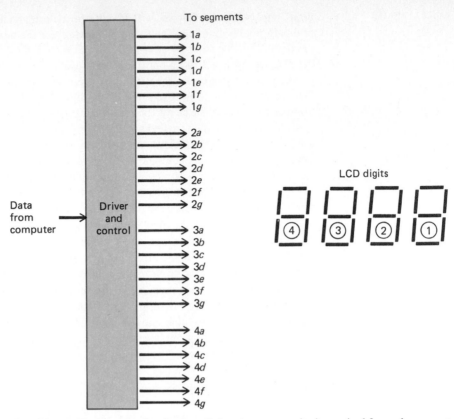

Fig. 4-21 LCD drive circuitry for four digits. A separate wire is required for each segment of each digit.

QUESTIONS FOR SECTION 4-5

1. What are the two main types of 7-segment displays?

2. What are the advantages and disadvantages of each?

3. Why is a display driver needed between the decoder and the display?

4. How is an LED segment driven?

5. How are the segment drivers combined to save on driver circuitry for a 7-segment LED?

6. How many transistors are required without the special circuitry?

7. Explain how demultiplexers are used to save on the number of driver transistors.

8. What is a troubleshooting drawback to the scheme in question 7?

9. How are LCDs driven? Can the demultiplexer scheme be used with LCDs? Why not?

PROBLEMS FOR SECTION 4-5

1. For a six-digit LED display, how many transistors are needed if the display is not multiplexed? If it is multiplexed? (Assume the input is BCD code).

2. For a 10-digit LCD display, how many drivers are needed?

4-6 A COMPLETE COUNTER AND DISPLAY CIRCUIT USING ICS

Now that we understand several types of IC circuits, we can combine them into a fully functional system. A good example is a counter circuit with readout. Look at the block diagram of a counter that is designed to count the number of input pulses that occur in a 10-s period (Fig. 4-22). In order to keep things from getting too complicated, restrict the counter values to a range of 0 through 9. (Actually, there are many practical applications where this range is useful, such as counting the rate at which people use a certain door, or measuring the frequency of very low frequency waveforms.)

The ICs required are:

- A 4-bit counter, which counts from 0000 to 1001. Input pulses from the event being counted go into this IC.

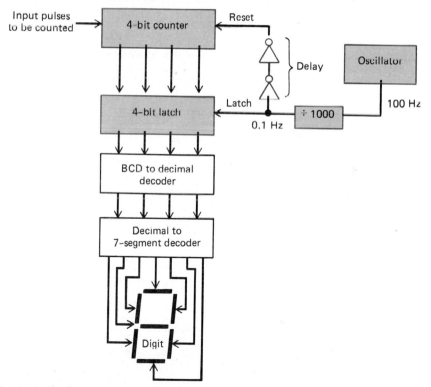

Fig. 4-22 Block diagram of a counter which counts the number of pulses (up to a maximum of 9) that occur in each 10-second period.

- A 4-bit latch register, which can store the value of the counter.
- A BCD to 7-segment decoder.
- A clock oscillator, which is providing a 100-Hz square wave to act as the system clock.
- A counter wired to act as a divide-by-1000 circuit, to provide a pulse every 10 s from the system clock (which is running at 100 Hz).
- Some NAND gates to provide some miscellaneous logic functions.

The system works this way:

1. Pulses from the event to be counted, such as a switch connected to the door, are counted by the counter.

2. While the counting is going on, the system Clock signal is being divided down to a 0.1-Hz (10-s period) signal.

3. Every 10 s, the divider produces a short pulse. This pulse goes to the 4-bit latch register and causes it to latch and hold the value of the counter at that instant.

4. The same latching signal goes through four NAND gates, wired as inverters. The logic state of the signal is not changed, but it is delayed by approximately 50 to 100 ns while going through the NANDS.

5. The delayed pulse goes to the reset line on the counter, so that the counter can reset to 0000 and begin counting over the next 10-s period.

6. The latched value of the count goes from the register to the BCD to 7-segment decoder, where it is decoded into the appropriate 7-segment information. It then drives the LED display.

There are a few things to notice about this system. While the counter is counting the input pulses, the displayed digit is not changing. This is because the latch register *holds* the count of every 10-s period while the counter *counts*. The delay (due to the NAND gates) is needed because the counter reset signal must not reach the counter until after the register has latched the count of the previous period. If there were no delay, the pulse from the divider would reach the counter and the latch register at the same time. The result would be that the register would, in many cases, be latching just as the counter was clearing to 0, and the count data would be lost. This situation, where a signal to one part of a system must be delayed slightly, occurs quite often in asynchronous circuits.

QUESTIONS FOR SECTION 4-6

1. What IC functions are required for a simple counter?

2. What is the function of each of these in the counter?

3. How does the counter work?

4. Why is the latch needed? Why are the NAND gates, and their delay, needed?

5. What method is used to make sure that the displayed count value does not change during the counting period, although the count itself is changing?

SUMMARY

In this section we have examined three types of ICs that are used in systems to provide specialized but very important functions. These are:

Decoders—used to identify a unique set of input conditions, and then enable an IC in a system based on an address and qualifiers, or provide a trigger based on these inputs.

Multiplexers—allow one of several input lines to be directed to a desired output line, under the control of a multiple-bit select code.

Demultiplexers-Decoders—used to direct data from a single line to one of a group of desired output lines, based on a select code. They can also be used to decode a binary value into the corresponding decimal value.

All of these functions are available in standard ICs. They are not complicated in operation because they are built up of basic gates, without flip-flop memory or internal feedback.

There is also a need for a wide variety of decoders and drivers, used to take one binary format and convert it to another, as well as provide the necessary physical signals for a display. Each of these types was developed because there was a need for them in different applications. ICs that perform these functions are available as standard parts. Some of the functions, such as decoding and driving a display, are used together so often that ICs are designed with both functions combined in a single package. This makes it easier to design a system at lower cost.

Troubleshooting decoders and drivers is straightforward because they are relatively simple combinations of gates and have a clear relationship between their inputs and outputs, defined by a truth table. The troubleshooting becomes more difficult when either the input signals or output signals, or both, are multiplexed to reduce the number of signal lines. For multiplexed systems, understand and relate what the multiplexed signals are doing, what is seen with the test instruments, and what is seen on the display.

REVIEW QUESTIONS

1. Explain the need for a decoder in a multiple-IC system.

2. What is an Enable signal? Where does it come from? What does it do?

3. What types of signals can be the inputs to the decoder?

4. What would be a limitation on a decoder and the ICs it was connected to if there were no qualifiers?

5. What is the reason for a multiplexer? What does it do?

6. Explain what the term eight-line to four-line multiplexer means.

7. What is the select code? The select lines?

8. How many select lines are used in a 1024-to-1 multiplexer?

9. What size multiplexer can be used with eight select lines?

10. How can a larger multiplexer be made of smaller ones?

11. What is a demultiplexer? How does it compare in function to a multiplexer?

12. How is the size of a demultiplexer described? Give two examples.

13. Where are demultiplexers used? Why?

14. What is the functional difference between a number format or data format decoder and an IC selection decoder?

15. What is the importance of the 7-segment format?

16. If 7-segment is so useful, why is it not used internally in systems?

17. How many unique values does the regular binary format indicate with four lines compared to BCD?

18. Why are two different 7-segment formats in use? What are they?

19. What is the difference between an LED display and an LCD display?

20. Give two good and bad points of LED and LCD relative to the other.

21. What is the role of the display driver?

22. Why is a multiplexed LED display necessary in most practical displays?

23. What is a problem with determining what is wrong with a multiplexed LED display?

24. How is the drive of the LCD different than it is for the LED?

25. Explain how the IC functions covered to this point are brought together in a counter.

26. What would happen if the counter did not have a latch? Would this be a useful function?

27. What decimal number is put on a 7-segment display to test the segments?

REVIEW PROBLEMS

1. Draw a decoder that provides a low output for addresses 1 1000 and 1 1001.

2. Repeat for a decoder that decodes addresses 1 1000 and 1 1010, but not 1 1001.

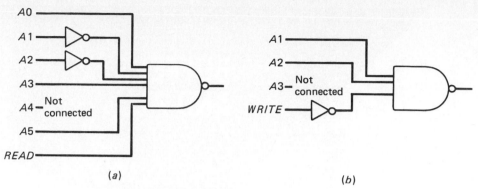

(a)

(b)

Fig. 4-23 Figure for Summary Problem 4.

3. Draw a decoder that is the same as the previous problem, but also decodes a Read (high) signal at 1 1000 and a Write (low) at 1 1010.

4. What address and qualifier are decoded by the decoders shown in Fig. 4-23a, and b?

5. A 256-to-1 multiplexer must select the one-hundredth input. What should the select code be? What about for the two-hundredth input?

6. Can a multiplexer with a select code of 5 bits allow selection from 1 of 34 bits? Why?

7. A demultiplexer has seven select lines. Can it steer an input to 1 of 128 outputs? To 1 of 256 outputs?

8. What code would be used on a 4-line to 16-line demultiplexer to direct an input to output 12?

9. Sketch the 7-segment display that would be shown for BCD 0110; for 0010.

10. Modify the counter of Fig. 4-22 to function as a totalizing counter, one which shows the total number of pulses received.

5

BUFFERS AND ALUs

A relatively simple IC called a *buffer* is often needed to provide a system with some specific characteristics and performance. The buffer can be designed for one or more roles: isolating one part of the system from another, protecting the circuit against various internal and external faults, allowing the output of a gate to provide a signal to many gate inputs simultaneously, and accommodating the voltage levels in the different parts of a system. There is a special buffer which provides both logic 0 and logic 1 output levels and can also be turned off so that it acts as if it is not in the circuit at all. This last kind of buffer is essential where many ICs share a common signal line, or bus.

A much more complicated IC is the arithmetic and logic unit (ALU). This IC can perform arithmetic operations on binary numbers, and perform boolean logic operations on groups of bits. The ALU is used whenever binary data must be used in calculations and decision making. It is available as an IC by itself or as part of larger, more complicated ICs. Some ALUs perform one function only, while others have signal lines that let the user select what ALU function should be performed.

5-1 BUFFERS

Digital signals in electronic systems travel from one point in the system to another, or from one system to another (from a computer to a printer, for example). Whenever the signals travel, buffers are needed for a large number of reasons. Without a buffer, the digital impulse would be reduced in strength by the time it reached the receiving end and be corrupted by noise picked up along the way. The result would be system malfunctions and errors as logic 1's appear as 0's or vice versa.

Buffers solve this problem and many other related ones. In this section we will discuss how buffers:

- Reduce the electrical load on the output of the IC, provide a known load, and provide more electrical drive power to the signal lines.
- Isolate and protect parts of a system from failures in other parts.
- "Clean-up" weak or noise-corrupted signals.
- Change signal voltage levels to meet the requirements of different parts of the same system or other interconnected systems.
- Allow more than one gate to use the same signal path in a "bus-oriented" system and allow the same path to be used for two-way signal flow.

The word *buffer* is the standard industry term for this IC. However, it is very *different* from the buffer studied in Chap. 3. That buffer used flip-flops to accumulate digital signals and then output them as needed, at various rates. The buffers of this section operate on a single digital line and accomplish very different goals.

Buffers make up an extremely important type of IC, and yet they are also the simplest type of logic function. A buffer does not change the boolean logic value of the signal passing through it (Fig. 5-1) (except for the special case of the inverting buffer which always changes the state of the signal). For noninverting buffers the boolean truth table is B = A; for inverting buffers it is B = A'.

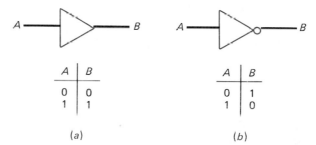

A	B
0	0
1	1

A	B
0	1
1	0

(a)   (b)

Fig. 5-1 Symbol and truth table for (a) Noninverting and (b) Inverting buffers.

To meet the various needs of circuits, many different buffer characteristics are needed, depending on the situation. Buffers are used at both the output of a gate, where the signal leaves the gate to go to the rest of the system, and at the input of a gate, where the signal arrives. Some buffers are called *line drivers* and *line receivers* because they are used for a special purpose in connecting one system to another. We will discuss all these applications of buffers and line drivers and receivers.

QUESTIONS FOR SECTION 5-1

1. What are three things that buffers do?

2. What is the difference between these buffers and the kind of buffer studied previously?

3. What does a buffer do to the logic state of a signal?

4. What is the truth table of a noninverting buffer? Of an inverting buffer?

5-2 BUFFERS AND LOADING

Any time the output of one gate sends a signal (drives) to the input of another gate, that input is called a *load* on the output. The load consists of the resistance, capacitance, and inductance of the wires and next input. There is a limit to how much of a load an output can drive properly with the correct voltage and current levels. This limit means that to drive a large number of loads, some help may be needed by the output of the gate. If an IC is connected to ICs of the same type, we can specify how many loads it can drive. This is called the *fan-out*. A logic IC can typically drive 10 to 20 similar types of ICs (has a fan-out of 10 to 20) (Fig. 5-2).

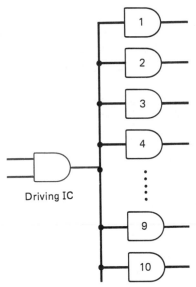

Fig. 5-2 An IC may be driving many similar loads. In this case, there are 10.

If the load is a different type, such as an IC driving a lamp or LED, then calculations are used to decide if the IC can drive the load directly.

If help is needed, a buffer is used. The buffer acts as a booster. It does not increase the voltage, but it increases the amount of current that can be supplied at that voltage. This is an important point: A signal is a logic 1 or 0, which refers to specific voltage levels depending on IC type. But at these voltages, there is also a value of current that must exist. If this amount of current does not exist at the voltage level, the system will not work properly. The buffer provides the extra current drive capability. The situation is similar to having a car with a trailer. When you do not have the trailer, you can do 55 miles per hour (mi/h) without any problem; with the trailer, you cannot go above 40 mi/h. You do not want more speed (voltage), but you do want more engine horsepower or torque (voltage and current) in order to reach 55 mi/h.

The buffer lets the IC output drive more loads, whether of the same type of IC or of another type. We saw in the previous section that an IC system used for a counter latch decoder could not drive the segments of a 7-segment LED directly. External transistors were used to supply the 10 mA that the segment required. Special high-current output buffer ICs may also be used to provide the current, at the same voltage that the decoder would have used.

Buffers are often used where the load on the gate output may vary. One effect of additional loads on the output of an IC is that the performance of the IC may get worse as it has more loads put on it. Just as in the case of the car with the trailer, the IC actually will slow down as the load capacitance increases. Even though the maximum load may be within limits, some critical applications require that the load not change. But the load may change if more circuit cards are plugged into the system, or more memory ICs are added to the empty sockets on a board to give the user more memory for other programs. In these cases, buffers are used to make sure that the output load seen by the IC does not change even though the amount of circuitry does (Fig. 5-3). The IC output sees only the constant load of the buffer input, while the buffer handles the varying loads and changing circuitry.

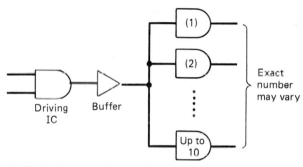

Fig. 5-3 The buffer isolates the driving IC from varying loads.

QUESTIONS FOR SECTION 5-2

1. What is the load on a gate output? What does it consist of?

2. Why is the amount and nature of the load important?

3. What is fan-out? What is a typical value for it?

4. What is the function of the buffer when the load is large? Is the output voltage changed?

5. How can large or varying loads affect performance of the IC that is driving them?

6. How does a buffer help the driving IC in the case of large or varying loads?

5-3 BUFFERS FOR ISOLATION AND PROTECTION

Most ICs are not electrically rugged. They will fail if they get too much voltage, either on the power-supply lines or on the signal lines. They also are very sensitive to short circuits; being connected to other ICs by accident; or if too much resistance, capacitance, or inductance is applied to their outputs. Even if the IC does not fail outright, the system it is in may malfunction as long as this unexpected or incorrect voltage or load exists.

It may seem that these unexpected things could not exist in a properly designed and functioning system. There are several ways that they could:

- A test technician could be probing around at the circuit board, looking for some other problems, and the probe may slip.
- A user could be attaching some other device to the system, such as a printer, and that device is bad or is the wrong type.
- One of the new memory chips installed to add more system memory may have a short circuit internally.
- Surges and other problems may occur on the power line. This is very common. The standard ac line power from the wall is not a clean sine wave, but is corrupted by unwanted fluctuations and electrical noise.

Buffers are used to protect the main part of the circuit from any damage. They also ensure that as much as possible of the system keeps working. For example, a system with buffers installed in the right places will make sure that a bad memory chip inserted lets the rest of the memory still be used, rather than "crashing" everything.

Buffers do this by acting as isolators between one part of the system and another, or one entire system and another. They isolate because their outputs are usually designed to be more rugged than a regular IC output. Just as important, the problems at the output of a buffer usually do not ripple back to the input of the buffer, which is the part of the buffer that the main part of the circuit sees. The effect of the buffer is to keep the external problems away from the internal part of the circuit. Consider the case of a large group of memory ICs connected to a microprocessor. Assuming the microprocessor had the drive capability to handle this load, a failure of any memory chip would appear on the signal lines shared by all the memory ICs and the system. Nothing would work properly. Now consider the same system but with a buffer between the microprocessor and each memory IC (Fig. 5-4). Any failure in the memory IC would be between that IC and the

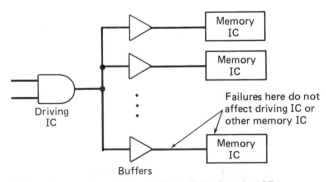

Fig. 5-4 A buffer protects the driving IC from faults in other ICs.

processor only. The other ICs would be unaffected, and the processor could still send signals to them.

The same situation occurs when a system has a connector for the user to attach to some external device. If this connector is driven directly from the processor or other internal circuitry, the wrong hookup could cause the whole system to crash. Instead, a buffer is used between the internal part of the system and the connector. If something is wrong externally, the external device will not work. However, the buffer has isolated the internal part of the system and so the system itself is both undamaged and unaffected.

QUESTIONS FOR SECTION 5-3

1. What are three things that could damage an IC? How do they occur?

2. Why do systems sometimes need isolation?

3. How do buffers provide isolation?

4. Explain how buffers minimize the effect of a localized failure in a system.

5-4 BUFFERS AND NOISE

As digital signals travel along a signal wire, they lose some of their voltage and often pick up noise. This noise may be from within the circuit or from electrical noise in the environment. The noise corrupts the digital signal, and if the noise is great enough, it may make a logic 0 look like a 1 and vice versa, despite the noise resistance (margins) inherent in digital ICs. Special buffers are available that have circuitry designed to clean up the digital signal before the noise reaches this level.

Consider a case where a logic 0 is represented by any value between 0 and 1 V, and a logic 1 is any value between 4 and 5 V. If less than 3 V is added to the signal, there is no danger of misunderstanding whether the digital signal is a 0 or 1. But if the noise increases beyond 3 V, there can be a mistake. Special buffers with what are called *Schmitt trigger* or *hysteresis* inputs are used to restore the digital signal to the proper levels (Fig. 5-5). This type of input has a threshold built

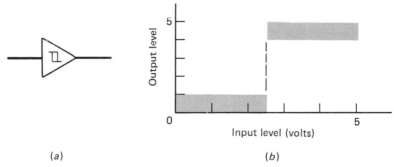

(a)   (b)

Fig. 5-5 Schmitt trigger input on a buffer. (*a*) Symbol; (*b*) Relationship between the input voltage and the output voltage for TTL levels.

in—any signal below 2.5 V going into the buffer is made into a logic 0 (below 1 V) and any signal above 2.5 V going into the buffer is made into a logic 1 (above 4 V). The effect of this type of buffer is that "dirty" logic signals are cleaned up before being passed on to the next part of the circuit or system. The result is that signals are not allowed to vary too far from what are acceptable values.

QUESTIONS FOR SECTION 5-4

1. What happens to a signal as it travels along a wire?

2. Where can noise come from? What can its effects be?

3. What is a Schmitt trigger? What does it do? What is another name for the function?

4. Why is it needed?

5-5 CHANGING VOLTAGE LEVELS

In many applications, different voltage levels are needed in different parts of a system or in interconnected systems. Some special buffers are designed to handle this situation easily and effectively. We will discuss two such applications.

Many digital circuits and ICs use relatively low voltage levels, such as 5 V. These same systems may have to control a device that needs higher voltages, such as an electromechanical relay which uses 24 V. The 5-V digital logic cannot be connected directly to a 24-V device—either the IC would burn up or it would not be able to provide the required voltage and current. Instead, a special driver buffer IC is used as the interface (Fig. 5-6). The driver buffer can accept digital logic signals that go up to 5 V and can itself be supplied by a 24-V power supply and thus provide digital signals at 24 V. The buffer makes the interface possible—it

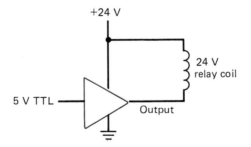

TTL "0" : Output at 0, current
flows through coil

TTL "1" : Output is at +24 V_1 so
coil has no current flow

Fig. 5-6 A buffer can act as a shifter of voltage levels. Here, acting as a shifter between a TTL signal and a 24 V relay coil.

allows control of 24 V, it protects the low voltage ICs from the higher voltage, and it isolates the two parts of the system.

Another very common interface is the one between a computer system and a peripheral device, such as a printer or terminal. The electrical standard which this interface uses is usually an industry standard called RS232. The RS232 specifications require signal voltages that are very different from those used by most ICs. While many ICs use 0 to 1 V as a logic 0 and 4 to 5 V as a logic 1, the RS232 standard calls for a voltage between -3 and -25 V for logic 1 and $+3$ to $+25$ V as a logic 0 (Fig. 5-7). In addition, because the device attached to the RS232

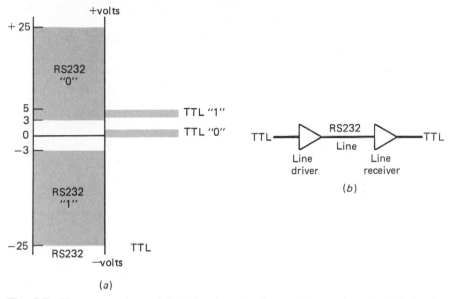

(a)

(b)

Fig. 5-7 The commonly used RS232 voltage levels are different than the TTL levels, shown in (a). Diagram (b) shows the use of RS232 drivers and receivers to send a signal from one system to another.

connection is often hooked up incorrectly by the user, the physical interface is required to withstand short circuits of various types, including being connected to $+25$ V when the system power is off. This voltage level and power situation would destroy most ICs, but special RS232 line driver and line receiver buffers are available. These drivers can take the lower IC logic voltages and provide signals at the required levels, and also withstand the power-off problem. The receivers are not damaged by the voltages and convert the larger voltage levels to values compatible with the ICs within the circuit. Since the RS232 interface is so common, many manufacturers provide the necessary interface ICs.

It has been shown that buffers serve a wide variety of needs and solve a large number of practical problems that are found in digital circuits. Many types of buffers are available as standard ICs, with each type specifically designed to do one or more type of buffering: more drive current, larger voltage levels, protection against failures, and so on. In some situations where the buffer is used only as a driver to allow more fan-out, the standard AND or OR gates that are otherwise unused in the circuit may be put to use as buffers. If 1 output can drive 10 other inputs, then 1 output can drive many more than 10 total inputs if the gates are used as inbetween drivers. This is why some circuit diagrams have gates which serve no logical purpose—they merely take the existing digital signal and pass it on. These apparently useless gates are actually serving as buffers within the circuitry.

If buffers do so many things, what are the disadvantages of using them, if any? There are several reasons why a system would try to use as few buffers as possible:

- Buffers add cost to the system.
- They are another component, and every additional component adds to the overall number of ICs in the system and so increases the chances of something going wrong or failing.
- Buffers, like any logic element, do not pass signals instantaneously. The few nanoseconds (billionths of a second) that it takes for the signal to go from the input of the buffer to the output (propagation delay) makes the overall system run slower. A complex system may have many buffers in the path of a single signal, and the small individual delays can add up quickly to a much larger delay.

QUESTIONS FOR SECTION 5-5

1. Why do not all systems or parts of a system use the same voltage value?

2. What is the role of the buffer when there are different voltages?

3. What is the RS232 standard? Where is it used?

4. What does a buffer do for the RS232 standard in terms of voltage levels and protection?

5. Why do some circuits have gates that serve no apparent purpose at first glance? Describe one such gate.

6. What are three disadvantages of buffers?

5-6 TROUBLESHOOTING BUFFERS

Buffers are among the easiest part of a system to troubleshoot. They do not do anything except pass the signal along. Noninverting buffers keep the same logic state between input and output, while inverting buffers (which are slightly faster for reasons that have to do with the internal aspects of IC design) invert the logic state. When troubleshooting, you have to make sure you know what type of buffer you are looking at. Then, compare the buffer input and output. The logic states should agree with the buffer type, and the voltage levels should also be compatible with the buffer type. Since one of the roles of buffers is to isolate a circuit from problems beyond the buffer output, you will find the most common problem to be that the input of the buffer is all right, but the output is not. This means that the output is being held at one level or the other by a short circuit or other external failure. Sometimes, the external failure is severe enough to damage or destroy the output stage of the buffer IC. In order to check this, disconnect everything connected to the buffer IC output. If the output is still stuck, the buffer has been damaged. If the output now follows the input properly, something connected to the output is at fault. In some unusual cases, the buffer has been internally damaged but still appears to follow the input properly when everything is disconnected, and not work when things are reconnected. In most cases, this means the problem is beyond the buffer, but in some instances the buffer has been damaged so that it can provide the correct voltages (which you see on your oscilloscope or voltmeter) but it cannot reach these voltages once a load (which draws current) is applied. It is a situation like a car engine that seems fine and smooth at idle, but has no power when the transmission is put into gear. The problem is not with the car power train, but with the engine itself.

QUESTIONS FOR SECTION 5-6

1. Why are buffers easy to troubleshoot?

2. What is the most commonly observed buffer problem? Why?

3. How is buffer operation checked?

4. What is the problem when the voltage levels on an input or output are good without a load but bad with the load?

5-7 BUFFERS IN BUS-ORIENTED SYSTEMS

Many of the applications for buffers discussed up to this point involve a single IC output connected to one or a few other IC inputs. Signals flow in one direction only. However, many systems and all microprocessor systems are bus-oriented systems, where the signal flow is more complicated.

A bus is a common path for the flow of logic signals (Fig. 5-8). A logic signal starts at one IC, usually the processor, and travels along this path. Many other ICs, often on different circuit boards, listen to this logic signal. These other ICs include memory ICs, timers, flip-flops used for handshaking, and special-function ICs to be studied later on. In a bus system, the part of the system that provides the digital signal has a buffer to drive the bus, and all the receiving parts of the system have

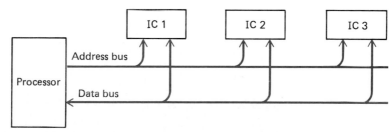

Fig. 5-8 A bus is a common path for signals between many ICs.

their own buffers to receive the signal and pass it on to their part of the circuit. There are two reasons for all these buffers:

- The driving part of the system will see only a number of loads equal to the number of receiving buffers. The driver does not see the number of loads beyond each receiving buffer. In other words, the load can be guaranteed by the design to be less than a certain maximum value.
- The buffers isolate the various parts of the system (usually separate circuit boards) so that failure on one board do not affect the entire system.

But buffers do much more than this. In a processor system, digital data flows to the processor from memory ICs, and from the processor to the memory ICs (as well as to and from other system ICs).

The typical system has perhaps a dozen ICs that may need to originate a digital signal and try to send it to the processor. This can be a problem. It is fine for many ICs to receive a signal on the bus, but it is more complicated for many ICs to send signals to the processor without causing confusion (Fig. 5-9). It is exactly like the difference between a group of people listening to one speaker and a group of people all trying to talk and be heard by one listener.

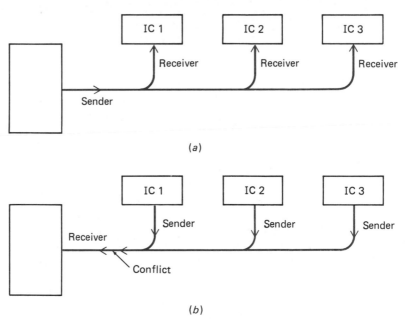

Fig. 5-9 A bus can have conflicts. (*a*) There is no conflict with one sender and several listeners; (*b*) There is a conflict with multiple senders at the same time.

The solution is to use a special kind of buffer called a three-state, or tristate, buffer (Fig. 5-10). The digital gates studied so far had two output states: 1 or 0 (also called high or low). In a three-state output, the output can be 1, 0, or *high-impedance*. The high-impedance state has the effect of turning the output completely off, and it puts out *no output at all*. It is as if the IC output is just not

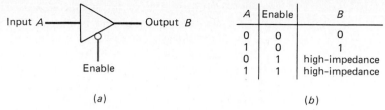

A	Enable	B
0	0	0
1	0	1
0	1	high–impedance
1	1	high–impedance

(a)   (b)

Fig. 5-10 The three-state buffer is activated by the enable line. (a) Symbol; (b) Truth table.

there. All that is needed is a way to control whether the buffer is in the normal (1 or 0) mode, or in the high-impedance three-state mode.

This is done by another signal line brought into the buffer called the *Enable line*. The Enable line does not set the output to a 1 or 0. It either turns the buffer output off (when in the disabled state) or lets the buffer present to the bus the logic output that the buffer has on its input (when enabled). Special circuitry in the system is responsible for controlling the Enable line itself. Most buffers use a logic 0 on the Enable line to turn the buffer active, while a 1 turns the buffer to high-impedance mode.

Now the bus situation is changed. There is one listener, the processor input, and many potential talkers, the outputs of the various three-state buffers (Fig. 5-11).

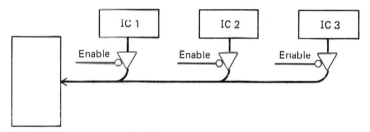

Fig. 5-11 The three-state buffer in a bus system can be used to allow only one talker at a time.

However, only one of these talkers is "authorized" to talk at any given instant, via the Enable line to that talker. The end result is that multiple sources of logic signals and data can share a single line, all of which need to talk to a single input at various times.

The development of the internal IC circuitry that made the three-state output possible was a major technical advance because it allowed all sorts of flexible designs and circuits to be used. Systems were no longer required to have a simple "end-to-end" link. Instead, common buses which allow efficient data flow could be used. Bus-oriented systems also save a great deal of circuitry.

Three-state buffers at the outputs of ICs also allow another very important system feature: bidirectional signal lines. These lines are most often used to carry data to and from the system processor. It is possible to use separate lines to carry the data to and from a processor and its memory ICs, but this is very inefficient and uses a lot of circuitry. Realize that data is stored in a memory IC, and that data storage location has to have two characteristics. The processor must be able to write to it, and the processor must be able to read from it. Even if two separate lines are used between the processor and the memory, they will rejoin at the flip-flop where the bit is stored. Therefore, using separate paths for the data being read and the data being written does not eliminate the need for bidirectional data paths—it only moves the problem further along.

The three-state buffer completely eliminates this problem and allows for a cleaner, easier to layout and troubleshoot design. The buffers are arranged as shown in Fig. 5-12. When the processor needs to send data to memory, it gener-

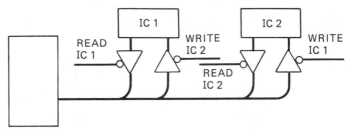

Fig. 5-12   The three-state buffer allows a bus to be used for bidirectional signal flow.

ates a Write control signal that is used to enable the buffers pointing in the direction of the memory. The buffers in the other direction remain disabled, and it is as if they were not there. When the processor needs to read data from memory, it generates a Read control signal that turns the other buffers on, so that data can flow from memory to the processor. As in the Write case, the buffers in the other direction remain off and so do not affect the operation.

Buffers and three-state operation are essential to nearly any functioning system. They are so useful that many of the ICs used with processors have their own three-state buffers built into the outputs of the IC chip itself to allow maximum flexibility in operation. This will be seen when these peripheral ICs are studied. The only drawback to three-state operation is that failures can make troubleshooting more difficult. If a buffer gets stuck, for whatever reason, in the enabled state, it will cause a failure of the bus line in both directions. Since there is another line involved, the Enable line, there is something else to go wrong: the Enable and three-state circuitry within the buffer IC, and the system circuitry that controls the Enable line. However, by examining the Enables, and the state of the buffer input and output when the Enable is active (IC output is on) you can usually narrow down the problem to the controlling circuitry, or the buffer itself.

QUESTIONS FOR SECTION 5-7

1. What is a bus? What is a bus-oriented system?

2. Why do bus systems need many buffers?

3. What is the three-state buffer? What bus problems does it solve?

4. What is the high-impedance state? What is the logic level when in that state?

5. What is the Enable line in a three-state buffer? What does it do?

6. What logic level is usually active on the Enable line?

7. Explain why bidirectional bus lines are an advantage in a system.

8. How does the three-state buffer make bidirectional lines practical?

9. Explain how bidirectional operation with three-state buffers works.

10. What is one drawback of three-state operation?

5-8 ARITHMETIC AND LOGIC UNITS

In previous sections, the binary system and the ICs that are used to develop binary numbers, convert them, and store them were covered. Next, the types of circuits that allow binary numbers and bits to be used to *make decisions* will be studied. There is a need to add binary numbers, compare them, and check certain individual bits within a binary group or byte. The circuits that do this are called *arithmetic and logic units* (ALUs). Since the bits of a binary group represent information, it is reasonable and necessary to make use of this information; ALUs help do this.

The arithmetic functions of the ALU are needed because any practical use of binary numbers involve calculations such as addition, subtraction, multiplication, and division. The ALU also provides logic functions. These include AND, OR, and comparisons on entire groups of bits. A group of bits, such as an 8-bit byte, can represent two different kinds of information. The entire group may be a number value, with each position of the group weighted twice the position to the right. Arithmetic operations are used for these groups. The group may also have bits which are not related to each other but are moving through the circuit and system together. A case might be where each bit shows if a door or window is open or closed. Each window or door is independent of the others, so the binary group really tells the state of the house entry points. These types of binary groups often need boolean logic functions.

The ALU is a very important part of a computer-based system. One common definition of a computer is that it is a device capable of performing calculations and making decisions based on data and information. The ALU makes this possible. The microprocessor, or "computer on a chip," is a very complicated IC that contains many of the functions of a computer, including the ALU, on the single piece of silicon in one IC package.

Arithmetic Functions

Arithmetic is a familiar subject. Addition, subtraction, multiplication, and division are used in many things each day. The same is true for computers. Arithmetic is one of the things computers need to do, and they can do it very well. The difference between a person doing arithmetic and a computer doing it is not just a matter of speed. Computers do their arithmetic using binary numbers. They also can internally do addition only (for most computer types).

If computers do addition arithmetic only, how can they do the other three functions of subtraction, multiplication, and division? The answer is that these three other functions can be made to look like addition with a little clever circuitry and preparation. Multiplication is just repeated addition: 3 times 4 is the same as $3 + 3 + 3 + 3$. In this way, any multiplication can be accomplished by repeated additions used as often as necessary. Subtraction is the same as adding the negative of the number you want subtracted: 10 minus 4 is the same as $10 + (-4)$. Finally, division is repeated subtraction, or repeated negative addition. The end result is that this fundamental operation of addition can be used for all arithmetic operations. This is important. It means that one good addition circuit in an ALU can provide all four arithmetic operations, which saves a great deal of IC design time and circuit space.

Binary Addition

Binary addition uses very simple rules, similar to the rules for decimal addition. In decimal addition, simply add up the numbers in each column. If the sum is greater than 9, carry the excess to the next column to the left. In binary addition, there are only 4 possibilities for a column:

First number	0	0	1	1
Second number	0	1	0	1
Binary sum	0	1	1	10

Note the carry for the last case, where $1 + 1 = 10$. The 1 of the 10 is the carry into the next column. See how this applies for binary numbers of 2 bits: The possible binary values to be added are 00, 01, 10, and 11. Here are examples:

$$
\begin{array}{cccc}
01 & 10 & 10 & 11 \\
+10 & +10 & +11 & +11 \\
\hline
11 & 100 & 101 & 110
\end{array}
$$

Binary addition can be expanded to add binary numbers with any number of bits, or columns. The same addition rule is applied to each column, starting from the LSB and working to the MSB.

QUESTIONS FOR SECTION 5-8

1. Why is there a need to add or compare numbers? Why does this need exist also for binary numbers?

2. What do the initials ALU stand for? What does an ALU do?

3. What are the two things that a group of bits can represent? Give an example of each.

4. How does an ALU perform subtraction, division, and multiplication using the addition function?

5. What is the rule for the addition of two binary numbers?

6. What is the carry bit in binary addition? What does it represent?

PROBLEMS FOR SECTION 5-8

1. What are the steps for an ALU to: (a) add 3 + 3; (b) do 6 minus 4; (c) do 6 × 4; (d) do 12 divided by 3?

2. Add these binary numbers: 01 and 01, 10 and 10, 11 and 10, 01 and 11.

3. Add these 3-bit binary numbers: 101 and 101, 011 and 110, 001 and 100, 001 and 010.

5-9 ADDITION CIRCUITRY

Since binary addition has a set of rules, circuits and ICs can be built out of standard gates that implement these rules. One circuit for building an adder for 2-bit numbers is shown in Fig. 5-13. This circuit takes two separate binary num-

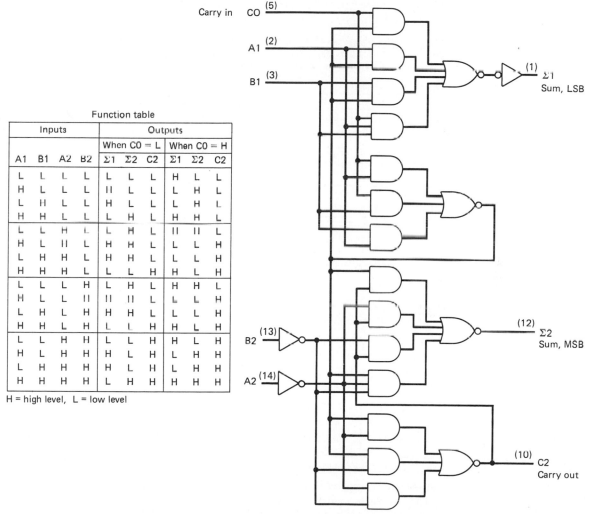

Function table

Inputs				When CO = L			When CO = H		
A1	B1	A2	B2	Σ1	Σ2	C2	Σ1	Σ2	C2
L	L	L	L	L	L	L	H	L	L
H	L	L	L	H	L	L	L	H	L
L	H	L	L	H	L	L	L	H	L
H	H	L	L	L	H	L	H	H	L
L	L	H	L	L	H	L	H	H	L
H	L	H	L	H	H	L	L	L	H
L	H	H	L	H	H	L	L	L	H
H	H	H	L	L	L	H	H	L	H
L	L	L	H	L	H	L	H	H	L
H	L	L	H	H	H	L	L	L	H
L	H	L	H	H	H	L	L	L	H
H	H	L	H	L	L	H	H	L	H
L	L	H	H	L	L	H	H	L	H
H	L	H	H	H	L	H	L	H	H
L	H	H	H	H	L	H	L	H	H
H	H	H	H	L	H	H	H	H	H

H = high level, L = low level

Fig. 5-13 A 2-bit adder, with carry-in and carry-out. (*Courtesy of Texas Instruments, Inc.*)

bers, each 2 bits long, and produces the sum of these two. There are also two additional connections in the circuit (besides power and ground): the *carry-in* and *carry-out*. The carry-out is the extra bit that results when the sum of the two-digit numbers requires a third bit for the carry to the next column. The carry-in is a special input that is used to bring a carry from another added IC into this circuit. The carry-in makes it possible to join together, or cascade, several small adders to make an adder for longer binary numbers.

Adders which can be combined (cascaded) are very useful functions and there are ICs available which perform the addition of 2 or 4 bits, along with carry-in and carry-out. There is one other feature of adder circuits which makes them even more useful. If many small adders are cascaded so that a larger binary number can be handled, the entire addition must wait for the carry (if any) from the first IC to go to the second IC before the second IC can begin addition, and so on down through all the adders. This *ripple-through carry* would result in slow overall operation because of all the internal propagation and carry delays adding up. Some additional circuitry can be built into the adder to provide what is called *look-ahead carry*. This is a carry that can be sent directly to the MSB part of the adder and does not have to go through all the adders along the way. The result is that a little more IC complexity gives much faster overall operation, which is usually needed in the application.

By themselves, arithmetic ICs such as adders do not do much. They need support circuitry to send them the numbers to be added and to take the result and put it somewhere to be used. But these arithmetic functions are the real heart of any computer-based system or processor and are also the building blocks for some advanced functions.

QUESTIONS FOR SECTION 5-9

1. In an adder circuit, what is the carry-out? The carry-in? Why are they needed?

2. What is the drawback of a ripple-through carry? What alternative is often used?

5-10   LOGIC FUNCTIONS

Adders are used to perform calculations on groups of bits that represent numbers. In computers, there is just as much importance in handling and manipulating bits that are not numbers but just represent individual bit states of 1 and 0. Logical manipulation is needed, and it is an extension of the basic boolean logic functions we have used.

Where are logic functions used? Go back to the case of bits representing open or closed doors and windows for several cases. Consider an 8-bit register which is used to store the data on the open or closed position of the doors and windows, one per bit (Fig. 5-14). These bits are called *flag* bits. When the flag is equal to 1, the flag is *set*. When the flag is equal to 0, the flag is *cleared* or *reset*.

The processor needs to know if a particular door or window flag bit is set. One approach would be to check the bit by comparing the binary value of the byte to

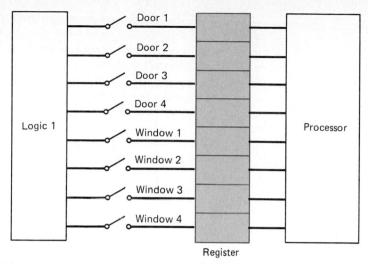

Fig. 5-14 An 8-bit register used for flag bits for doors and windows in alarm systems. (*Courtesy of Texas Instruments, Inc.*)

all possible binary values that could have this bit set. If the bit for the door of interest is the LSB of the byte, it would have to check 0000 0001, 0000 0011, 0000 0101, 0000 0111, and so on, up to 1111 1111. This means a total of 128 values out of the 256 possible values would have to be compared—this is very time consuming and inefficient. (Incidentally, the notation sometimes used is XXXX XXX1, where the X represents a bit that is "don't care—could be either a 0 or a 1.") A better way to do this is to use a *mask* and then logically combine it with the contents of the register (Fig. 5-15). A mask is a specific pattern of 1's and

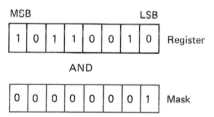

Fig. 5-15 A mask used to examine the LSB of door and window register.

0's. To examine the LSB, the mask would be 0000 0001 and the ALU would AND it with the register. This is done by taking each bit of the register and individually ANDing it with the corresponding bit of the mask, on a bit-by-bit basis. The AND function used with the mask causes the other 7 bits to become 0, since A AND 0 = 0. The result of this logical operation is that only the bit of interest has its value carried through to the result. If, after doing the AND operation, the byte has all 0's, then the bit for the door must have been 0. If, on the other hand, the byte has a 1, then the bit must have been a 1 (since A AND 1 = A). In this way, a mask and the AND function can be used to select and examine bits of interest within a group of bits. A mask can be used to examine only 1 bit at a time, or pick out several bits and ignore the rest.

Another powerful logic function on a group of bits is the OR function. Suppose there are two 8-bit registers, each representing eight doors and windows. To check the bits to see if *any* single door or window is open, the processor could examine each register separately. Another possibility is to OR both registers, by taking the OR boolean function of corresponding bits of both registers (Fig. 5-16). Then, if

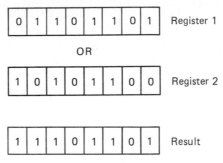

Fig. 5-16 The OR of two bytes to see if any bit is equal to 1.

any bit is a 1, something is open. (Recall that A OR 0 = A). The ORing of two registers or groups of bits has the effect of combining, not adding, contents and providing a summary. For example, if one byte is 0110 1101 and the other is 1010 1100, then the OR result is 1110 1101. Depending on the application, some other logic functions may be needed. These include the NOT function and the XOR function.

There is one other logic function that is often needed. The comparator function looks at two groups of bits and compares them to see if they are equal in value, or which one is greater than the other. The outputs of the comparator indicate if input A > input B, input A < input B, or input A = input B. This logic function is slightly different than the simple AND or OR function. For the comparator function, the individual bits cannot be looked at separately from each other, because the comparator is comparing the *value* of the groups. Practically, this is done by looking at the two groups at the same time, starting from the leftmost or MSB, and working toward the right, or LSB. When one group has a binary 1 in one column and the other does not, then the one with the 1 is larger. If all the bits are the same in each position, then the two values are equal. Here are some examples for 4-bit numbers:

$$0110 > 0100 \qquad 1000 > 0010$$
$$0101 = 0101 \qquad 0011 < 0100$$

A comparator can be built out of standard gates that compares each position of the two inputs and also carries the result over to the next position, or it can be part of a larger ALU.

QUESTIONS FOR SECTION 5-10

1. Give two examples where multiple-bit logic operations are needed.

2. What is a flag bit? What do set and reset mean with flag bits?

3. What is a mask? Where is it used?

4. How is the AND operation done with a mask? The OR operation?

5. What is the logical compare function? What does it indicate?

PROBLEMS FOR SECTION 5-10

1. A byte has the following pattern: 1110 0100. What mask would be used to check for set flag bits in the LSB and MSB positions? What would be the result of ANDing the mask and the byte?

2. Perform the OR function on these binary numbers: 101 and 100, 000 and 010, 110 and 010.

3. OR these binary numbers: 1010 and 0111, 0011 and 1001, 0000 and 0100.

4. AND these binary numbers: 101 and 100, 000 and 010, 1100 and 1001.

5. Perform the AND on the numbers of problem 3.

6. Perform the XOR function on the numbers in problem 3.

7. Which of the following numbers is larger: 100 and 011, 101 and 110, 0110 and 1011, 0011 and 0100?

8. Which of the numbers is larger in each pair of problem 3?

5-11 ALU ICs

The ICs that perform ALU functions come in many different forms with various capabilities. Simpler ICs are available that perform one arithmetic function only or one logic function only. More complex ICs that perform many arithmetic and logic functions are also available.

A typical logic function IC is the 4-bit comparator. This IC has four inputs for binary group A, four for binary group B, power and ground connections, and three outputs (Fig. 5-17). These three outputs indicate, on separate lines, if $A > B$, $A = B$, or $A < B$ by making that output line high and leaving the other two low. The comparator also has three cascade inputs, which are used if this 4-bit comparator is to be linked to another comparator so that binary groups of more than 4 bits can be compared. The three cascade inputs are $A > B$, $A = B$, and $A < B$, corresponding to the outputs of a previous comparator.

Another IC that performs an arithmetic function is the adder IC. These ICs are available in 2- and 4-bit packages. The adder has two inputs A and B (either 2 or 4 bits each), a 2- or 4-bit output, and power and ground. In addition, the adder has a carry-out bit, which has the carry (if any) from the addition, and provides for feeding a carry-in from another adder which may be cascaded to allow addition of larger numbers of bits. The carry may be handled internally either in the slow but simpler ripple mode, or the faster but more complicated look-ahead mode.

For systems which need more than one arithmetic or logic function, there are ICs which can do many different ALU operations. These chips are more complicated internally, of course. A typical ALU has 4 bits for input A, 4 bits for input

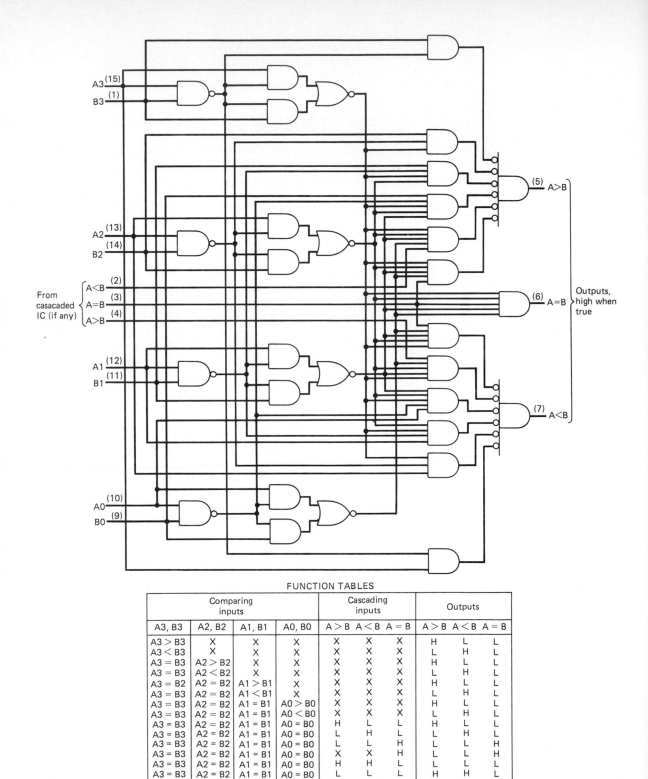

FUNCTION TABLES

	Comparing inputs			Cascading inputs			Outputs		
A3, B3	A2, B2	A1, B1	A0, B0	A > B	A < B	A = B	A > B	A < B	A = B
A3 > B3	X	X	X	X	X	X	H	L	L
A3 < B3	X	X	X	X	X	X	L	H	L
A3 = B3	A2 > B2	X	X	X	X	X	H	L	L
A3 = B3	A2 < B2	X	X	X	X	X	L	H	L
A3 = B2	A2 = B2	A1 > B1	X	X	X	X	H	L	L
A3 = B3	A2 = B2	A1 < B1	X	X	X	X	L	H	L
A3 = B3	A2 = B2	A1 = B1	A0 > B0	X	X	X	H	L	L
A3 = B3	A2 = B2	A1 = B1	A0 < B0	X	X	X	L	H	L
A3 = B3	A2 = B2	A1 = B1	A0 = B0	H	L	L	H	L	L
A3 = B3	A2 = B2	A1 = B1	A0 = B0	L	H	L	L	H	L
A3 = B3	A2 = B2	A1 = B1	A0 = B0	L	L	H	L	L	H
A3 = B3	A2 = B2	A1 = B1	A0 = B0	X	X	H	L	L	H
A3 = B3	A2 = B2	A1 = B1	A0 = B0	H	H	L	L	L	L
A3 = B3	A2 = B2	A1 = B1	A0 = B0	L	L	L	H	H	L

Fig. 5-17 A 4-bit comparator indicates if input A is greater than, less than, or equal to B. Auxiliary inputs allow comparators to be cascaded. (7485 IC) (*Courtesy of Texas Instruments, Inc.*)

B, 4 bits for the resulting output, carry input and carry output, and power and ground. In addition, there are control or select bits which allow the external circuitry and system to select exactly what function the ALU will perform (Fig. 5-18, page 132). A typical ALU IC has three select lines, which provide eight ALU functions: B minus A, A minus B, A plus B, A AND B, A OR B, clear all outputs, and preset all outputs. The outputs can represent numbers or logic results; they change as the operation performed by the ALU is changed.

Multiple-function ALUs are more versatile than single-function ICs; however, there is a need for both types. The multiple-function ALU can do many things, but it is more expensive and requires a great deal of support circuitry to prepare the inputs, select the mode of operation, and handle the outputs. In contrast, the single-function IC is less expensive and easier to interface because it always is called upon to do the same operation.

Another reasonable question is: Since the functions of the ALU are part of every microprocessor, why not use the microprocessor instead of an ALU IC? The answer is similar to the difference between a single-function IC and a multiple-function ALU—the microprocessor is more powerful but requires more support and interface circuitry, which may be excessive or too expensive for the application. Also, dedicated ALU ICs are usually faster than a microprocessor, and this may be an absolute requirement for the application.

Troubleshooting ALUs

The most important thing to have when checking ALUs is the specification and function table for the specific ALU IC type that is used in the circuit. This allows the ALU inputs and outputs to be compared for proper operation. Single-function ICs are relatively easy to check—for several input values, the outputs should agree with the calculations performed. Multiple-function ALUs are tested in a similar way, except that you have to be careful about what ALU mode or function is selected, and then check operation. If there is an internal fault in the IC, the ALU may work in one mode but not another. If the fault is in some external IC or connection to the ALU, then the fault will usually show up in many types of operations.

QUESTIONS FOR SECTION 5-11

1. What is in a typical comparator IC? What features does it have?

2. Repeat question 1 for an adder IC.

3. What type or types of arithmetic function does an ALU perform? What types of logical functions does it provide, typically?

4. How does the user indicate the function desired to the ALU?

5. Why are ALUs sometimes used instead of a microprocessor?

PROBLEMS FOR SECTION 5-11

1. For the ALU of Fig. 5-17, what is the logic level of A < B when the input bits are equal, except that A0 > B0?

Logic symbol

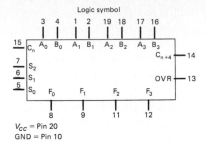

V_{CC} = Pin 20
GND = Pin 10

Function table

Selection			Arithmetic/Logic
S_2	S_1	S_0	Operation
L	L	L	Clear
L	L	H	B minus A
L	H	L	A minus B
L	H	H	A plus B
H	L	L	A $\oplus$ B
H	L	H	A + B
H	H	L	AB
H	H	H	Preset

H = high level, L = low level
See Truth Table for full description.

Definition of functional terms

A_0, A_1, A_2, A_3	The A data inputs.
B_0, B_1, B_2, B_3	The B data inputs.
S_0, S_1, S_2, S_3	The control inputs used to determine the arithmetic or logic function performed.
F_0, F_1, F_2, F_3	The data outputs of the ALU.
C_n	The carry-in input of the ALU.
C_{n+4}	The carry-look-ahead output of the four-bit input field.
$\overline{G}$	The carry-generate output for use in multi-level look-ahead schemes.
$\overline{P}$	The carry-propagate output for use in multi-level look-ahead schemes.
OVR	Overflow. This pin is logically the Exclusive-OR of the carry-in and carry-out of the MSB of the ALU. At the most significant end of the word this pin indicates that the result of an arithmetic two's complement operation has overflowed into the sign-bit.

Truth table

Function	S_0	S_1	S_2	C_n	A_n	B_n	F_0	F_1	F_2	F_3	$\overline{G}$	$\overline{P}$
Clear	0	0	0	X	X	X	0	0	0	0	0	0
B minus A	1	0	0	0	0	0	1	1	1	1	1	0
				0	0	1	0	1	1	1	0	0
				0	1	0	0	0	0	0	1	1
				0	1	1	1	1	1	1	1	0
				1	0	0	0	0	0	0	1	0
				1	0	1	1	1	1	1	0	0
				1	1	0	1	0	0	0	1	1
				1	1	1	0	0	0	0	1	0
A minus B	0	1	0	0	0	0	1	1	1	1	1	0
				0	0	1	0	0	0	0	1	1
				0	1	0	0	1	1	1	0	0
				0	1	1	1	1	1	1	1	0
				1	0	0	0	0	0	0	1	0
				1	0	1	1	0	0	0	1	1
				1	1	0	1	1	1	1	0	0
				1	1	1	0	0	0	0	1	0
A plus B	1	1	0	0	0	0	0	0	0	0	1	1
				0	0	1	1	1	1	1	1	0
				0	1	0	1	1	1	1	1	0
				0	1	1	0	1	1	1	1	0
				1	0	0	1	0	0	0	1	1
				1	0	1	0	0	0	0	1	0
				1	1	0	0	0	0	0	1	0
				1	1	1	1	1	1	1	0	0
A $\oplus$ B	0	0	1	X	0	0	0	0	0	0	0	0
				X	0	1	1	1	1	1	1	1
				X	1	0	1	1	1	1	1	0
				X	1	1	0	0	0	0	0	0
A + B	1	0	1	X	0	0	0	0	0	0	0	0
				X	0	1	1	1	1	1	1	1
				X	1	0	1	1	1	1	1	1
				X	1	1	1	1	1	1	1	0
AB	0	1	1	X	0	0	0	0	0	0	0	0
				X	0	1	0	0	0	0	1	1
				X	1	0	0	0	0	0	0	0
				X	1	1	1	1	1	1	1	0
Preset	1	1	1	X	0	0	1	1	1	1	1	1
				X	0	1	1	1	1	1	1	1
				X	1	0	1	1	1	1	1	1
				X	1	1	1	1	1	1	1	0

Fig. 5-18 The signal lines and corresponding function table for a typical 4-bit ALU (AM25LS2517 IC). (*Courtesy of Advanced Micro Devices, Inc.*)

2. For the ALU of Fig. 5-18, what are the select code bits that cause the ALU to perform A + B? To perform A plus B? What is the difference in the meaning of + versus plus?

The buffer IC has a very simple truth table; the output is in the same state as the input (for noninverting buffers) or the complement of the input (for inverting buffers). But within this simple boolean function is a group of ICs that are essential in many practical systems. Buffers do the following:

- Protect a system circuitry from external problems.
- Allow a gate output to drive more gate inputs than it normally could handle.
- Isolate one part of the circuit from another.
- Allow a low-voltage gate to control higher-voltage loads.
- Shift voltage levels as required between two systems.
- Eliminate noise on logic signals so that the noise does not cause logic 1's and 0's to look like each other.

A special category of buffer, called the three-state buffer, is used whenever multiple ICs may want to talk on a single bus line. The three-state function allows the IC to be disconnected, in effect, from the bus so that it does not interfere with the IC that is putting a signal onto the bus. These three-state ICs use an Enable line to turn the buffer output on, so that the buffer can then pass the logic level through.

ALUs are specialized ICs that perform the basic arithmetic operations on binary numbers. Arithmetic is the heart of a computer processor function and is a necessity for a practical system. The ALU can also perform boolean operations, such as ANDs and ORs, on groups of bits. This is used where bits are combined with a mask to identify the logic state of specific bits in a group.

REVIEW QUESTIONS

1. What is the need for buffers in three different aspects of system operation?

2. How are these buffers related to buffers that are built of registers?

3. If an inverting buffer on one circuit is connected to an inverting buffer on the next circuit, what is the output level of the second buffer when the input to the first is a 0?

4. What can excessive or varying load do to an IC?

5. Can most ICs drive 100 IC loads of the same type?

6. What are two causes of damage to an otherwise good IC?

7. Describe the meaning of the isolation that buffers can provide.

8. Explain how buffers can prevent a single failure on a bus from causing an entire system to be inoperative.

9. Why does hysteresis improve performance in a series of multiple gates and buffers?

10. What might be the function of one AND gate driving two other AND gates, and each of these driving 10 AND gates (if the AND gates have only one input)?

11. Give two disadvantages of buffers.

12. What is done to troubleshoot buffers? Is it usually simple or difficult?

13. Why are bus-oriented systems so common? What is the advantage of a bus?

14. What is the problem with bus systems?

15. What is a three-state buffer? How does it solve the bus problem?

16. What is the function of an Enable line on the three-state buffer? What controls the Enable line?

17. How does bidirectional operation occur with three-state buffers?

18. Why are numbers sometimes added or compared?

19. Where is an ALU used, whether by itself or as part of another function?

20. Explain how a group of bits can represent two completely different types of information.

21. How do ALUs perform the four basic arithmetic functions? Why is this approach used?

22. What is the carry rule for decimal addition? How does it compare to the rule for binary addition?

23. Why are a carry-out and carry-in needed?

24. Why is look-ahead carry used instead of ripple-through carry in most applications?

25. What is the importance of flag bits? What is the logical value of a set bit? A reset bit?

26. What does the expression ''clear the flag'' mean?

27. Where is a mask used? With what logical operations can it be used?

28. Give an example where the compare function might be used.

29. Are ALU functions available only by themselves, or are they sometimes part of another IC? Which one? Why?

30. What method is used to select the different ALU functions that are available?

31. What does the symbol X stand for in a binary number?

REVIEW PROBLEMS

Unless otherwise indicated, use these pairs of numbers for the problems: 10 and 01, 11 and 10, 100 and 010, 101 and 110, 1010 and 0111, 0101 and 1000, 1000 and 1000, 0110 and 0110, 0001 and 0010, 1100 1100 and 0101 0101, 0001 1101 and 1110 1100, 0110 0001 and 1000 1000.

1. Perform the OR function on these pairs.

2. Perform the AND function on these pairs.

3. Add these pairs.

4. Perform the XOR function on the pairs.

5. Do a compare on these pairs.

6. A system has the binary byte 0100 1101. What mask would be used to decide if the second LSB and second MSB were set? What logical function?

7. For the same byte, what mask would be used to see if any of the four LSBs are set?

8. Verify the operation of the adder of Fig. 5-13 for input values A1, A2, B2 = 1, B1 = 0. Assume the carry in C0 is 0.

9. Verify the operation of the comparator of Fig. 5-17. Use input values for A = 1010, B = 1100, and the cascade inputs are "don't care." The correct output is indicated by a low signal on the output, while the other two remain high.

THE MAJOR IC FAMILIES

6

This chapter discusses the types of ICs that are used to provide the logic functions presented in the previous chapters. The ICs belong to families which share common technical specifications to ensure that the ICs of the family can be interconnected to build larger circuit functions. There are two major technologies used to produce these ICs, and each one has a family (and several related families) associated with it. Each of the families provides a choice in operating speed, power used, ability to connect to other devices, as well as the actual voltages used by the IC. The choice of one family versus the other is made by the system designer based on the requirements of the design and the actual application of the equipment that will use these ICs. The techniques for troubleshooting these families have many similarities but there are also some differences that must be understood.

There is another category of IC which is important in IC circuitry systems. This category consists of ICs which are specifically designed for the application. They may be entirely customized, or they may be ICs which have many of the production steps complete, but with the last few steps deliberately left unfinished. When the logic function requirements of the product and design are finally defined, these last steps are done to meet the specification.

6-1 CHARACTERISTICS OF IC FAMILIES

As seen from the previous sections, there are many different standard ICs available. In a complete system, various ICs have to work together to provide the overall function and capability for which the product is designed. If each IC were designed with different power voltages, logic signal levels, and operating characteristics, the complete system would never work. To make the process of developing a working product more practical, IC families were developed by the companies that manufacture ICs.

Members of a family have some common characteristics:

- All the ICs in the family use the same power-supply voltage or voltages. This means that a single power-supply unit and single set of power-supply wires can be used for the entire system based on these ICs.
- The signal levels that the ICs require for their inputs, and the signal levels that they provide for their outputs, are the same throughout the family. This means that the various ICs can be interconnected and drive each other without any problems, since the input and output interfaces are compatible.
- The internal speed, or propagation delay, of the ICs is consistent across all members of the family. In this way, the components of overall systems will be able to operate in unison and not be slowed down or thrown out of step by one very slow IC.
- The functions provided by members of a family fit well together. A real electronic system is like a puzzle, and all the right pieces have to be there to develop a complete system. A family provides a wide range of IC functions and makes sure that all the necessary variations of functions are available. For example, a single family would have all the flip-flop types that we have seen available for use in the appropriate parts of the system.
- The wide selection of ICs within the family means that everything from simple boolean gates through much more advanced ICs, such as counters and ALUs, are available and compatible.
- Members of an IC family are designed with features of other members in mind. This means that an IC in the family might provide a special output signal which in itself is not useful but can be combined with a corresponding input line on another IC to provide a very powerful pair. This has the effect of allowing very powerful and flexible systems to be created by proper interconnection of the family members.
- Finally, members of the family have an orderly numbering plan for all the ICs within the family. This is a major practical convenience when examining a circuit schematic and trying to troubleshoot the system.

In this chapter two of the most common families of ICs used today are examined. New types of ICs which are becoming more important in systems and circuits will also be covered.

QUESTIONS FOR SECTION 6-1

1. Why is a family of ICs needed?

2. What are five characteristics of an IC family?

6-2   THE 7400 SERIES TTL FAMILY

The 7400 series of ICs is a family of over 400 small- and medium-scale ICs that has been developed since 1972. It has several related variations within the main family. These variations offer the user choices in speed of operation and power used by the ICs. The 7400 series of ICs uses a transistor to drive the transistors of the output stage, so the term *transistor-transistor logic,* or TTL, is often used to describe the type of IC technology that the 7400 series represents. Another common term for this technology is *bipolar,* since both plus and minus electrical charges are used within the internal IC transistors to carry the signals. The 7400 series was started by one manufacturer and has become a major logic family. Many IC manufacturers now make identical 7400 series parts, so a company can choose to buy the same part from multiple sources. This fact also makes the 7400 series attractive to users because they are not tied down to one supplier.

Many logic functions are represented by the 7400 series. Here is a listing of some of the types, starting with the first and simplest member, the 7400:

Gates	7400	Quad two-input NAND gates (has four independent NAND gates)
	7401	same as 7400, but with open-collector outputs
	7402	Quad two-input NOR gates
	7404	Hex inverters (has six separate inverters)
	7408	Quad two-input AND gates
	7410	Triple three-input NAND gates
	7411	Triple three-input AND gates
	7420	Dual four-input NAND gates

Flip-flops and latches	7470	*J-K* flip-flops with Preset and Clear
	7474	Dual *D*-type edge-triggered flip-flops with Preset and Clear
	74174	Hex *D*-type flip-flops
	7475	4-bit bistable latches

Counters	7490	Decade counter
	7493	4-bit binary counter

Registers	74170	4-bit, four-deep register
	7495	4-bit shift register
	74164	8-bit parallel-output serial shift register

Decoders mux, demux, code converters	74184	BCD-to-binary code converter
	74185	Binary-to-BCD code converter
	74150	1-to-16 data selector and multiplexer
	74154	4-line to 16-line decoder and demultiplexer
	74138	3-line to 8-line decoder-demultiplexer

Display decoders and drivers	74142	Decimal converter–4-bit latch–4-bit to 7-segment decoder-driver
	7445	BCD-to-decimal decoder-driver
	7446	BCD to 7-segment decoder-driver

Buffers	7406	Hex inverter buffer-driver with open-collector outputs
	74241	Octal buffer with noninverted three-state outputs
	74245	Bidirectional octal bus driver with three-state outputs

ALUs	74181	16 arithmetic operation–16 logic function ALU
	7482	2-bit binary adder

In general, the simplest ICs and functions have the lowest numbers, and the higher numbers indicate more complicated ICs in the family.

The 7400 series has been so successful that many non-TTL ICs are specially designed so that their input and output pins look to the rest of the circuit like TTL IC pins. Such inputs and outputs are called *TTL compatible*.

QUESTIONS FOR SECTION 6-2

1. What are three of the ICs in the TTL family?

2. What is the general trend in numbering the 7400 series family of ICs?

3. Is the 7400 series made by just one manufacturer? What are the other manufacturers called?

4. What does *TTL compatible* mean? What does TTL stand for?

5. What is another term often used for TTL ICs?

6-3 CHARACTERISTICS OF THE 7400 SERIES FAMILY

Power Supply

The power supply used for the TTL ICs is a single positive 5-V supply. A tolerance of ± 10 percent is allowed, so the actual voltage may be 4.5 to 5.5 V. There is no negative supply voltage required. All the 7400 series of ICs, as well as variations of the family, use this supply voltage. The amount of current required varies with the total number and type of specific ICs in the circuit. Typical values for a TTL system based on 7400 ICs are anywhere from 0.5 to 10 A. The power-supply voltage is often called V_{CC}; the return side of the power supply is called *ground*.

Logic Levels

This is the most important specification for digital ICs. The logic levels are the specific voltage levels that an IC must produce or see to be a valid logic 1 or 0. For all 7400 series ICs, and any derivative families, these values are:

A valid logic 0 (or low) exists if the voltage is between 0 and 0.8 V.
A valid logic 1 (or high) exists if the voltage is between 2.0 V and V_{CC}.

If a voltage value in a TTL circuit does not fall into either of these two ranges (Fig. 6-1), the circuit is either not working or barely working and will operate

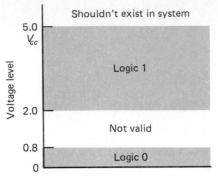

Fig. 6-1 TTL logic levels versus voltage. Note the band of acceptable and invalid voltages.

erratically. In a properly working system, these are the only voltage ranges that can exist.

Drive Capability

The voltages that are used by these ICs are not the complete characteristics of valid signals. The IC must also be able to produce enough current, at the specified voltage, to drive the next IC in the circuit. Voltage and current are the two aspects of any electrical signal. It is similar to water flowing through a pipe: There must be water pressure (voltage) and water flow (amps) to make use of the water. There can be lots of pressure but a very small pipe, so there is little flow through the pipe; there can also be a large volume of water flowing, but with so little pressure that it cannot reach the sink. Neither situation is useful; a proper combination of both is needed to do the job.

The same is true with ICs. The logic level is a voltage, but in order to have it be useful within the system it must also have enough current capability. The amount of current capability that is needed depends on the number of ICs connected to the output line that the IC is driving. For the 7400 series, there is a specification that defines how much current the IC must provide, both at logic 1 and logic 0 levels. When an IC is putting out a logic 1, or high voltage, it is actually supplying, or *sourcing,* current to the ICs that it is driving (Fig. 6-2).

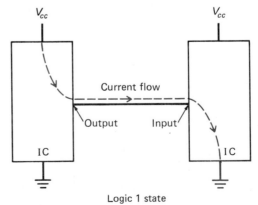

Fig. 6-2 A TTL IC at logic 1 sourcing current to another TTL IC.

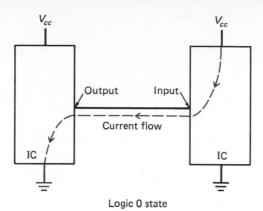

Fig. 6-3 A TTL IC at logic 0 sinking current from another TTL IC.

When it is putting out a logic 0, or low level, it is accepting current that is supplied to it by the next IC. This is called *sinking* current, because the driving IC is sinking the logic 0 current to system ground (Fig. 6-3). The terms source current and sink current are used quite often in describing the operation of ICs in circuits, and problems sometimes occur when an IC fails partially and cannot source or sink enough current for all the loads it has.

The number of loads that an IC can drive is called the *fan-out*. A 7400 series IC can handle a fan-out of 10, which means that it can supply enough current at high output voltage, and sink enough current at low output voltage, for 10 inputs of the same family. Some IC manufacturers make 7400 series ICs that have a fan-out of 20 loads. If the circuit requires greater fan-out, buffer ICs are required.

Speed and Power Consumption

Every IC consumes electrical power as it operates. This can be a design problem for several reasons. Sometimes the amount of available power is limited because the system is designed to run on batteries (as opposed to the powerline, which can supply much more current). But power consumption is also bad for the IC and system. As ICs have more and more transistors on the silicon, they may generate enough heat as they operate to actually cook themselves—to heat themselves up beyond the temperatures at which they can operate. A standard 7400 family IC is designed to work over the temperature range of 0 to 70° C (32 to 158° F), which may seem like a wide range but is not. The temperature in a system can rise if the outside (ambient) temperature is high, if the system is in the sun, and if the heat that is generated by the ICs (self-heating) is not drawn out of the system quickly. The high power consumption of an IC may mean that there may be a need for mechanisms like fans, air conditioners, or water cooling. Heat also stresses the mechanical connections in any system and contributes to failure and poor reliability.

The other requirement of a system is to perform the application—calculations, response to the operator, software program execution—within a certain time period. The speed at which the IC can operate is crucial here. This speed of the system is determined by the internal propagation delays within each IC, which add up as the signals flow through the system. In general, faster is better.

It turns out that there is a close relation between the speed of an IC and the power it consumes. In general, faster ICs use more power. In order to give engineers a range of choices for various applications needs, some additional TTL families are developed.

Each of these families offers a choice of operating speed and power consumption. They all share the same numbering plan as the basic 7400 series family. These other TTL families are:

74S00	High-speed (The S stands for Schottky, which is the name of the technique used to provide high speed in the IC)
74LS00	Low-power Schottky—Not quite as fast as S, but much lower power
74H00	High-speed—Not as fast as S and uses more power
74L00	Lower power but lower speed than LS
74AS00	Advanced Schottky that provides the highest speed of operation
74ALS00	Advanced Schottky that is slower than AS but uses less power

Figure 6-4 shows a graph of typical speed and power consumption for these TTL IC variations.

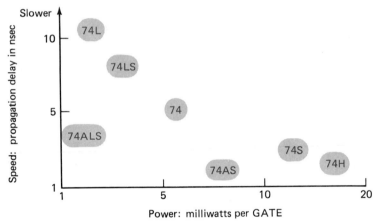

Fig. 6-4   The choices in TTL IC versus power consumed for the various TTL IC families.

There may also be the prefix 54 instead of 74. These ICs are identical to the 74 prefix ICs but are designed to work over a wider temperature range of −55 to 125° C (−67 to 257° F), as required for military and aerospace applications.

While it may seem confusing to have all these family variations, it really is not, for these reasons:

- All the family variations use the same power-supply voltage.
- They all use the same voltages to represent logic levels 0 and 1.
- The pinouts of the ICs are the same, regardless of family type.
- A given system generally uses only one family—mixing of 74S and 74LS, for example, is not commonly done within a circuit.
- The logical flow of signals is not affected by the family variation.
- The speed and power characteristics are choices for the system designer but do not affect the person who is trying to troubleshoot the circuit.

QUESTIONS FOR SECTION 6-3

1. What is the nominal power-supply value for TTL?

2. What is the actual range of the power supply allowable?

3. What is the power supply often called?

4. What voltage range represents logic 0? logic 1?

5. What are the two simultaneous requirements for an IC output to drive another IC?

6. When is a TTL output sourcing current? What does it mean?

7. When is a TTL output sinking current? What does that mean?

8. When is a TTL input sinking current? Sourcing current?

9. What is the fan-out capability of a typical TTL IC?

10. What is the operating temperature range of TTL? Why is this not as wide as it may first seem, at either the low or high end?

11. What are two things that can cause an IC temperature to rise? What are two consequences of this?

12. What is the relationship between TTL speed and power consumed?

13. Describe two other TTL families. Why do these exist?

14. What is the series 54?

15. Give three reasons why having all these families is really not confusing to someone working with these ICs.

PROBLEMS FOR SECTION 6-3

1. A TTL power supply provides 4.8 V and has noise on it of ± 0.4 V. Will the TTL work?

2. A TTL IC produces a 1-V signal for a logic 1. Is this good enough? How about the same voltage for a logic 0?

3. Show two ways to have a TTL NAND gate handle a fan-out of 30, using the same NAND gates.

4. A system is being used in air at 37°C (about 98°F). The self-heating from the ICs and power supply adds another 20°C to the temperature that the IC sees. Will this IC operate reliably?

6-4 CONNECTING ONE IC TO ANOTHER

A practical digital system uses more than one IC. There are some standard techniques that are used to interconnect ICs and ensure proper operation, protect the ICs from too much electrical noise (which would corrupt the signal), or connect the IC to external devices such as keyboards, displays, and relays.

Inputs

Many ICs have some input lines that are not used in a particular circuit configuration. For example, a system might require two inputs of a three input NAND gate. To make the gate work properly, one input of the NAND must be permanently set to a logic 1. It cannot be left unconnected. An open input is not allowed in digital systems because such an input cannot be guaranteed to be a 1 or 0. The IC then misunderstands the logic level, and at one instant may think it is a 1, and at another think it is a 0. This would cause the system to malfunction occasionally, and would be a difficult problem to solve.

Such unused inputs must be connected to a voltage in the system that is at the proper logic level, either 0 (for a logic 0) or V_{CC} (for a logic 1). Inputs that need to be at 0 V can be connected directly to the system ground (Fig. 6-5). The situation

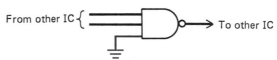

Fig. 6-5 The proper connection of an unused input that should be at logic 0. It is connected directly to ground.

for inputs that must look like logic 1 is a little more complicated. It may seem that connecting the input to the power-supply line will do the job—and it will. But it also has some drawbacks in that noise spikes in the power-supply lines can get directly into the IC and either cause malfunction or actual damage. The solution is to connect the input to the supply line via a resistor, usually of 2 to 10 kilohms ($k\Omega$) in value (Fig. 6-6). This is called a *pull-up*. It connects the input to a logic 1,

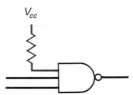

Fig. 6-6 The connection of an unused input that represents logic 1. It is connected to the power supply via a pull-up resistor.

but at the same time limits any sudden surges into the input. Pull-ups are very common in circuits and are often required in many situations. If an IC has several input lines that need to be connected to a logic 1, they can share the same pull-up resistor (Fig. 6-7). This reduces the number of components needed in the system.

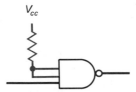

Fig. 6-7 Several unused inputs from the same gate, or even different gates, can share the same pull-up resistor.

Outputs

The output circuit ("stage") of an IC is as critical as the input. There are rules and guidelines that the circuit designer follows when connecting *anything* to the IC output, and these rules are derived from the nature of the output stage. These rules also affect the procedures that are used when troubleshooting a circuit.

The output stage is responsible for providing a logic 1 or logic 0 voltage, at the specified current level, to the next IC input or inputs. For the 74LS00 series of TTL ICs, the voltage and current values are:

- A logic 1 is at least 2.4 V, V_{CC}, while sourcing up to 0.4 mA. (*Note:* Sometimes, 2.6 or 2.8 V is specified, depending on the manufacturer.)
- A logic 0 is 0 to 0.4 V, while sinking up to 8 mA.

As shown in Fig. 6-8, the difference between logic 0 and logic 1 voltages for inputs and outputs guarantees some overlap and helps ensure that an output level will be within the proper input range at the next IC. Note that the sinking ability of the IC is much greater than the sourcing ability. This affects the way an IC can be connected to external devices such as lamps and displays.

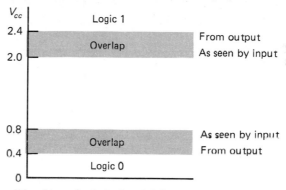

Fig. 6-8 The valid voltages for logic 0 and 1 from the IC output and the IC input have some overlap to insure that signals arrive in the valid region.

The output stage of a TTL IC usually has two transistors, arranged in a "totem-pole" fashion between the power-supply V_{CC} and the power ground (Fig. 6-9). When the output has to be a logic 1, the upper transistor is turned on while the lower one is off, and the output is then connected, through the upper transistor, to the power supply (Fig. 6-10). In this way, the IC can provide the necessary voltage level and current. Conversely, a logic 0 is provided by turning the upper transistor off and the lower one on. This connects the output to the ground point via the lower transistor and provides the logic 0 voltage, and the path to sink required current (Fig. 6-11).

The supply voltage is 5 V and ground is at 0 V. However, the logic 1 and logic 0 voltage values can never reach 5 or 0 V, because the output is connected to the supply or ground through a transistor. Even when a transistor is turned on, there is a small voltage loss, or drop, across it. This is why in practice the allowable specifications for the TTL levels are close to the supply value and ground but are not equal to it.

The totem-pole output is ideal for driving other TTL inputs. It offers a good

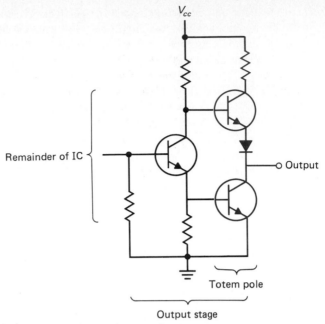

Fig. 6-9   Most TTL ICs have a totem-pole output structure.

combination of operating speed, ease of design and fabrication, and other operational characteristics. However, two totem-pole outputs from different gates or ICs can never be connected together (Fig. 6-12). If they were, the output-stage transistors from one gate could be in conflict with the transistors from the other gate, if one IC was trying to put out a logic 0 and the other trying to put out a logic 1.

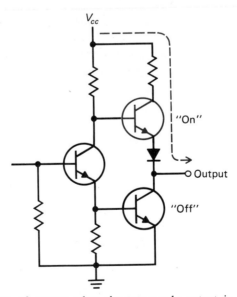

Fig. 6-10   The flow of current when the totem-pole output is providing a logic 1 output.

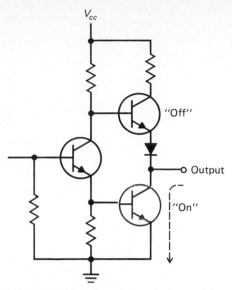

Fig. 6-11   The flow of current when the totem pole is providing a logic 0 output.

There are situations where the totem-pole output is not best for the application. A TTL gate may be driving a load that is not another TTL IC, such as a lamp or LED. In this case, the IC output needs to actually control the flow of current through the load. If a totem-pole output were used, the current would flow through both the load and the totem-pole transistor, and it would not be possible to control the amount of current flowing through the load (Fig. 6-13).

The solution is a special output stage called an *open-collector output*. In the

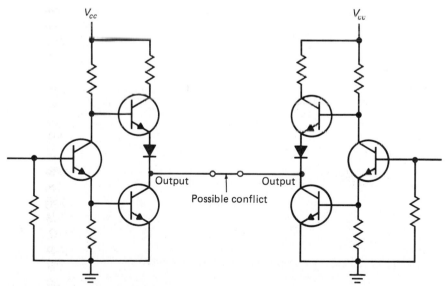

Fig. 6-12   When two totem-pole outputs are connected directly together, a conflict will exist.

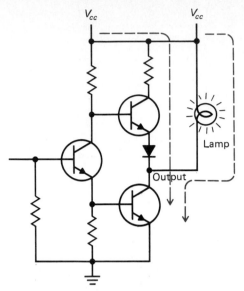

Fig. 6-13   The totem-pole output cannot be properly used for driving a non-IC load, since the amount of current shared between the output transistor and the load is uncertain.

open-collector output, the upper transistor of the totem-pole is removed, and the collector of the lower transistor is left open with respect to V_{CC} (Fig. 6-14). The

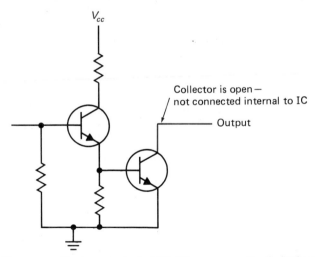

Fig. 6-14   The open collector output of a TTL IC overcomes the limitations of the totem pole in some applications.

circuitry outside of the IC must therefore complete the path for current flow to V_{CC}. The load is connected from the output of the IC to V_{CC}. An example is shown in Fig. 6-15. When the output of the IC is at logic 0, the lower transistor turns on, and current has a path from V_{CC} through the load, through the transistor, to ground. When the IC output is a logic 1, the lower transistor is off and there is no current flow from power source to ground. The load is therefore off.

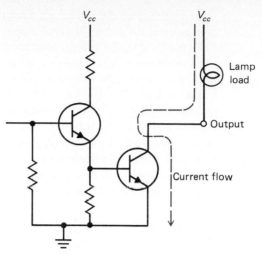

Fig. 6-15 The open collector output connected to a non-IC load. The amount of current flow is controllable and predictable.

Note that in the open-collector configuration the load is on when the logic output is 0, and the load is off when the logic output is 1. This is called *active low logic,* and it is commonly used because TTL ICs can sink more current than they can source. The sink current capability is used because it can handle more current and thus larger loads.

Open-collector outputs have another interesting characteristic: they allow the outputs of two ICs to be tied directly together, along with a shared resistor to V_{CC} (Fig. 6-16). This configuration is called *implied AND* because it creates a phantom

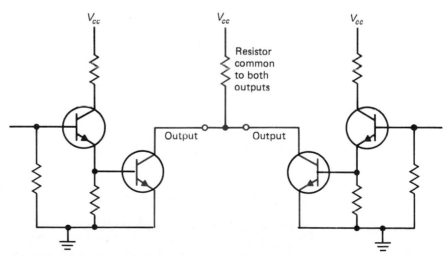

Fig. 6-16 Two open collector outputs can be wired together in an *implied-AND* configuration, often called (incorrectly) a *wired-OR* connection.

gate, with the two open-collector outputs as the inputs to this gate, and the common point as the output. If either output is low, then the combined output is low. The combined output is high only if both open-collector lines are high. The implied AND is sometimes called a *wired OR* configuration, although this is a misleading name since the connection performs the AND function for positive signals.

The implied AND (or wired OR) configuration is not considered good design technique, but it is used only when circuit parts count, cost, or circuit board space must be kept to as low as possible and every extra component eliminated. Because the two open-collector outputs are connected directly to each other, without a real logic gate, there is no separation of the signal from one open-collector output and the other. This lack of physical isolation means that any problem on one open-collector output will immediately affect both outputs and the implied AND function.

QUESTIONS FOR SECTION 6-4

1. What must be done to unused gate inputs? Why?

2. An OR gate with three inputs is being used with only two of the inputs connected to the circuit. What must be done with the third input?

3. What is a pull-up resistor? Where is it used? Why?

4. Which is greater for a TTL IC output: the sinking or the sourcing capability? What is the ratio of one to the other?

5. What is a totem-pole output? Explain the path of current for a load connected to a totem-pole output when the output is a logic 0 and logic 1.

6. Why can a gate output never reach 0 V or the power-supply voltage?

7. What happens when two totem-pole outputs are connected together?

8. Where is the totem-pole not the best solution? Give an example.

9. What is the open-collector output? How does it physically differ from the totem-pole output?

10. How is the open-collector output connected to the load? Where does the current flow at logic 0 or logic 1?

11. What is an implied AND output? What is it often called, incorrectly? What characteristic does it provide? What is the problem with it?

PROBLEMS FOR SECTION 6-4

1. Sketch a three-input AND gate with two inputs connected to some other part of the circuit. Sketch the connection of the unused third input.

2. Show three open-collector outputs connected together to a load. What happens when one output is 0? When all outputs are 1? When all outputs are 0?

3. The load of a TTL IC is a lamp which requires 2 mA. Can this be connected so that the source or the sink current drives the load? For each case, where is the other end of the lamp connected? Explain.

There are many types of functions within the TTL families, and an almost limitless number of combinations of these ICs. Therefore there are many different trouble-shooting techniques, and the difficulty involved can vary from very simple to extremely complex. However, there are general rules and guidelines, as well as some common problems.

First, always check the power-supply voltage V_{CC} at the circuit board and at the IC. For TTL it should be a nominal 5 V (from 4.5 to 5.5 V including 10 percent tolerance). It is a good idea to check the numerical value of the voltage with a digital voltmeter and also check the appearance of the supply voltage with an oscilloscope. The oscilloscope will show if the voltage is steady or has unwanted fluctuations and noise superimposed on it. The voltmeter will not show the variations, which can cause malfunctions and other problems. It must be checked at the power and ground of the IC itself because the power supply may be "clean," but noise can be picked up on the way to the IC and the power may be noisy when it reaches the IC (Fig. 6-17). Most of this noise comes from other ICs in the circuit.

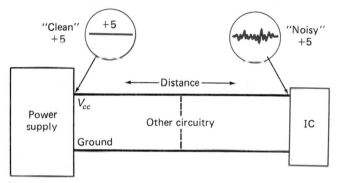

Fig. 6-17 Even if the power supply is perfect, the power supply lines to the ICs can pick up noise along the way.

As their outputs make the sudden change from logic 0 to 1, the ICs can generate noise spikes on the circuit power works. For this reason, a TTL IC often has low-value capacitors [0.1 microfarad (μF)] called *decoupling capacitors* near the IC itself, to shunt this noise away from the regular IC (Fig. 6-18).

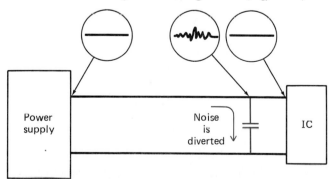

Fig. 6-18 A decoupling (bypass) capacitor steers this noise to ground before it reaches the IC and causes problems. This is often called *shunting* the IC.

The voltages of the logic signals should be in the allowable ranges: 0.2 to 0.8 V for a logic 0, and 2.4 V (nominal) to V_{CC} for a logic 1. If the circuit is static (nothing is changing, no clocks, all gate inputs and outputs "frozen") then a voltmeter will do. If the circuit is dynamic (signals changing state, as in normal operation), then an oscilloscope, logic probe, or logic analyzer will be needed. (These special instruments will be discussed in detail in the final chapter.)

Unused inputs that are pulled up via a resistor are no problem, except when a signal resistor is shared by several inputs. This technique is good from an engineering perspective, but it can make troubleshooting difficult. Since two or more inputs are connected together and to the resistor, it is impossible to access the inputs independently.

For some troubleshooting methods, however, it may be necessary to gain access to one unused input at a time in order to force that input to the low state (and thereby not have it pulled up to V_{CC}). In fact, a test instrument called a *logic pulser* is designed to force a gate input that is at one logic level to the opposite logic level. When several inputs are tied together, the pulser would change them all simultaneously.

For example, a flip-flop may have its Preset and Clear inputs pulled up through a single resistor, because the circuit never requires that they be used. However, for service purposes, it may be necessary to clear the flip-flop, and so it would be necessary to force the Clear line low (for flip-flops with active low Presets and Clears). It would not be possible to change only the Clear input and leave the Preset input high if both are tied together. The result is that the service engineer or technician is restricted from using certain routine procedures and techniques.

Open-collector outputs present a different problem. As long as the output is connected to V_{CC} through a load, the open-collector output can reach valid high levels. But when the load is removed, the open-collector output can sink current to ground (and so reach a valid logic 0), but it cannot source current (logic 1) because there is no longer a connection between the V_{CC} and the output. In many real applications the load is not located on the circuit board with the open-collector output. It may be a lamp on the front panel. When the cable to the front panel is disconnected as part of the service disassembly of the system, the open-collector output loses its load (Fig. 6-19). Any readings made with a voltmeter, oscillo-

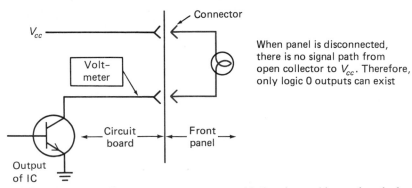

Fig. 6-19 The open collector output can cause troubleshooting problems when the load is disconnected as part of the test plan.

scope, or logic probe are meaningless at this time. The solution is to use clip leads and connect a resistor (usually 10 kΩ) from the open-collector output to V_{CC} to ensure the presence of a load, which substitutes for the disconnected load.

QUESTIONS FOR SECTION 6-5

1. What are three items to check first when checking a TTL-based circuit? What tools are needed?

2. What is a decoupling capacitor? What does it do?

3. What is the troubleshooting problem with one pull-up resistor common to several unused inputs? Give an example.

4. What is the possible misleading output indication that an open-collector output can provide? When does this occur in troubleshooting?

6-6 THE 4000 SERIES CMOS FAMILY

The TTL logic families studied in the previous section are extremely versatile and can be used to develop many digital circuits and systems. But the TTL family has some drawbacks:

- It requires a steady 5-V power supply that is free from almost all electrical noise. The tolerance specification of ±10 percent on the supply under all operating conditions may also be difficult to achieve in some practical situations.
- The TTL families consume considerable electrical power. This means that large supplies are required, that battery and portable operation is impractical, and that there will be a large amount of heat produced by the ICs as they operate. This heat, in turn, means system cooling must be provided and limits the complexity of the ICs themselves—if they generate too much heat (because of too many gates on the silicon) they could self-destruct.
- While all digital logic is resistant to unwanted noise voltage superimposed on the logic 1 or 0 levels, the amount of resistance to misinterpreting a logic 1 and logic 0 (called *noise immunity*) is not as great in the TTL series as some applications require. Electrical noise is a problem in systems that will be used with motors, engines, machinery, and large electrical equipment.

There is an IC family which overcomes these problems. This family of ICs does not use the bipolar circuitry of TTL, but uses another IC technology called *metal oxide semiconductor* (MOS). The transistors used are called *field-effect transistors* (FETs), and they are designed to use either positive electric fields (*p*-channel) or negative electric fields (*n*-channel) to control the flow of electrons within the transistor. The configuration used in the MOS ICs that overcomes the drawbacks of TTL ICs requires a complementary pair of *p*-channel and an *n*-channel transistors working together. This is called *complementary* MOS, or CMOS.

The 4000 series of CMOS ICs is a family that provides a wide range of boolean functions, roughly comparable to the 7400 series of TTL. These include:

Description	CMOS IC	Function
Gates	4000	Dual three-input NOR gates
	4001	Quad two-input NOR gates
	4002	Dual four-input NOR gates
	4011	Quad two-input NAND gates
	4012	Dual four-input NAND gates
Flip-flops and latches	4103	Dual *D*-type flip-flop with set-reset
	4027	Dual *J-K* flip-flop with set-reset
	4042	Quad *D* clocked latch
Counters	4017	Decade counter-divider
	4020	14-stage ripple carry binary counter–divider
Registers	4006	18-stage shift register
	4015	Dual four-stage shift register
Decoders	4028	BCD-to-decimal decoder
Display drivers	4026	Decode counter-divider with 7-segment display outputs
Buffers	4009	Hex buffer
ALUs	40181	4-bit ALU
	4008	4-bit full adder with parallel carry-out

CMOS ICs are not without disadvantages, however. They have less output current source and sink capability than TTL, which can be a drawback when driving loads besides other CMOS ICs. They are very sensitive to damage from static electricity, such as the spark from walking across a carpet on a dry day and then touching metal. This sensitivity means that CMOS ICs have to be handled under special conditions from the time they are fabricated through their actual installation in the electronic system. Finally, CMOS tends to be slower and have longer propagation delays than TTL—although recently the speed gap has been narrowed, and many applications do not need the maximum speed.

QUESTIONS FOR SECTION 6-6

1. What are three drawbacks to TTL?

2. What special transistors are used in MOS ICs?

3. What is CMOS?

4. What are three disadvantages of CMOS versus TTL?

6-7 CHARACTERISTICS OF THE 4000 SERIES CMOS FAMILY

Power Supply

The power supply used for CMOS ICS in this family can be anywhere from +3 to +15 V. The amount of current required at this voltage is very low, and a typical 4000 series IC consumes only a few microamperes. It is very practical to have a whole circuit board of CMOS ICs powered by a few batteries. The higher supply voltages result in faster IC operation, but more power is consumed. The higher side of the supply is called V_{DD}; the ground side is called V_{PP}.

Logic Levels

The exact values for a valid logic 0 and logic 1 depend on the value of the power-supply voltage used to supply the ICs. A logic 0 is a value near ground, while a logic 1 is near the power-supply value (typically within 0.5 to 1 V of power-supply voltage).

   The noise immunity of CMOS ICs is very high. Typically, the switching point between logic 0 and logic 1 is midway between power and ground, so the noise voltage would have to be almost half the power-supply voltage to be misinterpreted (Fig. 6-20). Noise of this magnitude is very uncommon.

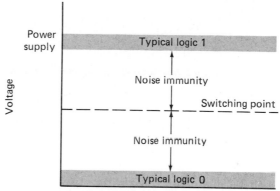

Fig. 6-20   The noise immunity of CMOS ICs is equal to nearly half the power-supply voltage because the logic 0 and 1 levels are near 0 and power-supply values.

Output Drive Capability

The ability of CMOS ICs to source and sink current is very low compared to TTL. Once again, the exact amount is a function of the power-supply voltage used. Typical source and sink values are 100 μA. This is not a problem in circuits built entirely of CMOS ICs, since an input current required by the next IC is very low, less than 1 μA. As a result, the fan-out of CMOS ICs is very large. There is a family of 4000 series CMOS that provides a signal buffer built into the IC which increases the drive capability to about 1 mA.

Speed and Power Considerations

In order to meet the needs of circuit designers, several CMOS family variations are available, just as there are variations of the 7400 series TTL family. These include:

CMOS	Characteristics
4000A	Moderate speed, with output buffer, power supply up to +15 V.
4000B	Moderate speed, with output buffer, can run on up to +18 V and so is capable of higher speed than the A suffix.
4000AU	Somewhat faster than 4000A since it has no output buffer and thus no extra propagation delay.
4000BU	Faster than 4000B since there is no output buffer to add delay.

   The IC pin numbers corresponding to gate inputs, outputs, resets, and other functions on 4000 series CMOS is different than on 7400 series TTL. Some manufacturers provide CMOS ICs that use the same pinouts as 7400 series TTL. These are: 74C00, similar to 4000B series, but "pinout function identical" to the equivalent 7400 IC; and 74HC00, a faster version of 74C00, also compatible with 7400 series ICs.

QUESTIONS FOR SECTION 6-7

1. What is the power supply range for CMOS?

2. How much current does a CMOS IC use?

3. What voltage is a logic 0 in CMOS? A logic 1? What does the exact value depend on?

4. What is the noise immunity value in CMOS? Why is this relatively large value an advantage?

5. What is the typical output source and sink value of a CMOS gate?

6. What are two CMOS families and their key features?

7. What is the 74C00 family and its relation to the standard TTL IC family?

8. What is V_{CC}? What is the equivalent designation for CMOS?

6-8 CMOS IC INPUTS AND OUTPUTS

As with TTL, unused logic inputs on CMOS gates must never be left open since this will result in an ambiguous or varying logic level voltage on the input. Inputs which should be low can be connected to ground. Inputs which should be at logic 1 can be connected to the power supply through a resistor. Several unused inputs can be connected through the same resistor if needed to save cost and board space.

Another reason for never leaving the input of a CMOS IC open (floating) is that CMOS inputs are very sensitive to static electricity. Normal handling of the IC can produce enough of a static voltage to burn out the internal input circuitry of an open CMOS input. CMOS ICs have some protection circuitry built in, but this circuitry is not able to protect against even the several-hundred-volt static electricity charge that a person can generate in routine handling. CMOS ICs are normally shipped in special bags and tubes as part of the overall static damage protection plan.

The output structure of CMOS ICs is similar to TTL, except that the TTL output transistors are replaced by FETs (Fig. 6-21). When the output is a logic 1, the upper FET is on and the lower one is off, so current flows from the power supply through the transistor and to the load. For a logic 0, the upper transistor is off while the lower one is on, and current flows from the load through the lower FET which acts as a current sink. The output FETs of CMOS ICs have little current drive capability, so specially designed CMOS buffer ICs are used where more sink or source current is needed.

There are applications which require that the CMOS IC be used in the same

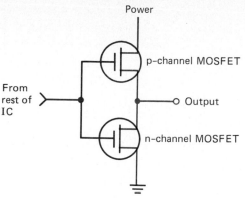

Fig. 6-21 The output structure of a typical CMOS IC uses FETs.

circuit or system as another IC family, such as TTL. The CMOS IC may have to drive the TTL IC, but it does not have the ability to source and sink enough output current for a TTL input. Conversely, a TTL output may have to drive a CMOS input, and there is adequate drive current available from the TTL IC, but the TTL IC is using +5 V for its supply and the CMOS may be between +3 and +15 (or +18) V.

In some special cases it may be possible to connect the two families directly or with a few simple intermediate components such as resistors. In general, however, the interface is handled by a special type of IC called a level shifter-driver. The level shifter-driver does several things (Fig. 6-22). It accepts signals with one set

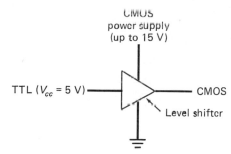

Fig. 6-22 A level shifter-driver acts as the interface between TTL and CMOS voltage and current.

of logic 0 and 1 levels and converts these to another set (such as required between CMOS and TTL). It also provides whatever current source and sink capability is needed for the input it is driving. The level shifter-driver can also be used between an IC and a load such as a lamp. The CMOS IC may be running at 10 V and is capable of providing only 1 mA of output current, while the lamp may require 24 V and 10 mA. The appropriately chosen level shifter-driver will act as the interface (Fig. 6-23).

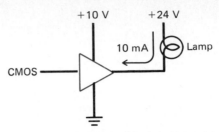

Fig. 6-23 The level shifter-driver can also interface CMOS (with its low sink and source capability) with higher voltage and current loads.

QUESTIONS FOR SECTION 6-8

1. What is done with unused (open) CMOS inputs? Why?

2. What does the output circuit of a CMOS IC look like?

3. How does current flow for a logic 0? For a logic 1?

4. What is a level shifter-driver? Why is it used?

6-9 TROUBLESHOOTING CMOS ICS

The general guidelines of troubleshooting digital logic ICs and their boolean functions applies to CMOS ICs. However, there are some special considerations.

The correct power-supply value for the CMOS-based circuit cannot be assumed. Unlike TTL, which always uses a +5-V supply, the CMOS ICs may be using a supply of +3 to +15 (or +18) V. The correct value must be obtained from the service documentation for the system. Since CMOS is much more noise resistant than TTL, the noise of the power-supply line is generally not a problem as it might be with TTL.

The other important consideration is that CMOS inputs are very sensitive to static electricity and must never be left disconnected. Sometimes the troubleshooting plan involves disconnecting the inputs from the outputs of the previous gates. If this is done, and there is no pull-up resistor on the input to act as a connection, the input will be floating and susceptible to damage (Fig. 6-24). Sometimes, this

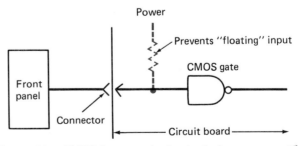

Fig. 6-24 The sensitive CMOS input can inadvertently become open (floating) during troubleshooting procedures.

pull-up is located outside the circuit board that the IC is on (such as on a front panel). When the external circuit is disconnected as part of the troubleshooting procedure, the CMOS input is left open inadvertently.

In general, CMOS ICs are less resistant to being connected, even momentarily, to voltages within the circuit that are beyond the normal specifications for CMOS ICs. Some systems have several different voltages for the different circuits and functions they are performing. A system may have +15 V for the CMOS ICs and +24 V controlling external relays, and the electrical interface may be performed by appropriate level shifter-drivers. However, when probing the circuit board with logic probes, voltmeters, and oscilloscopes, the service technician may inadvertently connect the +24 V to a pin on a CMOS IC or a circuit wire leading to one of these pins. This would ruin the IC. Small slips of the hand and misprobing are common occurrences when troubleshooting circuits, and CMOS ICs are more likely to suffer damage from this so-called probe poisoning than TTL ICs.

QUESTIONS FOR SECTION 6-9

1. What are the differences between the first rules of troubleshooting CMOS versus TTL?

2. What are the problems that may inadvertently occur when troubleshooting CMOS?

6-10 NONSTANDARD IC TYPES

The ICs of the various TTL and CMOS families are extremely useful and common in electronic circuits and systems. They provide the following features to engineers designing products with them, and to companies involved with manufacturing and servicing these products:

- These ICs are available from many manufacturers, so an adequate supply is normally available.
- They are produced in high quantities and are available at reasonably low cost. A 74LS00 IC sells for about 10 to 20 cents.
- There are many logic functions available in the IC families, and they can be interconnected to form an endless variety of circuits. The ICs are extremely versatile and flexible in application.
- These standard ICs are easy to service because they are familiar to many technicians.

However, the standard ICs have some drawbacks. A given circuit function or system may require a large number of these ICs (a large computer can have over a thousand), and the large quantity requires considerable circuit board space; high total power consumption; and the significant cost of inserting, soldering, and testing all the ICs. The equipment manufacturer also needs to keep a large inventory of all the different types of ICs used in the design, and this adds cost and paperwork to the manufacturing process. Finally, the use of standard ICs means that a competitor can look at the physical circuit boards and figure out (or "reverse engineer") the design (or critical portions of it). This would result in lost business opportunities if the competitor came out with a similar product.

There are three alternatives to using standard IC families that a company and its designers can consider to overcome these problems.

1. A completely new custom IC can be designed which precisely performs the circuit function needed. This custom IC would have only the gate functions needed for the application. This approach is attractive but has some very real drawbacks. Designing a custom IC and getting it into large-scale production can take several years. The design time and cost are very large, and even after the design is complete, there can be problems in the actual production phase. Recall from Chap. 1 that IC production is a very complex process, especially for a new design. A precise combination of temperature, impurities, and many other factors is critical to making the production yield high enough. Finally, the high initial cost can only be justified if the total number of ICs to be produced is very large, so that the cost per IC is reasonably low. A custom IC is justified most for high-volume, low-cost applications such as video games, electronic toys, or home electronic devices.
2. Semicustom ICs are another alternative. A semicustom IC is an IC wafer which is deliberately incomplete. The individual gates, flip-flops, and similar gate functions are already fabricated. However, the final step of interconnecting these with metallization is not done. Instead, the manufacturer of the semicustom IC completes the metallization to a specific pattern which provides the system function that the customer needs for a product (Fig. 6-25). The metallization acts as the wiring between the logic functions.

The advantages of semicustom ICs are that the first one can be produced in relatively short time of 2 to 6 months. The semicustom IC manufacturer keeps the

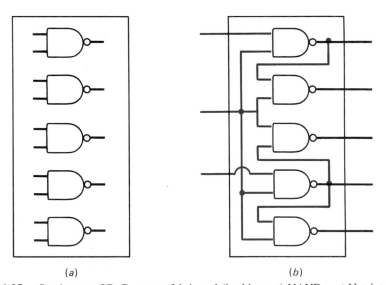

(a)   (b)

Fig. 6-25   Semicustom IC: Gates are fabricated (in this case) NANDs, *a)* Not interconnected, and *b)* Final step—interconnected to customer specification.

nearly completed wafers in stock, and most of the time is required only for getting the first correct metallization pattern produced. After that, the ICs can be made at a high production rate, and the most complex parts of production (the photographic steps, the high-temperature diffusion of impurities) are not involved since they were done beforehand.

The cost of semicustom ICs is usually low and acceptable for the products in which they will be used. The disadvantages of semicustom ICs are the 2 to 6 month design time, which may be too long for some products, and the fact that the IC is produced at another factory to a specific design plan. In the practical world of product design, the function that the final circuit has to perform may be constantly changing up to the last moment, and yet the exact specifications for the semicustom IC must be defined months in advance. If the user of the semicustom IC finds that he or she needs to add or change a function, the 2 to 6 month cycle must start over again. The semicustom IC provides design security because the final function of the IC is within the IC and very difficult to examine.

3. Gate arrays and programmable logic are another alternative to standard ICs. Sometimes called programmable gate arrays (PGAs), these gate arrays and programmable logic ICs are specially designed with large numbers of gates, registers, and other simple logic functions, *all* interconnected by the metallization. (This is the opposite of the semicustom IC situation.) To make the IC have the specific function for the application, some of this metallization must be removed (Fig. 6-26). This is done by deliberately destroying some of the metallization using

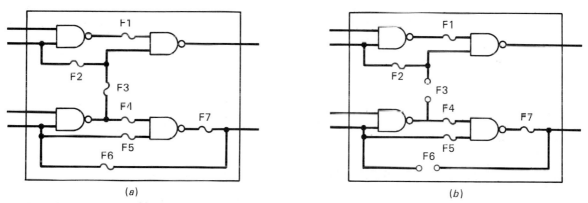

(a)   (b)

Fig. 6-26   Programmable gate array—a) Interconnected, with fuse intact, and b) After some fuses are blown.

special equipment. It may seem impossible to physically destroy the metallization in some places without destroying it by accident in other places. However, a special IC design and fabrication technique is used. Part of the metallization is made of tungsten and behaves like an electrical fuse. If too much current passes through this fuse part, the tungsten heats up and burns itself out, leaving an open connection where previously there was a complete circuit. This is exactly how a fuse in a house, car, or electronic product works; excess current causes the fuse to "blow" and open the circuit. It is called *programming* because it sets a pattern, but it has no relation to computer software programming. The special instrument used for blowing fuses in PGAs puts higher-than-normal current on specially designated leads of the IC, and this blows the desired internal fuses. The result is an IC with a remaining circuit pattern that interconnects only the necessary gates and registers in the required pattern. Both semicustom ICs and PGAs are important alternatives to standard ICs. The semicustom IC requires more planning on the part of the designer and larger production quantities to justify the initial costs. The customization is done on the silicon wafer itself in a special clean room as the final step of the IC fabrication process. By comparison, PGAs are practical for quantities as small as one, since the desired array pattern is blown on the completely packaged IC, using a typewriter-sized machine at the factory of the company that is actually using the IC. It can be done in a regular factory environment.

The PGAs are used for address decoders, for example. A single piece of electronic equipment may use several gate arrays, each with a different pattern but made from the same unprogrammed gate array. This is a major practical advantage: a manufacturer can keep a large stock of unprogrammed arrays and then program them as needed for various functions of the system for a particular production batch.

QUESTIONS FOR SECTION 6-10

1. What are four advantages of using standard TTL and CMOS IC families?

2. What are two disadvantages?

3. What is a custom IC? Where is it practical to use one? What are its good and bad features?

4. What is a semicustom IC? When is it practical? Why is it not a good idea in some cases?

5. What are gate arrays and programmable logic? How are they implemented?

6. What does the term *programming* for gate arrays and programmable logic mean? How is it done?

6-11   SEMICUSTOM AND PROGRAMMABLE ARRAY ICS

In a semicustom IC, the silicon die that will become the heart of the final IC has many complete boolean logic gates, registers, flip-flops, and similar functions already fabricated. These are called *uncommitted* because they are not yet dedicated to any other function on the silicon but are available for use by the desired overall connection pattern. The final metallization layer of the fabrication process connects the required functions to each other, and to the points that will become the external leads of the IC. The IC technology can be TTL, MOS, or a combination of the two, depending on the technologies that are offered by the semicustom fabrication company.

In the example shown in Fig. 6-27, there is a block diagram of the uncommitted functions on a MSI semicustom IC. This IC has flip-flops, gates, and registers available for use. Figure 6-28 shows the simplified diagram of how these would be connected in order to form a frequency meter similar to the one studied in Chap. 3.

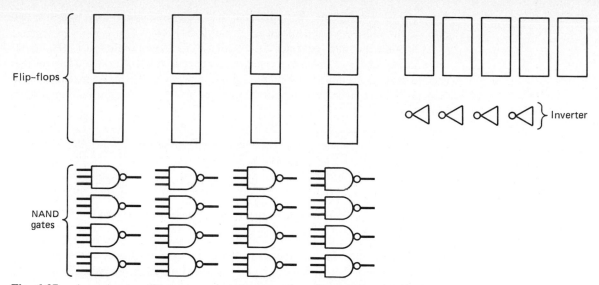

Fig. 6-27 A semicustom IC may contain a large number of uncommitted logic functions.

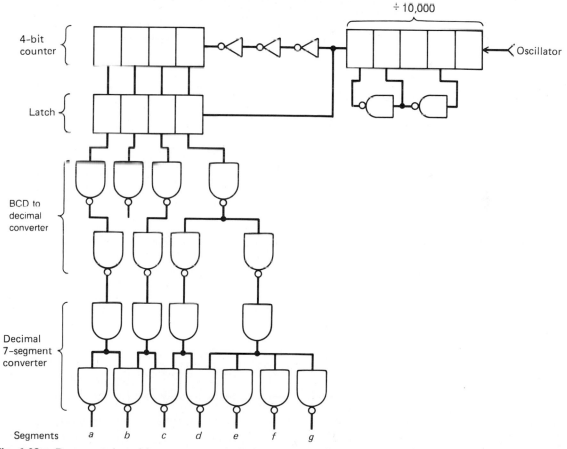

Fig. 6-28 Representation of frequency meter built from a semicustom IC. (Note: Details of gate interconnections are not shown, for clarity.)

A single IC can replace the individual ICs that were previously required and provide the same overall usefulness.

As another example, consider a die which has a large number of NAND gates only. These can be arranged by proper interconnections to form a decoder which is used to develop the active low enable signals for various ICs in a computer system (Fig. 6-29). The decoder is required to accept binary inputs representing address

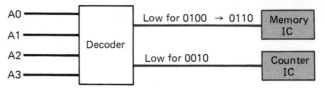

Fig. 6-29   The function of a decoder for address lines A0 through A3.

values and enable a memory IC when the address pattern is 0100 through 0110. It also must recognize the input pattern corresponding to 0010 and enable a counter IC. The required interconnection pattern is shown in the schematic of Fig. 6-30.

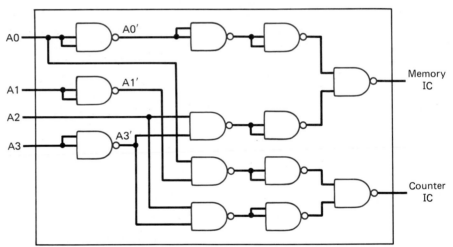

NOTE: NAND's with one input only have 2nd input
pulled up, so NAND functions as inverter.

Fig. 6-30   Two input NANDs of semicustom logic are connected to perform the decoder function. Many of the NANDs are used simply as inverters.

The single semicustom implementation replaces four NAND gates, which require 7400 ICs, shown in Fig. 6-31.

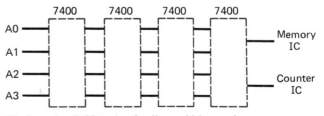

Fig. 6-31   ICs from the 7400 series family could be used.

Programmable Arrays

In contrast to the semicustom IC, the PGA has many electrical paths already completed between the gates and functions on the IC die. The die is packaged in one of the regular IC packages discussed in Chap. 1. The user of the PGA puts the IC into a special fuse-burner instrument and uses a keyboard to enter a sequence corresponding to the desired final interconnection pattern. The fuse burner puts a special voltage on a pin of the IC that is called a *control line*. This voltage tells the PGA that the signals on the IC pins are not normal operational logic levels, but special signals identifying which internal fuses are to be burned. Once the fuse burning is completed, the special control voltage is removed and the PGA behaves like any logic IC, except that the internal functions are more complex than a simple 7400 series IC.

One popular type of PGA is called *programmable array logic* (PAL) by its manufacturer. The PAL consists of a large collection of input buffers, inverters, AND gates, and OR gates. These are all connected in an organized pattern. Figure 6-32 shows a small PAL. Each input to the PAL logic array is available nonin-

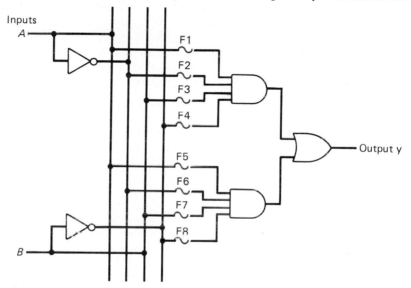

Fig. 6-32 PAL with 2 inputs, 1 output, 8 fuses, and NOT, AND, and NOT gates.

verted or inverted, then is ANDed with all the inputs, and finally the AND results are ORed together. The fuse links F1 through F8 let the PAL user decide which signals should be combined in the AND circuit, and so implement any boolean function once it has been expressed as the OR result of AND intermediate results. This is called the *sum-of-products form* (since the OR is sometimes referred to as a boolean sum, and the AND is called a boolean product), and this form can be used to express *any* boolean logic function.

Consider a circuit which requires the boolean function

$$Y = [A \text{ AND } (\text{NOT } B)] \text{ OR } [(\text{NOT } A) \text{ AND } B]$$

This would be realized by blowing fuses F2, F3, F5, and F8, so the final interconnection would be as shown in Fig. 6-33.

The PAL IC can implement many useful boolean functions and is limited only

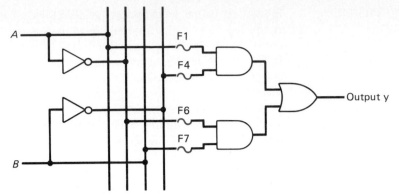

Fig. 6-33 The PAL of Fig. 6-32 shown with fuses blown to implement: Y = [AND (NOT B)] OR [(NOT A) AND B]

by the number of AND and OR gates that are available on the IC. To meet the needs of user applications, the manufacturer of PALs provides a complete series of PALs with different numbers and combinations of AND and OR gates on the IC.

Troubleshooting Semicustom ICs and PGAs

Most semicustom ICs and PGAs use TTL level inputs and outputs as well as TTL V_{CC}. Internally, the ICs may use various IC technologies, depending on the technical and production characteristics that are required. As far as a user is concerned, these ICs are like TTL parts. Some semicustom and PGA ICs use non-TTL technologies with non-TTL signals and voltages, but these are less common.

There is another important characteristic common to both semicustom and PGA ICs: service and documentation are more difficult. Because they are special, semicustom and PGA ICs cannot be numbered in a logical, orderly pattern as 7400 series or 4000 series standard family ICs are labeled. This can cause confusion when the service technician is trying to determine the function of the IC in the circuit. Also, it is no longer possible to look at the numerical designation on the package and then find the function of the IC in a reference handbook. The same IC can have different functions, depending on the customization or the programming burned into the IC. The electrical schematic cannot show the extremely complex functions within these ICs, and in many cases the reason for using these ICs was to hide their function from competitors. The result is that special troubleshooting techniques and procedures must be specified for these semicustom and PGA ICs by the manufacturer of the equipment in which they are used. Without this information, it is nearly impossible to determine the proper function (if there is a fault within the IC) or to get a replacement, except from the original manufacturer.

QUESTIONS FOR SECTION 6-11

1. What is the difference between a semicustom IC and a PGA?

2. What is a PAL? Why is the internal logic structure of inverters, ANDs, and ORs so useful?

3. What are three unique troubleshooting problems that semicustom and PGA ICs introduce?

1. What are the characteristics of an IC family?

2. What are two other names for the 7400 series family?

3. What is the nominal power-supply voltage for 7400 series ICs? What tolerance? What is the power return called?

4. What are the logic 0 and logic 1 voltage levels for the 7400 series?

5. What is source current? Sink current? Fan-out?

6. Why is power consumption by ICs a problem in some systems?

7. Why is IC speed needed in some systems?

8. What are the characteristics of the 74LS00 and 74AS00 designations? The 5400 designations?

9. In 7400 series ICs, what is done with unused inputs that must go to logic 0? To logic 1?

10. Which is greater for 7400 series ICs: the source current or the sink current? By what ratio?

11. How is a logic 1 level provided by a 7400 series IC output? How is a logic 0 level provided?

12. Why is logic 0 never at 0 V and logic 1 never at the power-supply voltage?

13. When is the totem-pole output not useful? What is an alternative?

14. What is implied AND function? How does it work? What is the drawback?

15. What are the troubleshooting steps for implied ANDs? Where should power be checked? Why?

16. What is the troubleshooting problem with shared pull-ups?

17. What is the troubleshooting caution with open-collector outputs?

18. What are the drawbacks of TTL logic?

19. What technology does CMOS use?

20. What are the drawbacks of CMOS?

21. What is the CMOS power-supply voltage? How much current is used? What affects the speed of CMOS IC operation?

22. What is the output drive capability of unbuffered CMOS? Buffered CMOS?

23. What can happen to a floating CMOS input?

24. How are CMOS ICs connected to TTL ICs?

25. Compared to TTL, what is different about troubleshooting CMOS ICs?

26. When troubleshooting, what situations that are potentially damaging to CMOS can occur?

27. What is the danger when probing CMOS IC circuits?

28. What are the drawbacks of standard family ICs?

29. What are the advantages and disadvantages of custom ICs?

30. What is a semicustom IC? How is it made? What are the drawbacks? The features?

31. What is a PGA? How is it made?

32. What are the advantages and disadvantages of PGAs?

33. Why are semicustom ICs and PGAs hard to troubleshoot?

34. What is sum-of-products form? Why is it useful?

35. What is different and more difficult about troubleshooting semicustom ICs and PGAs?

REVIEW PROBLEMS

1. Which of the circuits shown in Fig. 6-34 will function as inverters?

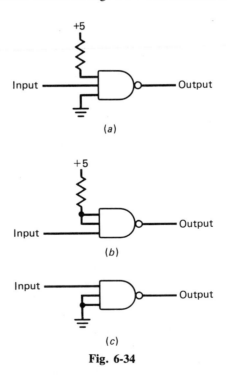

Fig. 6-34

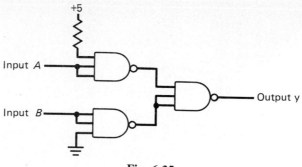

Fig. 6-35

2. What boolean function is implemented by the circuit shown in Fig. 6-35?

3. An open-collector NAND is connected to an LED. Is the LED on or off when input A = B = 0? When A = B = 1? When A = 0 and B = 1? (A and B are the NAND inputs.)

4. A semicustom IC has 20 two-input NOR gates. Show how they could be connected to produce the AND function.

5. For the PAL of Fig. 6-32, which fuses should be blown to implement the function Y = A OR B? For Y = A OR (NOT B)?

6. Complete this chart: (use general terms such as fast, good, high, low, medium)

Characteristic	TTL	CMOS
Power used		
Speed		
Source capability		
Sink capability		
Noise immunity		
Static immunity		

7. Complete this chart:
   (Use general terms as in problem 6)

Characteristic	Custom IC	Semicustom IC	Programmable IC
Initial cost			
Cost per IC			
Time for first IC			
Flexibility			
Difficulty to test			

MEMORY ICs

An essential part of nearly all digital systems is memory. Memory is used to store and retrieve bits. These bits can represent letters, numbers, or the steps the system processor program must follow. The memory falls into two broad classes: memory which can be written to and read by the system processor; and memory which can only be read (which would be used for storing the program and critical numerical values).

Within each category, there is a wide variety of ICs that can be used. The basic principles of operation are the same for all types, but each type has certain operational features which may be required in a special application.

Some of the ICs are normally permanent, but can be erased if needed with special equipment.

For systems that need more memory capacity than a single IC can provide, smaller memory ICs can be combined. The system decoder enables the correct IC by decoding the processor address lines.

7-1 THE NEED FOR MEMORY

Every computer system needs memory. Memory is used by the central processor of the computer for storing all of the information (instructions and data) that the computer will use as it performs its functions. The memory is needed for three purposes:

1. To store numerical data such as numbers to be processed and results of calculations, as well as 1-bit data such as flag bits.
2. To store alphabetic information (often called alphanumeric), which can be part of names, words, and messages to be displayed to the computer user.
3. To store the actual sequence of instructions that the computer will go through as it performs its intended application. This sequence is called the *computer program*.

Depending on the application, the memory may be used for the three types of information in different amounts and proportions. In all cases, the information is stored in memory as collections of 1's and 0's. The processor is able to determine whether the 1's and 0's represent numbers, letters, or instructions by the way the processor is designed and programmed. To understand how memory works, there is no need to understand what the contents of the memory represent.

Different types of memory ICs are required for different applications, and often for specific memory functions within a single system. There are also design reasons why one specific type of memory IC may be used rather than another, even though the memory function performed is the same.

The amount of memory in a computer system can vary depending on the system processor and the system application. Memory system size is usually measured in bytes, or groups of 8 bits. This is because many computer systems use a single byte to represent an alphanumeric character, or a part of a number. Program steps may occupy several bytes. A small computer used for a single dedicated function such as motor control may have only several hundred to a few thousand bytes of memory. A desktop computer can have from 64,000 bytes to several hundred thousand bytes. A larger minicomputer may have millions of bytes available.

The number of bytes is often expressed in kilobytes (K bytes), which is $2^{10} = 1024$ bytes. A typical specification may read: "64K memory" which means that a total of 64×1024 bytes is available. This memory capacity may be physically provided by many smaller-capacity ICs or a few larger-capacity ICs (Fig. 7-1). All memory ICs are built out of combinations of the boolean gates, registers, and decoders studied in previous sections. The different amount of memory available in a single IC is related to the number of registers, the number of bits per register, and the overall interconnection scheme that is used within the IC. Larger-memory ICs generally are more expensive than the smaller ones, but they are less expensive per bit, use less power per bit, and are more reliable (since reliability is a function of the number of ICs as well as their size). The general tendency in systems is to use fewer larger-capacity memory ICs to get the total amount of memory needed for the application.

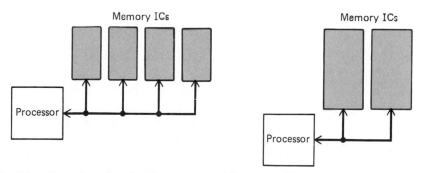

Fig. 7-1 Several small or few large memory ICs can provide the same system memory capacity.

QUESTIONS FOR SECTION 7-1

1. What kind of data is stored in memory ICs?

2. How is the data stored?

3. What unit is used to measure memory capacity? How many bytes is a kilobyte?

4. How much memory is used in a typical dedicated-function computer? In a desktop computer?

5. What are the advantages of larger-capacity memory ICs?

7-2 THE PROCESSOR-MEMORY INTERFACE

An understanding of the operation between a computer processor and its memory ICs is important because many different types of memory ICs are available and designed to work with processors in various specific ways. ICs also have specific performance characteristics which are critical if the IC is to function properly and reliably with the processor.

Memory ICs are connected to the processor via three groups of signal lines. These are the processor address lines, the data lines between the processor and the memory, and the various control lines required by the memory ICs to interact properly with the processor (Fig. 7-2). Each of these groups has a necessary role in allowing the processor to write data to memory and retrieve it later.

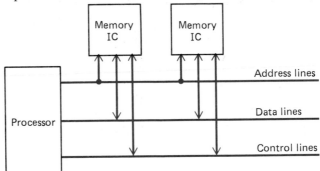

Fig. 7-2 Three groups of control lines, called buses, connect the memory ICs to the processor.

When the processor wants to store data, it must put this data in a known place so it can be retrieved later. This known place is a location in memory and is referred to by its address. Just like the address on a letter to be mailed, the memory address refers to a single specific location out of all the possible locations in the memory. The processor knows, through its program, at what address it wants to store the data. It indicates this address to the memory ICs by putting the address value, in binary code, on the group of address lines (the address bus) that go from the processor to all the memory ICs. This address pattern is decoded by the memory circuitry to point to and enable a single location in memory, which can be any number of bits. Usually, a processor is addressing a group of 1 or 2 bytes at that single address. Since the address is specified by a binary pattern, the number of address locations that can be used is determined by the number of address line bits.

A 2-bit address can identify $2^2 = 4$ addresses, while a 16-bit address can specify any one of 65,536 addresses. (Why?)

Once the processor has put the address on the address bus, it can then write the data pattern, consisting of 1's and 0's, on the group of data lines (data bus) that interconnect the processor and memory ICs. This data is received by the memory ICs and stored at the location indicated by the address bus. The processor then puts a new address on the bus so memory at a different location can be addressed.

To read the contents of memory, the same cycle as was used for writing to memory is repeated. The processor puts an address on the bus, the address is decoded within the memory IC, and the IC then puts the contents of that address on the data bus for the processor to read. Since the cycle is very similar, there must be a way to indicate to the memory ICs whether the processor intends to write data to memory or read data from memory. This is done by the control lines.

The exact type and number of control lines differs with the specific type of memory IC and processor used. The most important control line is the Read-Write line. When the processor intends to write to memory, it sets the Read-Write line high (or low, depending on the specific processor type). This enables the various three-state buffers between the processor and the memory to the proper signal flow direction, from processor to memory. It also indicates to the memory ICs that they should look at the data bus and transfer the binary pattern on the bus to the address location specified by the address bus. A timing diagram shows the state of the various buses, versus time, as the memory cycle takes place (Fig. 7-3).

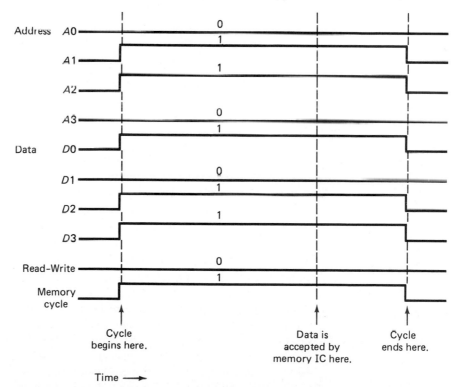

Fig. 7-3 The processor-memory write cycle that occurs when the processor is writing data to the memory. Here data 1101 is written to address 0110.

For reading the contents of memory, the processor sets the Read-Write control line to the opposite state. This enables the buffers that let data flow from the memory to the processor and directs the memory IC to transfer the contents of the specified address to data bus (Fig. 7-4).

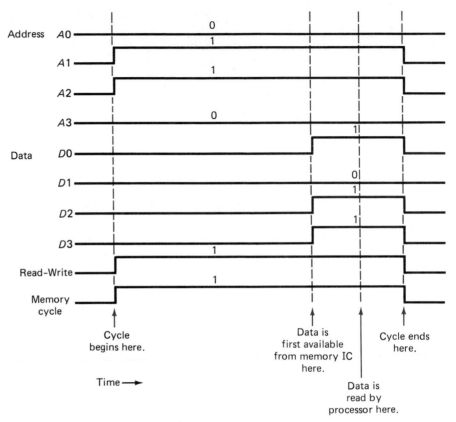

Fig. 7-4 The processor-memory read cycle as the processor reads data from address 0110. In this case, 1101 is the data.

Another control line that is available with some processor types is a memory cycle indicator. This is used to indicate to the other system ICs whether the processor really wants to read or write data (a memory cycle), or if it is putting out sequences of addresses for some other system purpose. The most common use is for another part of the system to access large blocks of memory at a high speed, without tieing up the processor itself. This technique is called *direct memory access* (DMA) and is often used where large amounts of data, such as from another computer system, need to be stored in the memory of the processor. If the processor had to signal the other computer system for each byte of data and then do a memory write with each byte, the overall system performance would be slow. Instead, special circuitry is built into the system which handles a transfer directly from the other computer into the memory of the processor. All the processor has to do is indicate the addresses where the data is to be stored. In effect, Read or Write cycles can be accomplished without the processor itself putting data on the bus or

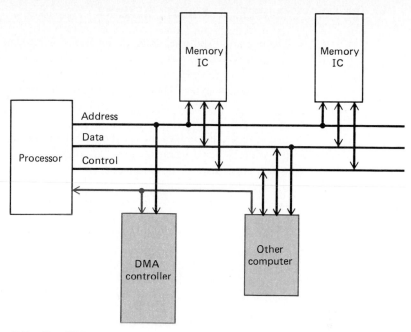

Fig. 7-5 In a DMA transfer, data flows from one device to another without intervention of the processor, after the initial setup.

reading data from the bus (Fig. 7-5). However, the system and memory ICs must somehow know that the bus to the processor is not to be used, but instead the other data path is to be used. The memory cycle indicator does this. If it is in one logic state, a regular processor Read or Write cycle is indicated. In the other state, the DMA operation is occurring.

The speed of operation of memory and processor interaction is determined by the speed at which the system and component ICs are designed to operate. The typical range of operation is from 100 to 500 ns to do a processor and memory Read or Write cycle.

QUESTIONS FOR SECTION 7-2

1. What are the three types of signal lines between a processor and memory ICs? What do they do?

2. What is the relation between the number of address lines and the amount of memory that can be used?

3. What are the steps of writing data to memory? Reading data from memory?

4. What is the function of the Read-Write line?

5. What is a memory cycle indicator? Why is it needed?

6. What are the typical speeds of a memory cycle?

7. What is DMA? Why is it used? How is it done?

The operation and interaction of a processor and a memory IC will illustrate how the internal circuitry of the memory IC makes the Read and Write activity possible. In order to keep the discussion to a manageable size, the memory IC will be very small, 4 bytes. The processor needs to write data to, as well as read data from, any 1 of the 4 bytes, as required by the program that is running in the processor.

The block diagram of the memory IC is shown in Fig. 7-6. The diagram shows the address lines which receive the address from the processor, the data lines which allow data to flow to and from the processor, and the control lines. The memory IC also requires power. The voltage required may be $+5$ V, or $+5$, $+15$, and -15 V, or some other combination. While single-supply memory ICs are easiest to use, some memory ICs use technology that requires multiple voltages, while providing characteristics and features that the specific application may require, such as extremely high speed operation.

The block diagram does not show the actual internal circuitry of the memory element, the location where the binary 1 or 0 is actually stored. Usually, this storage element is some type of flip-flop, but many unique configurations are used depending on the IC technology and memory characteristics required. The user of

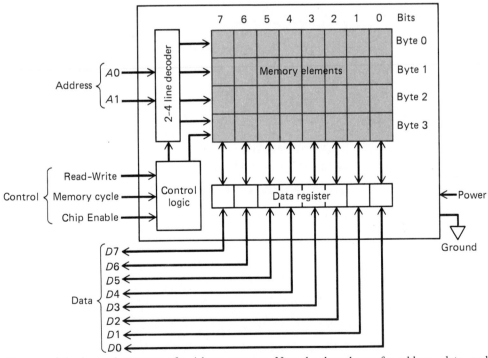

Fig. 7-6   Block diagram of the internal structure of a 4-byte memory. Note the three buses for address, data, and control.

the IC, or any service technician troubleshooting a circuit that uses the IC, is not directly concerned with the circuitry within the IC but mainly with the way the IC integrates with the rest of the system. It is important to understand how the memory IC works in conjunction with the processor.

The processor begins the memory sequence by putting the address of the desired byte on the address bus. Since this is a 4-byte memory, only two address lines are used for the address bus. These two address lines go to a two-line to four-line decoder inside the memory IC. The output of the decoder then activates the appropriate register where the data is stored. These registers are designed to act as parallel-in–parallel-out registers. The processor also sets its Read-Write line to indicate to the memory IC if this is a processor Read or processor Write operation. If it is a processor Write, the registers are set to accept the data from the data bus (parallel in), and the memory IC circuitry between the data lines and the registers lets data move into the registers. If it is a memory Read, the registers put the data that they contain onto the data bus (parallel out) so that the processor can read it. At the end of the operation, the processor removes the memory address from the address bus.

It may seem that a 4-byte memory is too small to be useful in any system. Certainly, for storing numbers, results of calculations, and alphanumeric characters, it is a very limited amount of memory. However, such small memories are in fact quite common when a processor needs to communicate with another complex IC such as a specialized interface to a screen and keyboard (Fig. 7-7). These specialized ICs, which will be discussed in detail in a later chapter, often perform complicated operations independent of the main processor. However, the processor is responsible for setting up these ICs with directions on the nature of the overall operation desired, and the processor must also read back into its program

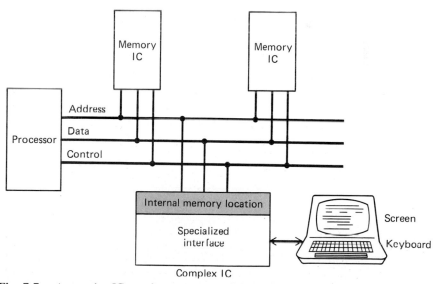

Fig. 7-7 A complex IC may have a very small internal memory which is used for communicating with the system processor.

the results of the operations that these specialized, complex ICs have performed. This processor-to-IC communication is often accomplished by having a small, multibyte memory built into the IC, and this memory area acts as a ''mailbox'' for instructions from the processor and results from the IC as the processor.

QUESTIONS FOR SECTION 7-3

1. What are the functional blocks inside a memory IC?

2. What is the usefulness of a small (several byte) memory? Where is it also used?

3. What power-supply voltages do many memory ICs require? Does the supply voltage affect the sequence of operation?

4. What is the most common storage element in a memory IC?

5. How is the address of the desired memory location indicated by the processor? What happens to it when it reaches the memory IC?

6. What kind of register does the memory location look like?

7-4 MULTIPLE-IC MEMORIES

In the preceding example, the entire memory used by the processor was physically located on a single IC. While this makes the interface easy to understand, many systems use more than one memory IC. The processor-to-memory interface must be able to accommodate several memory ICs on the address, data, and control line buses.

Consider the case of a system which needs twice as much memory as the preceding example. The first possibility is to use a larger memory IC, one with 8 bytes. However, the capacity of memory ICs is sometimes determined by the technology used to fabricate them, and larger ICs may not be available. The only other choice is to use two identical ICs to double the total amount of memory available (Fig. 7-8).

In order to use two ICs, some additional interface features must be provided by the design of the circuit, as compared to the single memory IC situation. First, each memory IC needs a Chip Select (CS) input, a single line that indicates to the IC that it is being selected and activated. When the CS is inactive (usually a logic 1), the IC ignores all signals from the processor on the address, data, and control lines. When the CS is active (usually a logic low) the IC responds normally to the activity on these buses. If there were no CS, both memory ICs would respond to processor bus activity at the same time and the entire situation would be chaos.

The second characteristic of a system with multiple memory ICs is the use of three-state buffers on the data lines between the memory ICs and the processor. This is to ensure that the data lines of a nonselected IC do not interfere with the data lines of a selected IC. The three-state buffers can be external to the IC or may actually be part of the memory IC itself.

A key element in any system with multiple memory ICs is the address decoder. This decoder is responsible for activating the appropriate memory IC chip selects

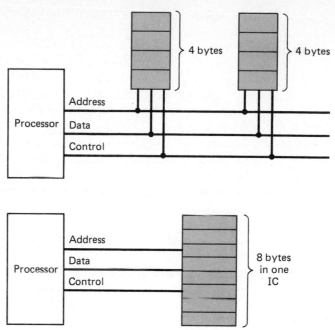

Fig. 7-8 The total memory needed can be provided by one large or two small memory ICs. The processor cannot tell which is being provided.

(CSs) and buffer enables, corresponding to the memory address the processor is accessing. The decoder takes the processor address lines and control lines as inputs and generates the necessary CSs (Fig. 7-9). The decoder can be built from a combination of logic ICs or it can be a semicustom IC such as a PGA. The processor does not know how may ICs are physically used for the memory—it is the decoder which manages these ICs for the processor. As such, the decoder is a critical element in a memory system.

Now the operation of a multiple memory IC system can be studied. This system

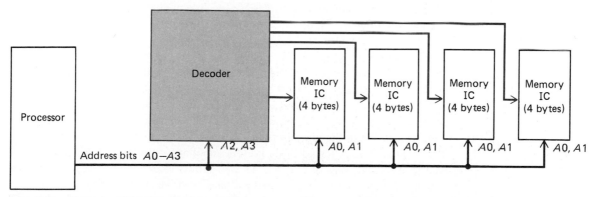

Fig. 7-9 The role of the decoder in a multiple memory IC system. More significant address bits do not go to the memory ICs. They go to the decoder which uses them to select the corresponding memory IC.

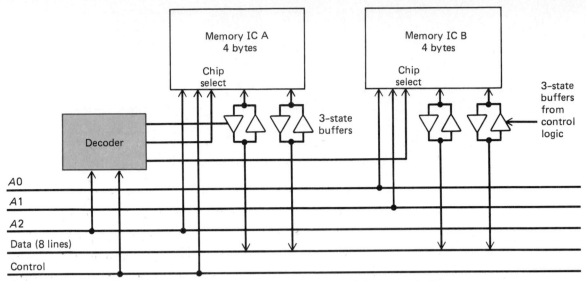

Fig. 7-10   A system with two memory ICs uses the Chip Select and three state buffers.

will have twice the memory of the 4-byte example, yet it will use the same 4-byte memory IC that was used before. The multiple-IC system is shown in Fig. 7-10. The overall sequence of activity between the processor and memory is similar to the single memory IC system, where the processor goes through the same steps and does not know if one large or two smaller ICs are being used to provide its memory needs.

The Write and Read sequences begin when the processor puts the desired memory location address on the address bus. Since eight locations can be addressed, three address lines are required. However, the memory IC itself can only use two address lines, A0 and A1, since the IC is a 4-byte chip. The third address line, A2, of the processor goes to a decoder which is used to provide a CS that indicates which of the two memory ICs should become active. When the address is in the range from 000 to 011, one IC is selected; when it is in the range 100 to 111, the other is selected. The processor Read-Write line goes to both memory ICs since there is no problem with a processor signal going to an IC that is not selected—that IC effectively ignores the signals.

When the selected memory IC receives the 2 bits of address, it is used to specify the corresponding address location within the chip. If the processor is writing to memory, the memory IC puts the data on the data bus into the memory location specified (Fig. 7-11). For processor Read cycles, the situation is more complicated. Even though IC A is selected and IC B is not, the outputs of IC B are still at logic 1 or 0. When the processor goes to read the data bus, it would find that the data outputs of IC A and IC B are present simultaneously, even though the output data of IC A only was of interest. As was shown in previous chapters, two gate outputs cannot be directly connected and used to drive a single input (in this case, the processor data bus), because system malfunctions would result.

The solution is to use three-state buffers on the data bus to control the flow of

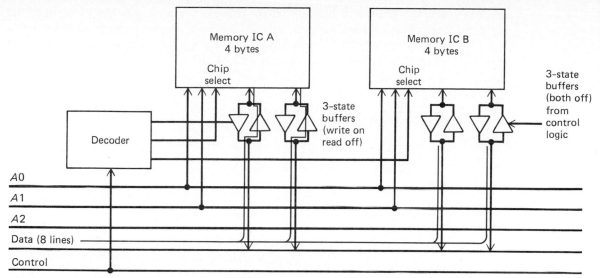

Fig. 7-11 The Write operation to IC A (A2 = 0), showing the signal flow of the data.

outputs from IC A and IC B (Fig. 7-12). The three-state buffers are controlled by a combination of the Chip Select line and the Read-Write line. The control circuitry is designed to provide the following operation: When the processor wants to read data from IC A, the buffers of IC A that allow data to flow from IC A to the processor are enabled. All other three-state buffers are disabled, so that data lines from IC B are not seen by the processor. In this way, interference is avoided and multiple memory ICs can share the same bus. The method that was used to have two memory ICs connected to a single processor can be extended to allow more

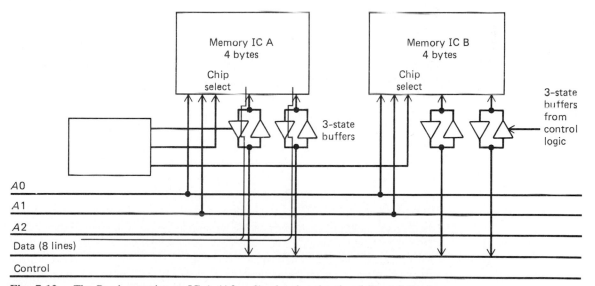

Fig. 7-12 The Read operation to IC A (A2 = 0), showing the signal flow of the data.

ICs and thus more memory on the processor bus, despite the capacity of the individual memory ICs. As long as the processor has enough address bits to uniquely specify the memory location to be addressed, the system can be designed to provide individual CSs to each memory IC.

QUESTIONS FOR SECTON 7-4

1. What are the two ways of getting more memory in a system?

2. What is a CS input? What effect does it have on IC operation?

3. Why are three-state buffers needed in multiple memory IC systems?

4. How does a multiple memory IC system operate?

5. What is the role of the address decoder in multiple memory IC systems? Does the processor know what the decoder is doing?

7-5 READ-ONLY AND READ-WRITE MEMORY

The processor and memory interaction shown in the previous sections involved writing data to memory and reading data from memory. For some types of data, the processor only has to be able to read from memory—there is absolutely no need to write to memory. For example, a system with a processor may be dedicated to a single application, such as controlling the motion of a motor which positions a robot arm. The program used is therefore always the same. Although the program itself may have to handle different inputs values from the arm position transducers and use different numbers in the calculations, the steps performed on these inputs and numbers do not change. In this situation, the processor program could be located in a memory IC that can be read but not written to by the processor.

To meet the needs of this type of application, memory ICs which can only be read are available, called *read-only memory* (ROM). It may seem that there is no advantage to using ROMs, since they are more limited in what they can do than a memory that is both readable and writeable. However, there are many reasons to use ROM when only a processor Read cycle is needed:

- ROM ICs need less interface circuitry to the processor.
- Internally, ROMs need less circuitry and thus more memory locations can be built onto a given size of die. ROMs are thus more "dense" than Read-Write memory. They are also easier to design and fabricate.
- ROMs are nonvolatile. This means that the contents of the ROM are built into the IC itself, and the memory is usable as soon as power is applied to the circuit. Thus the processor program is always available without any "reload" time. If power should fail, the system may stop, but the contents of the ROM are not affected. When a computer program (software) is stored in ROM, it is usually called *firmware* because it can no longer be changed.

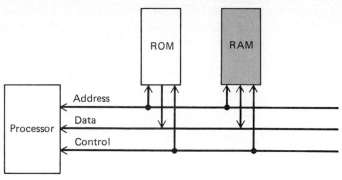

Fig. 7-13   ROM and RAM on one processor bus. Note that data can flow only from ROM to processor, but in both directions for RAM.

Of course, many applications need memory that can be written to as well as read from. The electronics industry usually calls this kind of memory *random access memory* (RAM) because any memory location can be accessed by the processor as needed to store data or intermediate results (Fig. 7-13). Many computer systems use some combination of ROM and RAM ICs on the processor bus. The ROM contains the program for the application and the RAM contains the data. The amount of ROM as compared to the amount of RAM depends on the intended purpose of the system. A system which is used for a specific, dedicated purpose, such as a video game or a motor controller, will have a larger amount of ROM for the program and a smaller amount of RAM for the data storage memory (Fig. 7-14a). A personal computer, in contrast, is used for a wide variety of different programs. Therefore, the largest proportion of the memory is RAM, for the pro-

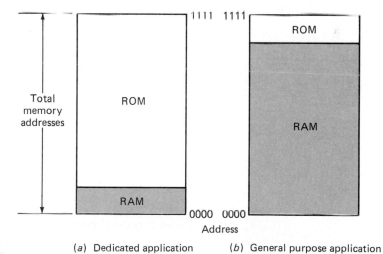

(a) Dedicated application   (b) General purpose application

Fig. 7-14   Different applications have different relative amounts of RAM and ROM.   *a)* Shows a dedicated, fixed program application while *b)* Is for a system which is constantly getting new applications.

gram and data, and there is usually a small amount of ROM which contains the instructions that tell the processor how to begin operation and find the rest of the program, which is usually located on a floppy diskette (Fig. 7-14b). The diskette can provide different programs to the RAM.

In both the RAM and ROM category, there are several technologies and implementations available to system designers. Each type provides different characteristics and features, and some require special circuitry to interface properly with the processor. The troubleshooting techniques used for servicing will also vary between RAM and ROM as well as between some of the various technologies.

QUESTIONS FOR SECTION 7-5

1. What is ROM? When is it needed?

2. What are the advantages of ROM versus Read-Write memory?

3. What is firmware?

4. What is RAM? Where is it used?

5. What is the relation between the amount of system ROM and RAM in different applications?

7-6 READ-WRITE MEMORY ICS

Read-Write memory ICs, usually called RAMs, are available in many memory capacity sizes to meet the needs of different applications. The entire category of RAMs can be divided into two types, regardless of size, static RAM and dynamic RAM. Each type has specific advantages and disadvantages, and while both are Read-Write memory IC types, the dynamic RAM requires additional interface circuitry and special troubleshooting techniques.

Static RAM ICs are constructed of a large number of flip-flops. Each flip-flop acts as the storage element for a single bit. The flip-flops are arranged so that they can be used for writing or reading when the IC is selected, and the address lines and Read-Write lines determine the operational mode and the actual memory bit or bits to be used. Static RAMS are available which hold as few as 64 bits and as many as 16K bits.

A typical static RAM is the 2114, made by several IC manufacturers. The block diagram of the IC is shown in Fig. 7-15, along with the physical location of the signals on the pins of the IC (called pinouts). The 2114 contains a total of 4096 bits. It is arranged internally to provide 1024 groups of 4 bits each. Therefore, for a given address, the processor can write or read a 4-bit group. In order to address 1024 groups, 10 address bits are needed. (Why? The size of individual memory ICs is usually designated by the number of address locations and number of bits in each location. Thus, the 2114 RAM would be called a 1024 × 4 or 1K × 4 RAM.) Operation of the static RAM is straightforward. The processor puts the address of the desired bit group on the address bus and also indicates if it intends to read or write the memory location by using a Read-Write line called Write Enable

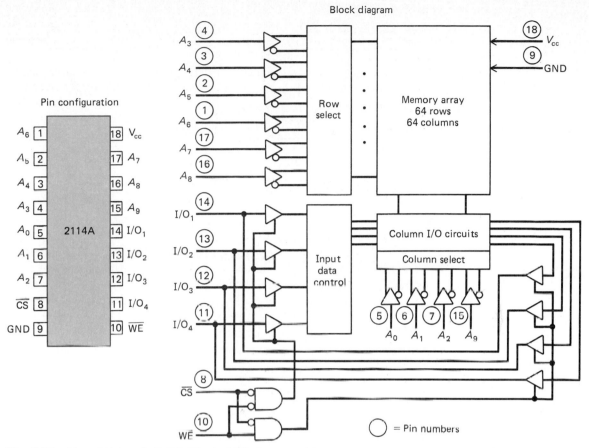

Fig. 7-15 Type 2114 static RAM: Physical IC pinouts and internal block diagram. (Courtesy of Intel Corp.)

(WE) (when the WE is low, it is a Write cycle; if it is high, it is a Read cycle). The RAM receives the address pattern and the WE, and also a CS (also active low). The CS is generated by decoder circuitry between the processor and the RAM, which looks at the more significant address bits as well as some other processor signals.

When the static RAM receives the CS going low, it then begins to look at the address lines and decodes them to point to the group of bits that the address specifies. The IC also looks at the WE line to see if it should accept data or present data back to the processor. If the line indicates a Read cycle, the IC puts the 4 bits of data on the data bus by transferring the data from the flip-flops and by enabling the three-state output buffers that are built into the IC. The buffers then allow the data to flow from the flip-flops to the data bus. The sequence or timing of the Read and Write cycles is shown in Fig. 7-16.

The data does not appear instantaneously on the data bus after the address, CS, and Write signals reach the IC. This entire operation can take from 100 to 250 ns, depending on the speed of the 2114 used. This time is called the *access time*, and must be compatible with the speed of operation of the processor. The RAM is said

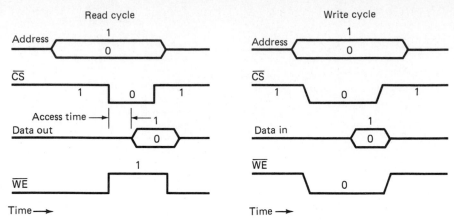

Fig. 7-16   Type 2114 static RAM: Read and Write cycle timing.

to be static because it will present the data outputs to the data bus as long as the input signals are present. This makes troubleshooting easier because the static RAM can be run at different speeds, including very slowly and even stopped in order to verify its operation. Static RAMs can also be used with processors that are much slower than the access time of the IC, since the RAM will keep the data present for as long as the processor needs to get to it.

For Write cycles, the RAM is ready to transfer data from the data bus into the flip-flops soon after the IC has been selected and the WE line goes to the active (low) state.

QUESTIONS FOR SECTION 7-6

1. What is static RAM?

2. Describe a static RAM Write cycle and Read cycle.

3. What is access time?

4. Why is RAM said to be static?

5. What is another name for RAM?

6. What is the memory bit storage element in static RAM?

7. What is the meaning of access time? What are typical values for static RAM?

7-7 DYNAMIC RAM

While static RAM is very easy to use and requires little additional circuitry in the system, it does have two drawbacks as the amount of memory needed in the product becomes large:

- Since six transistors are required for the flip-flop that stores each bit, and there are practical limits to how many transistors can be fabricated reliably on a silicon die, it is difficult to get static RAMs that have very large capacity.

- The power consumed by the transistors of the flip-flops becomes excessive as the static RAM becomes large. The system may not be able to provide this much electrical power, nor can it dissipate the heat that results from the power consumption.

The solution is another type of RAM called *dynamic RAM*. In dynamic RAM, the operation of Read and Write appear nearly the same to the processor as for static RAM. However, the storage mechanism is not a flip-flop, but rather just a single transistor and capacitor, fabricated on the die itself (Fig. 7-17). The result is that dynamic RAMs are available which can store 16K to 256K bits, and use much less power than static RAM.

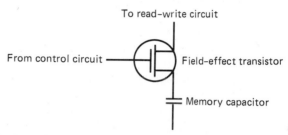

Fig. 7-17 Simplified diagram of a dynamic memory cell. A logic 1 or 0 is stored by the presence of positive or negative voltage on an internal capacitor.

Memory using a transistor and a capacitor works because the electrical charge representing a 1 or a 0 can be stored on the capacitor via the transistor. The transistor allows the charge to be put on the capacitor or removed, as required, to convert the stored bit from a 1 to a 0. It also allows the state of the capacitor to be read back to the data lines. Unfortunately, this simple scheme has a drawback. The capacitor can only hold the charge for a few milliseconds, since the charge is always draining off because of the fact that the transistor is not a perfect off switch but instead allows a small leakage of current. To overcome this problem, which would result in all memory locations losing their stored contents, the memory locations must be recharged, or refreshed, every few milliseconds. This refresh is done by activating every bit location using special signal lines which the IC provides. This means that the dynamic memory IC inputs cannot be held at fixed values indefinitely, since the contents of the bit locations will be lost if the refresh is not done. If an oscilloscope or logic probe is used on the pins of a dynamic memory IC, there will be considerable signal activity even if the processor itself is not reading or writing the IC. This complicates troubleshooting and debugging of the system.

A typical dynamic RAM IC is the 2164, which provides 64K bits (65,536) of storage in a 64K × 1 configuration. The physical pinout is shown in Fig. 7-18. Note the new signals Column Address Strobe (CAS) and Row Address Strobe (RAS), both used for the refresh operation. Some systems use the system processor to control refresh, which requires less additional circuitry but takes some of the processor time; other systems use special interface circuitry to manage the refresh.

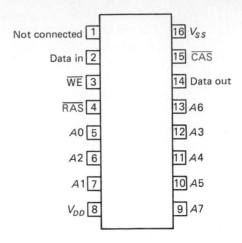

Not connected [1]
Data in [2]
$\overline{WE}$ [3]
$\overline{RAS}$ [4]
A0 [5]
A2 [6]
A1 [7]
V_{DD} [8]

[16] V_{SS}
[15] $\overline{CAS}$
[14] Data out
[13] A6
[12] A3
[11] A4
[10] A5
[9] A7

RAS = Row Address Strobe
CAS = Column Address Strobe

Fig. 7-18 Physical pinout of the type 2164 dynamic RAM.

QUESTIONS FOR SECTION 7-7

1. What are the disadvantages of static RAM?

2. What is the storage scheme of dynamic RAM? What is its drawback? How is it overcome?

3. What is a characteristic of dynamic RAM observed when troubleshooting?

4. What special signal lines exist for dynamic RAM? Why are they needed?

7-8 TYPES OF ROM

ROM ICs are intended to provide memory for the processor where the memory contents need to be permanent, unaffected by power loss, and immediately available when the computer is started up. There are several types of ROM available to system designers. Each type allows the ROM user to set (load) the memory contents into the ROM at a different point in the product manufacturing cycle. Depending on type, the ROM can have its memory contents set:

- At the IC manufacturing plant.
- Just before the ROM is installed in the actual product, at the factory which is assembling the circuit board and its components.
- After the ROM is installed in the final product itself.

Regardless of the ROM type, the performance of these ROMs in the final circuit is the same. The troubleshooting strategy used is also the same for the three types.

1. A ROM which has the memory contents built into the design of the IC is called a *mask-programmed ROM*. The 1 and 0 pattern is part of the design of one of the photographic masks used to expose the silicon die. Mask-programmed ROMs take the most time and money to develop but are least expensive per IC in very large production runs. A handheld calculator, which is made in quantities of several hundred thousand or millions, would use a mask-programmed ROM to contain the instructions to tell the processor in the calculator how to accept numbers, perform the arithmetic, and display results.

Mask-programmed ROMs have a disadvantage similar to custom ICs which makes them unsuitable for some applications; if the memory contents need to be updated or changed because of an error or bug in the sequence of steps that the processor is supposed to execute, then the entire ROM design cycle must start again, at considerable cost in time and money. Mask-programmed ROMs are thus not practical where the memory IC is needed quickly or may be subject to some changes. Products that use such ROMs are usually checked out and tested thoroughly before the ROM design is committed to the IC, since a change would be extremely aggravating and could keep the product off the market.

(Note: the word *program* when used with ROMs has two meanings: One is to set the 1 and 0 pattern into the ROM, by whatever method is used for that type of ROM; the other meaning is what the contents of the ROM represent, which is a program (sequence of instructions) for the processor that is used with the ROM. The correct meaning must be understood from the context, since the program is the word used by the electronics industry for both.)

2. ROMs which allow the product manufacturer to program the binary memory pattern just before the ROM IC needs to be used in the product are very popular because of their flexibility in application. These ROMs are called *programmable ROMs* or PROMs. The technology used is very similar to the programmable logic arrays discussed in the previous chapter. The PROM contains a tiny fuse that makes every location in memory equal to a logic 1 (or every location equal to a 0, depending on the manufacturer of the PROM). To program the desired contents, the PROM is put on a special piece of equipment called a *PROM burner* (Fig. 7-19), which blows the fuses of the memory locations which need to be changed from the unburned logic state (Fig. 7-20). In this way, a blank PROM (which contains either all 1's or all 0's) is converted into a memory IC with the correct pattern of 1's and 0's. Once a fuse is blown, there is no way to change the memory location back to the unburned value.

Despite the fact that PROMs are more expensive than mask-programmed ROMS, they are very popular for several reasons. The manufacturer can avoid the long development time and cost associated with mask-programmed ROMs and can in fact use the same blank ROM for several different products. By burning the different binary pattern into the blanks, the PROM can serve many applications. Finally, the PROM allows the manufacturer to quickly fix bugs and provide updates to the product, because the PROMs can be burned just before they are needed for manufacturing assembly.

3. There are situations where the product manufacturer needs the flexibility that a PROM gives but cannot afford the possible cost of throwing out any PROMs that are already installed when a product change or upgrade is added, or a bug is found. The solution is to use a PROM that can be burned at the factory and has the nonvolatility of a ROM, but also can be somehow erased and made into a blank

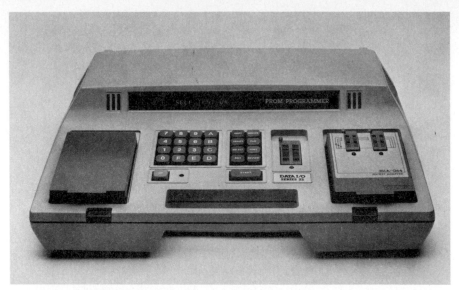

Fig. 7-19 A PROM burner is used to burn the desired memory pattern into the memory IC. *(Photo Courtesy of Data I/O Corp.)*

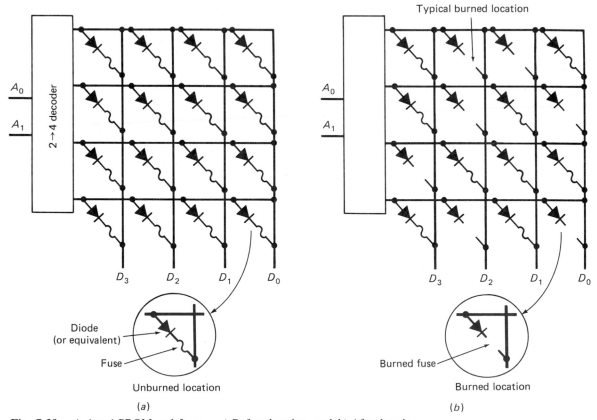

Fig. 7-20 A 4 × 4 PROM and fuses. *a)* Before burning, and *b)* After burning.

Fig. 7-21 A typical EPROM. The quartz window of the package, over the silicon die, allows ultraviolet (UV) light to pass through the package and energize electrons in the die. *(Photo Courtesy of Motorola, Inc.)*

memory IC again. The PROM that does this is called an *erasable PROM*, or EPROM (Fig. 7-21).

In an EPROM, the programming is achieved without using the fusible links that a PROM contains, since fuses cannot be restored once they are burned. Instead, the IC contains special latching circuitry for each memory bit. The latches of an unburned EPROM are all set to logic 1. The PROM burner applies special voltage levels to pins of the EPROM and sets the required bit latches to logic 0. At this point, the EPROM can be used in a circuit like any other ROM.

Erasing the contents of the EPROM is simple for the user and makes use of the special latches. To erase the contents, the user shines an ultraviolet (UV) light through the clear quartz window of the EPROM for about 1 hour. The quartz window allows the UV light to reach the silicon die. The energy of the UV light causes the electrons in the latches to become energized and actually move back to their original state, where the latches were at logic 1. In this way, the user can convert a programmed EPROM back into a blank EPROM for reuse.

The EPROM offers an excellent combination of good performance, flexibility, and reusability. EPROMs have been very popular and an entire family of EPROMs is available (from several IC manufacturers) where each member of the family is very similar to the others but has a different amount of memory capacity, ranging from 8K bits to 256K bits.

4. The EPROM has two drawbacks which have led to another type of ROM. First, the package of the EPROM is very expensive because of the quartz window, and the hermetic sealing required between the window and the DIP IC package. The package must be ceramic in order to bond to the window properly and form the hermetic seal, but ceramic packages do not work well with automated equipment which inserts ICs into circuit boards (the ceramic corners may chip and break). Finally, the IC must be removed from the circuit board to be erased and reprogrammed, and then reinserted back into the board. This involves manual labor and also means that IC sockets must be used which adds to cost and reliability problems.

To overcome these objections, IC companies developed ROMs that can be programmed *and* erased by special electrical signals on the pins of the IC package. These *electrically eraseable PROMs* (EEPROMs) function as regular ROMs when they are in use but can be easily erased and reprogrammed without being removed from the circuit. Since they require no window, they can use an inexpensive plastic package. Despite this, EEPROMs are presently more expensive than EPROMs because of the more complicated internal circuitry required, and they do not come with memory capacities as large as EPROMs.

A comparison of the key electrical characteristics of masked ROMs, PROMs, EPROMs, and EEPROMs is shown in Table 7-1.

TABLE 7-1
COMPARISON OF TYPES OF ROMS

	ROM	PROM	EPROM	EEPROM
Size available (bits)	16K–512K	8K–256K	4K–256K	2K–8K
Who programs	IC manufacturer	Product manufacturer	Product manufacturer	Product manufacturer or product user
How programmed	Mask at IC design	"Zap" fuses with PROM burner	Set latches with PROM burner	Set latches via actual final circuit (or PROM burner)
How erased	Cannot be done	Cannot be done	UV light	Electrical signals in final circuit (or PROM burner)

QUESTIONS FOR SECTION 7-8

1. What is a mask-programmed ROM? When is it used? What are the drawbacks?

2. What does the word program mean with respect to ROMs? How does it differ from a computer program?

3. What is a PROM? Why are they popular? How are they programmed?

4. What is an EPROM? How is it erased?

5. What is an EEPROM? How does it differ from an EPROM?

The leading EPROM family is the 27XX series, available from many IC manufacturers such as Intel, Hitachi, Texas Instruments, and National Semiconductor. The family consists of different-capacity EPROMs, all of which have many pinouts and functions in common. The EPROMs are designed to provide an 8-bit byte of data for each address of memory. The family includes the models listed in Table 7-2. All the EPROMs in the family use a single +5-V power supply and have TTL-compatible inputs and outputs.

TABLE 7-2 THE 27XX EPROM FAMILY			
EPROM Model	Total number of bits	Number of bytes	Configuration
2708	8K	1K	1K × 8
2716	16K	2K	2K × 8
2732	32K	4K	4K × 8
2764	64K	8K	8K × 8
27128	128K	16K	16K × 8
27256	256K	32K	32K × 8

Figure 7-22 shows the physical pinouts of two members of the family, the 2716 and 2732. Since the 2716 has 2K (2048) bytes, it requires 11 address lines, called A0 through A10. The 2732 has 4K (4096) bytes, and it requires 12 address lines

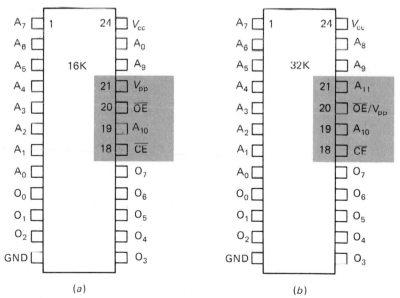

Fig. 7-22 Pinouts of pin-compatible EPROMs. *a)* Type 2716 EPROM, 2K × 8 and *b)* Type 2732 EPROM, 4K × 8. (Courtesy of Intel Corp.)

A0 through A11. In operation, the processor generates the address of the desired byte of memory, and the EPROM receives the address, along with CE (same function as CS) which activates the EPROM to look at the address. The CE is generated by boolean logic operating as a decoder on other address bits and processor lines. The EPROM then puts the 8 bits of data, called output 0 through output 7, onto the data bus. There is a separate control line to the EPROM called Output Enable (OE). This line, when set to logic 0, activates three-state buffers built onto the EPROM so that the data bits can go to the data bus. Because this EPROM family has the three-state buffers built in, no additional three-state buffers are needed when multiple EPROMS are used in a system to get larger memory capacity.

The EPROM can be obtained with various speeds. The data can take anywhere from 100 to 450 ns to appear at the data bus after the address and CE are sent to the EPROM, depending on the speed ordered from the manufacturer. Once the data appears, it will stay there indefinitely as long as the address and CE signals remain. This is called *static operation* and means that fast EPROMs can be used in a slower circuit.

Note that all the pins of the 2716 and 2732 are identical in function except pins 20 and 21. The function V_{PP} is the special pin used to allow the PROM burner unit to program the EPROM. On the 2716, it is brought out separately as pin 21. On the 2732, however, this pin is needed for the additional address bit A11, so V_{PP} is shared with OE, pin 20. This is not a problem in operation because programming the EPROM is a separate operation from actually using it in a circuit.

The 27XX family of EPROMS has been so popular that pin-compatible masked ROMs, PROMs, and EEPROMs are available. This lets a product manufacturer use the exact same board layout and circuit to accommodate whichever ROM technology is best. In initial production, for example, EPROMs may be used, but as the system is fully debugged the manufacturer may switch to PROMs or masked ROMs to reduce cost. The operation of these various memories in the system is identical from the standpoint of the processor and related circuitry.

QUESTIONS FOR SECTION 7-9

1. What is the 27XX EPROM family? What is common in this family?

2. What is the range of speeds available? The range of sizes?

3. What is the OE function? The V_{PP} function?

4. What is identical from one size of 27XX series memory IC to the next?

5. What has to be changed as the memory capacity of the 27XX increases? Why?

7-10 MEMORY CONFIGURATION AND ORGANIZATION

Memory ICs can be designed to provide a specific number of bits in different ways. For example, the 2114 static RAM has a total of 4096 bits, organized as 1024 groups of 4 bits (1K × 4). The 2164 dynamic RAM provides 64K bits, organized as 64K locations of 1 bit each (64K × 1). The reasons for the different organizations are partially due to technical and IC design reasons and partially to allow several memory ICs to be combined, or configured, into large blocks of memory to fit the application need.

Consider a system that needs a total of 1024 eight-bit bytes. This could be provided by two 2114 type RAMs, with one RAM for the first 4 bits and the other RAM for the second 4 bits. The processor would provide the same 10 address lines to both RAMs, but the data bus to the processor would be split, with four data bits for each RAM (Fig. 7-23).

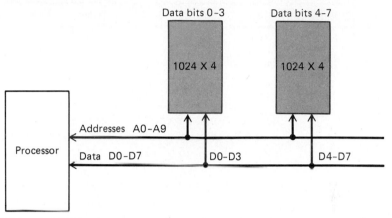

Fig. 7-23 A 1024-byte memory made up of two 1024 × 4 RAMs. (Note: Buffers not shown.)

Now consider a situation where a system requires a total of 2048 four-bit groups. Once again, two 2114 type RAMs could be used, but this time the first 10 address bits would go to both RAMs, and an eleventh bit would go to the circuitry that provides the CE (Fig. 7-24). When the eleventh bit was 1, one of the RAMs would be selected; when it was 0, the other would be selected. The four data bits from each RAM would go directly to the data bus.

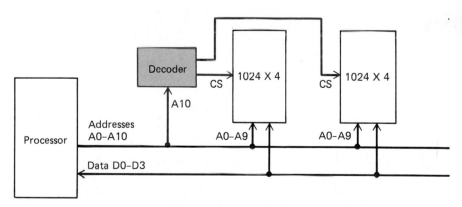

One IC: Processor addresses

Fig. 7-24 A 2048 × 4 bit memory, made up of two 1024 × 4 RAMs. (Note: Buffers not shown.)

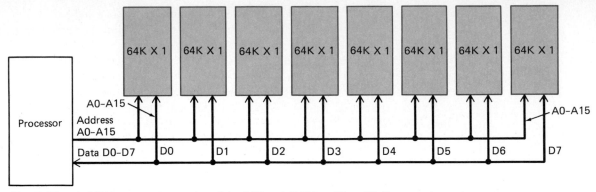

Fig. 7-25   A 64K-byte memory, using eight 64K × 1 RAMs. (Note: Buffers not shown.)

Dynamic RAMs are used in a similar way. Most dynamic RAMs provide a large number of bits, but each bit is independent of the others. To provide a processor with 64K bytes, a system could use eight of the 2164 type dynamic RAMs, as shown in Fig. 7-25. All 16 address bits would go to all the RAMs, and each RAM would provide one data bit for the 8-bit data bus.

Memory systems used with processors often require additional components between the processor and the memory ICs. Figure 7-26 shows a memory system of 2048 bytes (2048 × 8) made up of four 1024 × 4 ICs. The RAM IC used in this case does not have three-state buffers for the data built into the IC itself, so external buffers are required as shown.

It may seem that understanding the organization of memory systems is really a

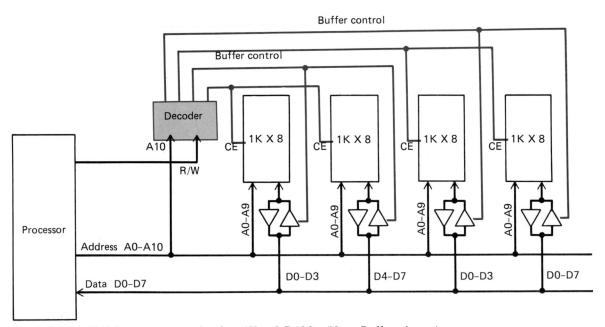

Fig. 7-26   A 2048-byte memory, using four 1K × 8 RAMs. (Note: Buffers shown.)

design function, and once a system is designed, the configuration is not important. This is true from the viewpoint of the system user, who sees only the total available memory and does not see how it is implemented. However, troubleshooting memory systems requires running tests to see exactly what memory locations are working and which are not. By looking at the groupings of the good and bad locations, and also looking at the organization of the overall memory, it is possible to determine which individual memory IC is bad, or which signal buffer or decoder that works with a group of memory ICs is malfunctioning. The repeated, regular structure of memory systems built of smaller memory ICs means that there are standard logical tests to run and conclusions to make, which will be studied in the next section.

QUESTIONS FOR SECTION 7-10

1. How are memory ICs combined to give larger capacity to a system?

2. Why is understanding the memory system configuration important?

3. Explain how memory ICs can be combined to provide more address locations with the same number of bits per location.

4. Repeat for more bits per location, with the same total number of locations.

7-11 TROUBLESHOOTING MEMORY ICS AND SYSTEMS

The regular, repeated pattern of memory ICs and systems means that clearly planned tests can be run on memories and good and bad areas can be identified. By looking at the types of problems and the system configuration, it is possible to identify what part or component has failed. In testing most systems, the assumption is made that only a single IC has failed. Multiple IC failures are very rare and usually the result of some larger system problem, such as a system power-supply malfunction which suddenly provides too much output voltage. The goal of testing is to look at the pattern of bad memory locations and bits and compare this to the memory system block diagram. Based on this, the goal is to see what single component failure would have caused this apparent pattern of bad memory locations.

How RAM Memory Tests Run

Memory is tested by writing known patterns of 1's and 0's to the various memory locations, reading them back, and comparing the results with what was written. Various patterns are used to verify the proper operation of the memory. These include all 1's, all 0's, and alternating 1's and 0's. Note that if only a single pattern were used, a memory location that was "stuck" at that pattern might appear good. Therefore, different patterns must be tried at the same locations.

The memory tests can be initiated several ways. Many processor-based systems contain the test routines as a special part of their program. The routine is executed when the system is started or when the user requests it via a keyboard. The processor runs the test patterns on all the memory locations and then shows the bad memory address on the user screen. The service technician uses this information to determine which memory IC is bad. While a new product is being developed, however, these built-in routines are usually not available. The new product instead often has a way for a technician to write and read individual memory locations via a keyboard, or write a short program which will run memory tests and display the results.

Consider the memory system built up of four RAM ICs, each 1024 × 4, in order to have a total memory of 2048 × 8 (see Fig. 7-26). What can fail?

- An individual memory cell within an IC, which means that one bit would be bad
- Some internal circuitry used by the memory IC, such as the address decoder, which would affect most of the IC
- The decoder circuitry that generates the CEs and controls the data buffers, which could interfere with operation of one or more ICs
- Address line buffers, which could prevent proper writing and reading of some addresses only
- A data buffer, which would affect a bit position throughout the memory

After the memory tests are run and the bad memory locations identified, the following questions must be answered:

- What physical component of the memory system is common to all the failures?
- Is any memory working that would require that component to also be working?
- What single IC failure would cause the effects recorded by the memory test?

Here are some examples, based on the figure:

1. If all memory addresses worked, except for the same bit in each location, then the data buffer for that bit position is probably bad.
2. If bits 0 to 3 were bad for *all* addresses, either the buffers for that group of bits (they may be a single IC) or the decoder circuitry which controls them is bad.
3. If bits 0 to 3 of addresses 0 through 1023 were bad, then the IC that contains those bits is bad or the control signals are not reaching the IC.
4. If all bits of addresses 0 through 1023 are bad, then either the most significant address bit A10 or the decoder is bad.
5. If nothing works, then the decoder is probably at fault.

Testing ROMs

The organization of ROMs is similar to RAMs, but testing them requires a different approach. This is because the system cannot write to the memory locations and then read back and compare. Several different approaches are used to check ROM systems:

- The contents of the ROM can be printed out, under processor control, and then compared to a listing of what the ROM should contain. This is very time consuming if the ROM contains thousands of bits.
- A special code can be placed in one memory address location. This pattern would, by itself, only tell the system if that one location was good. However, the code is usually derived by adding up all the contents of the memory IC. This

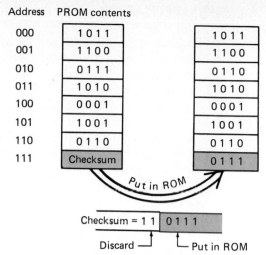

Fig. 7-27 The check-sum generation and location in an 8 × 4 ROM. The check sum is calculated with the data that will go into the ROM, and then the check sum is put in at the last address.

is called a *check sum* (Fig. 7-27). The check sum can be read by the processor. At the same time, the processor reads back all the memory locations, adds them up, and develops its own check sum. The two check sums are then compared and they should agree.

- Special test ROMs with known contents are installed in place of the usual system ROMs. These ROMs are then read back and should have the correct pattern. This technique requires that the memory ICs be in sockets, not soldered, and that the special test ROMs be available.

Regardless of what technique is used for identifying the bad memory locations, the logical process for deciding which component in the memory system is bad is the same for the RAM and ROM systems. RAM systems are easier to work with because more detailed tests can be run to additionally verify the suspected memory area.

QUESTIONS FOR SECTION 7-11

1. What is the key assumption of memory testing?

2. How are RAM tests run? What patterns are used? Why?

3. How are the memory tests initiated?

4. What can fail in a system? What is the effect of these failures?

5. Why is ROM testing different from RAM testing? What approaches are used to test ROMs?

6. What is a check sum? How is it determined? How is it used?

Although the majority of memory IC applications involve computer processors, memories are very versatile and inexpensive components that can be used for applications that require no processor. The ability to store large numbers of binary patterns lets memory ICs be used in obvious ways.

One example of this is called *code conversion*. Binary data in one format may need to be converted to another format. The 7-segment display digit studied in Chap. 3 is an electronic component that needs 7 bits of information to control each of the segments. The 4-bit input which represents the number to be displayed must be converted to the appropriate 7-bit pattern, which is similar to doing a word-by-word translation from English to French.

A ROM can be used for this code conversion (Fig. 7-28). Four input lines to the ROM are used. For each 4-bit input pattern, there is a desired 7-bit output pattern. The input bits specify an address in the memory IC, and the contents of that address are the 7-bit pattern that the display digit requires. These 7 bits are read out to the display digit via the data outputs of the ROM. To convert binary to 7-segment, then, a $2^4 \times 7 = 112$-bit memory is needed. (Why?) Similarly, there are many 8-bit patterns that represent information in one format and that need to be converted to another 8-bit format for another device. A 256×8 bit ROM can be used here. (Why?) ROMs are available commercially which convert many of the more common 8-bit patterns, such as ASCII, into other patterns, such as BCD. Of course, less common patterns can be converted by using a PROM or EPROM.

In some test situations, it is necessary to generate long patterns of 1's and 0's

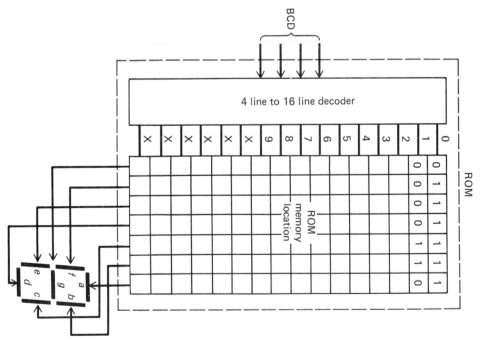

Fig. 7-28   A ROM used for code conversion of BCD format to seven segment digit format. (Note: Not all contents shown, for clarity.)

over and over again in order to test a communications channel or link. The known binary pattern is sent over the channel, and it is checked at the receiving end against the same pattern to see if all the bits were correctly received. This pattern generator function is also suited for ROMs. The desired pattern is stored in a ROM at each end. A pattern generator for 32 bits is shown in Fig. 7-29 using a 4 × 8 ROM. The sending end uses simple counters to step the addresses through every

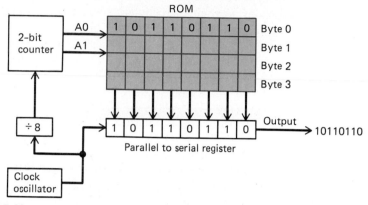

Fig. 7-29 A pattern generator, using a ROM to store the desired binary pattern. The counter causes the ROM to be stepped through all the addresses, and causes each byte of the address to be clocked out of the parallel-to-serial register.

bit of the byte of the ROM. The bit counter takes a byte of stored data and enables one bit at a time to reach the test line by using a parallel-to-serial converter. Every eight counts, the divide-by-8 circuit sends a pulse to the second counter which is generating the address of the byte in ROM for the parallel-to-serial register. As the counters go through each value, each bit of each byte is sent onto the communications link. When the counter reaches the maximum value, it automatically goes back to 0 and the whole process starts again. Using this technique, desired patterns of 1's and 0's can be generated repeatedly and consistently.

QUESTIONS FOR SECTION 7-12

1. What is code conversion? Where is it used? Give two examples.

2. What is a pattern generator used for?

3. How is a pattern generator made using a ROM?

SUMMARY

Memory ICs are a vital part of digital systems. They come in many sizes and with differing features to meet the specific needs of the wide range of their applications. Memory ICs include static RAMs, dynamic RAMs, ROMs, PROMs, EPROMs, and EEPROMs.

When used in actual systems, memory ICs connect to the processor address, data, and control busses. Some additional support circuitry, such as decoders and three-state buffers, is usually required. Regardless of the specific memory IC used and the physical configuration, there are basic principles of memory operation and troubleshooting that can be applied.

RAM memories are used when the system processor must both read data and write data, while ROM memories are used where the processor only needs to read, such as program instructions. Memory ICs are available in families where each member offers similar pin configurations but has the capacity for storing more bits. This storage capacity is often expressed using the letter K, for 1 kilobit, which is equal to 1024 bits.

REVIEW QUESTIONS

1. What three kinds of information can be stored in memory ICs?

2. Explain why the same memory ICs can be used for all the kinds of data. How is the data stored?

3. What is the difference between the letter k used to represent 1000 and the letter K as in kilobyte? How many bytes in a kilobyte? Why is this a convenient number?

4. Explain two of the pros and cons of using smaller memory ICs versus larger ones for the same total memory capacity.

5. What are the three buses between the processor and system memory?

6. What is the role of each bus? Which bus is bidirectional on each of its signal lines? Why?

7. What are two important lines of the control bus? Where are they used in a memory Read cycle? A memory Write cycle? A nonmemory cycle?

8. Explain how the various functional blocks discussed in previous chapters—registers, decoders, flip-flops, and buffers—come together in a memory IC.

9. How does the processor point to the desired location in the memory IC?

10. How can a system memory be increased without using larger-capacity memory ICs?

11. What is the CS input on an IC? Why is it important with memory circuits?

12. Where does the CS signal come from?

13. Explain the role of a memory decoder when used with a multiple IC memory system.

14. What happens to processor address lines that do not go to a memory IC directly in circuits with multiple memory ICs?

15. Explain the difference between ROM and RAM. Where would one be used versus the other? Why?

16. Is a computer program stored in ROM considered software, hardware, or is another term used?

17. What is the ratio between RAM and ROM in a memory circuit? What factor affects the ratio?

18. What is the difference between static RAM and dynamic RAM?

19. What are the pros and cons of each?

20. What is the meaning of *access time* for RAM? What is a typical range of values?

21. Why does dynamic RAM require refresh? What does refresh involve?

22. Explain the key differences and similarities between mask-programmed ROM, PROM, EPROM, and EEPROM.

23. Does the system processor realize the kind of ROM it is using? Explain.

24. What is the importance of a family of EPROMS such as the 27XX type? What does each successive member of the family provide?

25. Explain how memory ICs can be combined to provide more address locations; more data bits per location.

26. Why is it important to understand how the memory ICs are configured to provide the necessary total memory, even if the processor does not care?

27. What is the idea behind testing RAMs? Compare this to the idea behind testing ROMs.

28. Explain how the pattern of failures in memory tests can indicate whether a memory IC, a buffer, or a decoder has failed.

29. What does a check sum do? Does it indicate any problem in the ROM contents, or just some types of problems?

30. Why are ROMs used for code conversion? Give two examples of one digital code that might have to be converted to another kind.

REVIEW PROBLEMS

1. In a memory of 64K × 8 bits, how many address lines are there? How many data lines? How many address and data lines in a 16K × 4 memory?

2. (a) Sketch a block diagram of 4K × 4 memory system using 2114 type RAMs. (b) Sketch a block diagram of a 1K × 16 bit memory system using 2114 type RAMs.

3. Sketch a block diagram of a 64K byte memory system using: (a) 64K × 1 RAMs (b) 8K × 8 RAMs.

4. A small ROM, 8 × 4, contains the following 7 bytes: 0100, 1000, 0011, 0111, 0101, 1001, and 0110. What is the check sum value that should go into the last byte? Repeat the problem for these 7 bytes: 0100, 1011, 1000, 1100, 1001, 1001, and 0101.

5. The 2048 × 8 memory of Fig. 7-26 is tested by writing and reading each location. What component is probably bad for each of these symptoms?
   a. Bad memory in addresses 0 to 511, all bits
   b. Bad memory in addresses 0 to 511, bits 4 to 7
   c. Bad memory in address 0010, bit 3
   d. Bad memory in all addresses, bit 3
   e. Bad memory in all even addresses, all bits
   f. Bad memory at all addresses, all bits

6. A processor has both RAM and ROM in its memory. Sketch the timing of the address, data, and control lines for a ROM Read cycle followed by a Write cycle to RAM.

7. A code converter for converting BCD format to 7-segment format is connected as shown (Fig. 7-28). What should the ROM contents be in each location? (A logic 1 in a location turns a segment on.)

8. Show the relative timing of the counter lines and parallel-to-serial register of the pattern generator of Fig. 7-29.

9. A ROM can be used as an address decoder for a memory system. Consider a 32-byte memory system, made up of four 8 × 8 ICs.
   a. Sketch the block diagram of this system, including address lines and data lines.
   b. Use a 4 × 4 ROM for the decoder that develops the CEs for the memory ICs. What should the contents of the ROM be? (*Hint*: The decoder must decode addresses 0 0000 through 0 0111, 0 1000 through 0 1111, 1 0000 through 1 0111, and 1 1000 through 1 1111 to enable the correct 8 × 8 memory IC.)

10. Consider a 4-byte ROM using a check sum in the last byte. If the contents of the ROM are 0000 1011, 0000 1000, and 1000 0000, what is the check sum? Now repeat for bytes 0000 1010, 0000 1001, and 1000 0000. Is the check sum the same or different? In general, what categories of errors will be caught by the check sum?

SUPPORT ICs

The processors of digital systems are powerful and flexible, but they are sometimes not able to handle efficiently all the tasks they need to accomplish. This is especially true in the area of reading signals from the outside or generating output signals. To help the processor, support ICs are used which are specifically designed for a particular interface. The support ICs have the necessary signal lines for interfacing to the outside and for connecting to the processor. They are commonly used for handling groups of bits simultaneously (parallel) or a continuous stream of bits (serial). The support ICs communicate with the processor via some registers designed into the IC. Through these registers the processor is able to set the support IC to the desired mode of operation, pass data to the IC, and read data back from the IC.

8-1 THE ROLE OF SUPPORT ICS

A typical digital system has a microprocessor as the central "brain" which is called upon to perform the necessary operational steps for the application. These activities may include responding to user input from a keyboard, putting messages on a video screen, communicating to another computer, sending information to a printer, keeping track of time, checking other inputs, and controlling outputs (Fig. 8-1). All of these activities can keep the processor quite busy, to the point where it cannot complete all the tasks in the required time. For example, a processor that is busy may not respond to a keyboard input for several seconds. This would be annoying to the user and actually might be dangerous, since some keystrokes might be completely missed.

As computer systems get more complex and are required to do more and more, there is a point at which the designer cannot guarantee the required performance under all circumstances.

The solution to this problem is to use auxiliary ICs in the system. These ICs support the main processor by handling many of the activities that the processor

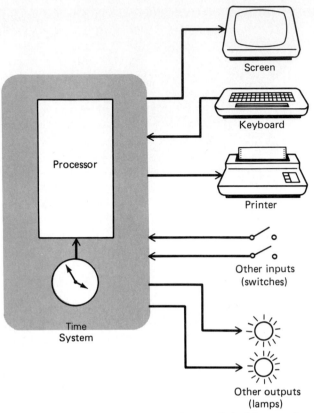

Fig. 8-1 A processor in a practical system is occupied with many internal and external tasks, such as handling a screen, a keyboard, a printer, and other inputs and outputs.

would otherwise have to do by itself. The tasks that the support ICs can handle include:

- Repetitive activities, where the processor must perform the same operation over and over, and yet in most cases there is no new data or information. For example, the processor must check the keyboard several times per second to see if a new key has been pushed, but most of the time there is no new keyboard activity by the user. Another case would be in the alarm system, where the processor would constantly check to see if a door or window switch had opened. The checking must be done frequently, about once per second, and most of the time the entrance switches are unchanged.
- Some external devices (peripherals) have special signal lines that the processor cannot handle directly. The signal levels may be different, or the number of signal lines may be greater than can be wired directly to the processor or its immediate interface circuitry. Here, the support ICs must provide additional lines, or lines of the type that the interface requires.
- Some applications have critical timing where the processor must perform some action within a very short time, and there can be no extension beyond this time.

This is typically the case where the motion of a mechanical object is being controlled, and any delay beyond the specified time would be dangerous. A motor that is controlled by a processor to move a robot arm must be controlled with precision to make sure that it moves the arm to the right spot at the right time. Otherwise the arm could move too far, or too fast, and not complete the action properly.

The role of support ICs is to reduce the workload on the main processor, called *offloading*. These ICs allow the processor to operate more efficiently by taking care of repetitive tasks, routine activities, and special interface situations. They also make the programmer's job easier because they perform functions that would otherwise have to be programmed into the processor itself. A support IC that costs a few dollars can let the designer develop a system that provides better performance and at the same time save many hours of programming effort. The functions of a support IC could be implemented by typically 50 to 200 small-scale ICs. However, the space, cost, and power would be excessive. The support function in a single IC is an example of large-scale integration (LSI).

QUESTIONS FOR SECTION 8-1

1. How can support ICs help a processor with:
   a. repetitive tasks?
   b. special signals to peripherals?
   c. critical timing?
   Give an example in each case.

2. What do support ICs do in terms of programming effort required?

3. Can the support IC be replaced by small-scale ICs? How many, on average?

4. What is the level of integration of most support ICs?

8-2 GENERAL STRUCTURE AND OPERATION

Support ICs vary in the range of activities they can handle. As a result, the internal complexity of these ICs also varies from simple to complex. The simpler support ICs are a collection of gates, flip-flops, registers, and components that are interconnected to provide a specific function. The more complex ones are actually microprocessors in themselves and capable of handling complicated situations and providing the central processor with only the really necessary information.

Regardless of the complexity of the different support ICs, their structure as seen by the rest of the system is fairly similar. One side of the support IC is connected to the processor address, data, and control bus. The other side is connected to whatever device or part of the system the IC is designed to help the processor with. The processor communicates with the support IC via registers which are internal to the support IC. These registers are used as mailboxes between the main processor and the actual operating part of the support IC (Fig. 8-2):

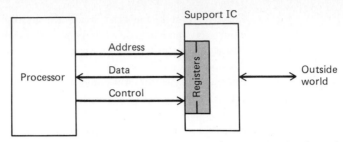

Fig. 8-2   Interfacing between the processor and the support IC is via registers in the support IC that act as "mailboxes" for commands, status, and data.

- By writing to these registers the processor can direct what activity it wants the support IC to perform, and select some specific choices from the variety the IC is capable of providing.
- By reading these registers the processor can find out the results of the activity of the support IC. For example, it can find out that a character from another computer device has been received and begin to process the character.

Interrupts

If the processor had to constantly check the register to find out the results, the overall system operation would be more efficient than without the support IC, but not as efficient as many applications require. What is needed is a way to alert the processor to the fact that something new and worthwhile is in the register, so then and only then does the processor need to check. To make this sort of operation possible, a special signal is usually used between the support IC and the processor. This signal is called an *interrupt*.

An interrupt is a signal line leading to the processor. When the interrupt line goes active (usually logic 0), the processor stops executing the routine of program steps it is currently in, and goes to handle (service) the interrupt. It does this by going to a special program routine which provides instructions on what to do if this interrupt occurs. After the processor has finished servicing the interrupt, it then resumes executing the main program from the point where it stopped (Fig. 8-3, page 209).

The concept of an interrupt is very important. Interrupts are what let the processor make maximum use of the power of the support ICs. People often use interrupts in their daily activities. Suppose someone is expecting a friend and has things to do until the friend arrives. The person could go to the door every few seconds and see if the friend has arrived, but would spend so much time doing this that they would be unable to get much else done. Instead, they continue doing the things that need to be done, and the friend interrupts by ringing the doorbell. This causes the person to stop what he or she is doing and answer the door. If that person is in the middle of something critical, such as cooking, he or she can ask the friend to come in and wait while the host goes back and finishes cooking or at least turns off the oven.

By properly balancing the use of hardware support ICs and the software program which supports both the system application and the support IC, the system is able to meet the required level of performance and cost. Support ICs allow a system to perform activities which would be difficult or impossible for the processor to perform alone.

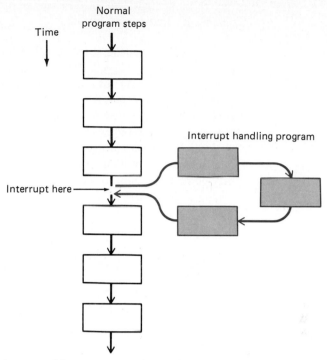

Fig. 8-3 An external interrupt causes the regular program to stop and a special interrupt program to run; the regular program then resumes.

QUESTIONS FOR SECTION 8-2

1. Why are support ICs needed? Why can't the main processor do the task by itself in many cases?

2. What types of tasks do support ICs handle?

3. What is *offloading*?

4. What does the support IC look like to the processor? How is it connected?

5. How does the processor communicate with the support IC?

6. What is an interrupt? Describe the operation of the system when interrupts occur. Give examples of interrupts in everyday situations.

8-3 SERIAL INPUT–OUTPUT SUPPORT ICS

A practical computer-based system must have some way of receiving information from the outside world and a way to present information to the system. Presenting information to the system from the outside world is called an *input*, while signals going to the outside are called *outputs*.

There are many formats that inputs and outputs (I/O) can have. Digital inputs and outputs are used when the information can be represented by one or more digital signals. The inputs and outputs can be a single bit on a single line, a stream of pulses on a single line, or they can be groups of 1s and 0s on several lines. Examples of digital input include:

- Open or closed switches from doors and windows in an alarm system with a single line per switch, and each switch is independent of the others.
- A group of eight signals, which indicate which buttons on a coffee vending machine have been pushed. All signals are checked at the same time.
- A single line with a continuous stream of pulses from a motor, with each pulse indicating that the motor has completed one revolution.
- A single line which carries data from another computer, where each alphanumeric character is represented by a burst of 1s and 0s. The characters themselves may come in a group, or may be widely spaced, depending on how the other computer is sending them.

Digital outputs are for information to the outside:

- Individual signal lines which are used to turn on lamps and indicators on a control panel.
- Groups of lines, which are used to simultaneously turn LED segments on and off in a multiple-digit 7-segment display.
- A single line which is used to send a stream of characters to a display or readout such as a terminal.

When a group of input lines is read at the same time by the processor, or a group of output lines is set at the same time, it is called *parallel* input or output (Fig. 8-4*a*). If a single line is used for carrying a stream of 1s and 0s, and the pulses are in a specific time relation to each other (for example, 1000 pulses per second) it is called *serial* input and output (Fig. 8-4*b*).

Support ICs are available to handle either parallel or serial I/O in a way that uses the fewest processor pins and requires the least amount of processor involvement. The result is an efficient system which can accomplish the required application.

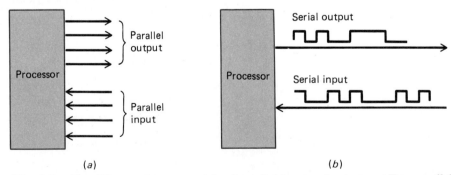

Fig. 8-4 The difference between serial and parallel input and output: *a*) Four parallel inputs and four parallel outputs; *b*) One serial input line and one serial output line.

QUESTIONS FOR SECTION 8-3

1. What is an output? An input?

2. What is serial I/O? Parallel I/O?

3. Categorize each of these as input or output, serial or parallel:
   a. Several people talking to you at the same time
   b. A meeting where each person takes a turn speaking
   c. Pulse from a control box to control a motor
   d. A processor reading data from a tape storage device
   e. Several letters being mailed at once to invite people to a party
   f. Three phones ringing at the same time

4. How can serial output be made parallel? How can parallel output become serial?

8-4   SERIAL I/O

Serial I/O is used when a system needs to communicate with an external device that may be located anywhere from a few feet to several hundred feet or more away from the system. The advantage of serial I/O over parallel I/O is that less circuitry is needed at each end and fewer wires are used in the interconnecting cable. The devices that may be connected to the serial "port" on the computer system include terminals, printers, bar code readers, scales, and electronic test equipment.

The most common serial interface is the RS232, which is an industry standard that defines the types of signal used and the signalling rate. Nearly all types of equipment used with computers have an RS232 port for communication from the device to the computer. A pattern of 1's and 0's is used to represent each of the alphanumeric characters (Fig. 8-5). This pattern is sent at a fixed bit rate called *baud*. The character can be sent at any time, so it is preceded by a start bit (logic 0) which tells the receiving end of the interconnection that a new character follows, and the character is ended with a stop bit (logic 1). Sometimes an additional bit is added to the group of bits that represent the character itself, in order to act as a checking bit for errors in transmission. This parity check bit allows the receiving

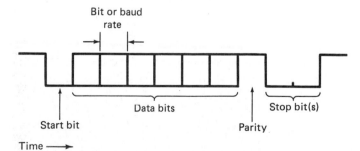

Fig. 8-5   A serial ASCII character is one formed by start, data, parity (optional), and stop bits, sent at a known rate.

end to determine if there were any errors caused by noise in the received signal.

In order to send a single character, the system processor would have to do several things:

- Determine the 1 and 0 pattern for the particular character (this information could be stored in its memory).
- Determine the value of the parity bit if one is required.
- Know the desired baud and set up the proper internal timing for this rate.
- Send a start bit.
- Send the 1s and 0s which represent the character.
- Send the parity bit, if required.
- Send a stop bit (some systems require two stop bits).

All these things can be done by a processor, but the bits must be sent at precise time intervals so that the receiving end can determine what was sent. To make sure that the time intervals are precise, the processor would have to stop doing anything else and devote all of its resources to this one task of sending a character.

Receiving a character is even more difficult. The character can come at any time, so the processor would have to constantly check, or poll, to see if a start bit had appeared. If this bit is present, the processor would have to read the state of the input line at exactly the right intervals to see each bit of the character. If the timing was not right, the processor might miss a bit, or read a bit twice, which would mean the character would be incorrectly interpreted (Fig. 8-6). If a parity bit is used, the system would have to check if the parity bit was correct for this character. Finally, the processor would have to look for the stop bit or bits, to know that the character pattern was complete.

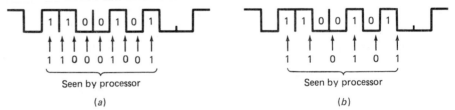

Fig. 8-6   The effect of a processor reading the serial bits representing 1100101 with the wrong timing: *a)* Too often *b)* Not often enough

Since sending and receiving a character requires exact timing and constant checking, most systems use a special IC for the purpose. This special IC works with the processor to handle the details of sending and receiving, and also provides some additional features that many practical systems require. The industry name for this type of IC is the *universal asynchronous receiver-transmitter* (UART).

The Signetics 2561 UART is typical of the UART ICs used. This 28-pin IC provides flexibility in use, offloads the processor, and also provides handshake lines to the external device connected to the system (Fig. 8-7). These handshake lines are used to let each end of the communication link know the status at the other end. If one end is busy for whatever reason, and cannot accept other charac-

PIN CONFIGURATION

D_2	1	28	D_1
D_3	2	27	D_0
RxD	3	26	V_{cc}
GND	4	25	$\overline{RxC}$
D_4	5	24	$\overline{DTR}$
D_5	6	23	$\overline{RTS}$
D_6	7	22	$\overline{DSR}$
D_7	8	21	RESET
$\overline{TxC}$	9	20	BRCLK
A_1	10	19	TxD
$\overline{CE}$	11	18	TxEMT/DSCHG
A_0	12	17	$\overline{CTS}$
$\overline{R/W}$	13	16	$\overline{DCD}$
$\overline{RxRDY}$	14	15	$\overline{TxRDY}$

PIN DESIGNATION

Pin. No.	Symbol	Name and function	Type
27, 28, 1, 2, 5–8	D_0–D_7	8-bit data bus	I/O
21	RESET	Reset	I
12, 10	A_0–A_1	Internal register select lines	I
13	$\overline{R/W}$	Read or write command	I
11	$\overline{CE}$	Chip enable input	I
22	$\overline{DSR}$	Data set ready	I
24	$\overline{DTR}$	Data terminal ready	O
23	$\overline{RTS}$	Request to send	O
17	$\overline{CTS}$	Clear to send	I
16	$\overline{DCD}$	Data carrier detected	I
18	$\overline{TxEMT/DSCHG}$	Transmitter empty or data set change	O
9	$\overline{TxC}$	Transmitter clock	I/O
25	$\overline{RxC}$	Receiver clock	I/O
19	TxD	Transmitter data	O
3	RxD	Receiver data	I
15	$\overline{TxRDY}$	Transmitter ready	O
14	$\overline{RxRDY}$	Receiver ready	O
20	BRCLK	Baud rate generator clock	I
26	V_{cc}	+5V supply	I
4	GND	Ground	I

Fig. 8-7 *a)* The IC pin configuration and *b)* designations for the 2651 UART. **Note:** some of the pins are used for special, nonsynchronous communication. *(Courtesy of Signetics Corp.)*

ters, it indicates this by setting the handshake line low. The sending end checks this line before sending each character and will not send if the handshake line indicates that the receiving end cannot accept another character. The result of this handshake scheme is that the receiving end is always ready for the character sent, and therefore no characters are lost. Without handshake lines, the rate of characters sent would have to be very slow, to ensure that each end had time to process the previous character before the next one was sent. A system with handshake lines might operate at almost 20,000 baud, while one without handshake lines might have to limit itself to about 1000 baud.

The UART functions as a specialized serial-to-parallel and parallel-to-serial converter, with some additional functions and features. A block diagram of the 2651 is shown in Fig. 8-8, page 214. A single UART contains the necessary circuitry for both sending and receiving characters at the same time. The serial-to-parallel port is used for receiving incoming characters and presenting them to the processor, while the parallel-to-serial port is used for getting characters from the processor and sending them out.

QUESTIONS FOR SECTION 8-4

1. What are the advantages of serial I/O over parallel I/O?

2. What is baud?

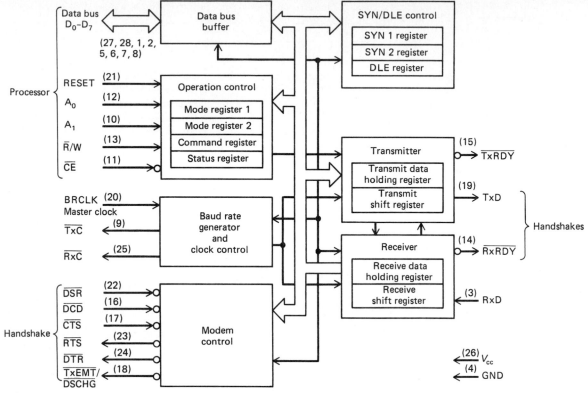

Fig. 8-8 The block diagram of the internal functions of the 2651 UART. A wide range of detailed functions are provided internal to the IC. *(Courtesy of Signetics Corp.)*

3. How is a character sent in RS232 serial communications?

4. What is a parity bit? What function does it serve?

5. Why is timing critical in sending serial bits? In receiving them?

6. What are handshake lines? How are they used?

7. What is polling? How does it differ from the interrupt method?

8-5 INTERCONNECTION AND OPERATION OF THE UART

The UART is connected to the processor address, data, and control buses. It is usually designed in the circuit to appear to the processor as a group of memory locations. The processor can then write to the UART as it would write to any memory IC. Internally, the 2651 UART has four registers, each of 8 bits. It is through these registers that the processor communicates with the UART. Three are mode and command registers, to allow the processor to direct the UART operation, and to set up the UART for the desired operational features. The fourth register is the location where the processor writes the code corresponding to the character to be sent. Figure 8-9, pages 216 and 217, details the many allowable modes of operation that can be set into these registers.

Mode registers MR1 and MR2, figures 8-9*a* and 8-9*b,* are used to indicate the desired operating specifications for the UART and therefore for the serial communication that the UART provides. In mode register MR1, the two MSBs (MR17 and MR16) are used by the processor to define the desired number of stop bits. MR15 and MR14 select the type of parity desired (odd or even) and determine if parity is used at all. The number of bits per character is set by MR13 and MR12, since different serial interfaces require either 5, 6, 7, or 8 bits per character. The second mode register, MR2, is used to select the transmit and receive clocks via MR25 and MR24. Some systems need the UART clock set by circuitry within the system, while others require that the clock be provided by an external source. Finally bits MR23, MR22, MR21, and MR20 indicate which one of 16 baud rates the UART should use. These baud rates are generated by the UART, using internal "divide-by-*N*" counters within the UART which divide the UART clock. The Command Register (CR) is used to set up some of the more advanced operating conditions that a serial interface requires.

The registers also allow the processor to read back the setup conditions and operating status of the UART and the external lines connected to it (Fig. 8-9*d*) via the status register (SR). The processor can tell if the UART is busy, and if there is any change in the state of the external handshake lines. Status register bit SR1 indicates if the register in the UART which receives characters from the serial interface is full (a 1 indicates that it is). If it is full, a character has been received by the UART and must be read by the processor before another character can be received. Similarly, SR0 shows whether the character that the processor put into the UART for transmission over the serial link has been sent. (It may not have been sent because the handshake lines indicate that the receiving end is not ready for it). If the transmit register is full, the processor cannot enter another character for transmission. These critical bits SR1 and SR0 can be checked by the processor in one of two ways. The processor can regularly check (poll) the bits to see what their status is, or the bits can be wired to the processor to cause an interrupt.

The interrupt would occur when the receiving register goes from empty to full (signifying that a new character has been received) or when the transmitting register goes from full to empty (indicating that the previous character was sent out and the UART is ready to accept the next one from the processor). The processor program would see the interrupt, halt its regular activity, and either read the newly received character or present the next one to be transmitted to the UART. Another register contains the code of the character received, which can then be read by the processor.

The part of the UART that faces the external device contains a line for the serially transmitted data, a line for the serially received data, and the handshake lines. The simplest operation is the asynchronous operation with an external clock, using eight transmitted bits. To transmit a character:

1. The processor sets up the desired conditions of baud, number of bits, whether parity is required, and other factors by writing appropriate bits to the registers. (There are 16 baud values from 50 to 19,200 to choose from, for example.)

2. The processor then parallel loads the UART register with the character to be sent.

3. The UART checks the handshake lines. (For a variety of reasons, most systems have more than one handshake line in each direction.) If the handshake lines indicate that the external device is ready to accept a character, the UART sends a start bit, takes the character in its register, and transmits it one bit at a time at the already specified rate, beginning with the LSB. (This is the parallel-to-serial process.) If a parity bit is needed, the UART figures out if the bit should be a 1 or 0 (based on the bits of the character) and then sends it.

4. The UART then sends the stop bit or bits.

5. If, however, the handshake lines indicate the other end is not ready, the UART

MODE REGISTER 1 (MR1)

MR17	MR16	MR15	MR14	MR13	MR12	MR11	MR10
		Parity Type	Parity Control	Character Length		Mode and Baud Rate Factor	
Asynch: stop bit length 00 = Invalid 01 = 1 stop bit 10 = 1½ stop bits 11 = 2 stop bits		0 = Odd 1 = Even	0 = Disabled 1 = Enabled	00 = 5 bits 01 = 6 bits 10 = 7 bits 11 = 8 bits		00 = Synchronous 1X rate 01 = Asynchronous 1X rate 10 = Asynchronous 16X rate 11 = Asynchronous 64X	
Synch: number of SYN CHAR 0 = DOUBLE SYN 1 = SINGLE SYN	**Synch: transparency control** 0 = Normal 1 = Transparent						

(a)

MODE REGISTER 2 (MR2)

MR27	MR26	MR25	MR24	MR23	MR22	MR21	MR20
		Transmitter Clock	Receiver Clock	Baud Rate Selection			
Not used		0 = External 1 = Internal	0 = External 1 = Internal	0000 = 50 baud 0001 = 75 0010 = 110 0011 = 134.5 0100 = 150 0101 = 300 0110 = 600 0111 = 1200		1000 = 1800 baud 1001 = 2000 1010 = 2400 1011 = 3600 1100 = 4800 1101 = 7200 1110 = 9600 1111 = 19,200	

(b)

CR7	CR6	CR5	CR4	CR3	CR2	CR1	CR0
Operating Mode		Request to Send	Reset Error		Receive Control (RxEN)	Data Terminal Ready	Transmit Control (TxEN)
00 = Normal operation 01 = Asynch: automatic echo mode Synch: SYN and/or DLE Stripping mode 10 = Local loop back 11 = Remote loop back		0 = Force $\overline{RTS}$ Output high 1 = Force $\overline{RTS}$ output low	0 = Normal 1 = Reset error flag in status reg (FE, OE, PE/DLE detect)	**Asynch: force break** 0 = Normal 1 = Force break **Synch: send DLE** 0 = Normal 1 = Send DLE	0 = Disable 1 = Enable	0 = Force $\overline{DTR}$ output high 1 = Force $\overline{DTR}$ output low	0 = Disable 1 = Enable

(c)

STATUS REGISTER (SR)

SR7	SR6	SR5	SR4	SR3	SR2	SR1	SR0
Data Set Ready	Data Carrier Detect	FE/SYN Detect	Overrun	PE/DLE Detect	TxEMT/ DSCHG	RxRDY	TxRDY
0 = $\overline{DSR}$ input is high 1 = $\overline{DSR}$ input is low	0 = $\overline{DCD}$ input is high 1 = $\overline{DCD}$ input is low	**Asynch:** 0 = Normal 1 = Framing error **Synch:** 0 = Normal 1 = SYN CHAR detected	0 = Normal 1 = Overrun error	**Asynch:** 0 = Normal 1 = Parity error **Synch:** 0 = Normal 1 = Parity error or DLE CHAR received	0 = Normal 1 = Change in $\overline{DSR}$ or $\overline{DCD}$, or transmit shift register is empty	0 = Receive holding REG empty 1 = Receive holding REG has data	0 = Transmit holding REG busy 1 = Transmit holding REG empty

(d)

holds the character and keeps checking the handshake lines. As soon as the line indicate ready, the UART transmits as in steps 3 and 4.

6. When the entire character, parity bit, and stop bit(s) have been sent, the UART does two things: It sets a bit in its status register indicating it's done, and it also generates an interrupt to the processor indicating the same thing. Depending on the design of the system and the program, the processor may respond to either done indication: It may see the interrupt and prepare the next character for the UART; alternatively, the system may be designed to not use the interrupt but just check the status bit in the register on a regular basis to see if the UART is done. Either way, the UART has relieved the processor of the critical checking and timing required to send the character.

7. The next character can be sent to the UART beginning with step 2. The processor does not have to set up the desired operating mode again, unless it wants to change them.

To receive a character:

1. The processor must instruct the UART to set up the initial conditions, similar to receiving a character.

2. The processor then enables the UART, in effect saying that it is ready for the UART to begin receiving.

3. The UART continuously checks its serial input line for the presence of a start bit. When this bit is detected, the UART begins reading the input line for the actual character bits. This reading is done at time intervals corresponding to the specified baud.

4. After the character bits are received, the UART detects the stop bit(s). The received character bits are in the UART register (serial-to-parallel conversion), and the UART then sets the handshake lines to the external device to indicate that the register is full and that no additional characters should be sent. The UART also determines if the parity bit is correct for the character.

5. Simultaneously with step 4, the UART sets the status bit indicating to the processor that a new, complete character has been received. An interrupt is also generated to the processor.

6. Either by polling or by responding to the interrupt, the processor sees that a new character is in the UART buffer. It reads this character, which causes the UART to set the handshake lines to indicate that the buffer character has been taken by the processor, and so the UART is ready for the next character, if any. The processor also checks the status register to see if there was a parity error or not. The operation resumes at step 3.

Once the UART is set up, the interaction between the processor, the UART, and the external device is rapid and continuous. The processor provides new characters for transmission to the UART and picks up characters that have come in. The processor is not involved in directly managing the handshake lines or the sending or receiving of the characters, or the generation or checking the start, stop, and parity bits. Depending on the application, the processor may periodically check the status bits of the UART to see if the last character was sent and a new character received, or it may wait for the interrupt indicating the same things.

Troubleshooting UARTs

UARTs such as the 2651 are fairly complicated internally, but their operation as seen from the processor and external serial communication port is much simpler. The UART has many modes of operation that can be selected via the processor and software instructions. Failures in operation are divided into two categories:

1. A problem with the software that sets up the UART or manages it. This should never occur in a properly designed and tested product. It can only happen during the debug stage of a new product while it is in engineering development. Finding the problem at this point requires specialized, complex equipment and a detailed understanding of the program and hardware.

2. A hardware failure that prevents the proper signals from reaching the UART, or prevents the signals for the UART from reaching either the processor or external device. Things to check include:

- The CE should go to its active (low) state when the UART is being addressed by the processor, either to set up the operating conditions or to read or write a register.
- The Read-Write line should be high when the processor is trying to read from the UART, and low when the processor is writing to the UART.
- The interrupt line should go active (low) when the UART is finished either sending a character or receiving a new one.
- The master clock to the UART should be running all the time. This is a clock signal which is used by the UART to develop the necessary timing for the various baud values. (*Note:* In some cases, the clock signal is supplied by the incoming data, so there is no external clock, and the UART is instructed by the processor not to use the external clock but instead use the data itself.)
- The signal lines for transmitted and received data, and for the handshake lines, should be working. These lines are almost always buffered between the UART and the external device, but signals from the UART pins should reach the external device and vice versa, through the buffers. This can be checked easily with a logic probe or oscilloscope.
- The UART may have failed internally.

Since the data signals at the external side of the UART are constantly changing, it is difficult to use an ordinary oscilloscope to analyze a pattern of 1's and 0's and see if it is correct. Special serial data analyzers, which capture a serial data stream and present it to the viewer as a single group, must be used. Many systems with UARTs allow a test technician to use special software routines from the system keyboard and specify which patterns and characters should be sent. The serial data analyzer is then used to verify that the proper pattern was indeed sent. Verifying the proper reception requires even more test equipment.

Another approach is to make use of the two channels of the UART. Since UARTs are used for communication to external devices, their signals are usually brought to connectors on the system. A special cable can be used to connect one channel of the UART to the other. The system itself is then used to send characters from one channel to the other, and the results compared. This loopback test is very powerful and involves no additional equipment besides the loopback cable.

QUESTIONS FOR SECTION 8-5

1. How is the UART connected to the processor?

2. How does the processor get data to the UART and from the UART?

3. What do the mode and command registers do? How are they used? What is the status register? Who reads it?

4. What are the steps for the transmitting of a processor character via the UART? What does the UART do in the sequence?

5. What is the sequence of steps for receiving a character? What does the UART do that reduces the work of the processor?

6. How does the processor use the interrupt and status bits when it is using the UART?

7. What are four things that can go wrong with a system using a UART?

8. What should be checked if a serial port which once worked no longer does?

9. What is a loopback test for serial I/O? How is it implemented?

PROBLEMS FOR SECTION 8-5

1. The processor wishes to send the letter A, which is represented by 0100 0001 in standard ASCII code. The UART is set up for one start bit, eight character bits, no parity, and one stop bit. What does the serial pattern coming out of the UART look like, beginning with the start bit?

2. The processor wishes to send the number 1, represented by 0011 0001, using the same conditions as above, and two stop bits. What is the serial pattern?

3. Parity is a method of checking if any bit in a group has changed. Parity can be set to either an odd or even mode. The parity bit is determined by counting the number of 1's in the group of bits. For even parity, this number plus the parity bit should be even. For odd parity, the number of 1's plus the parity bit should be odd. (*Note:* Zero is considered even.) What is the value of the parity bit for these binary groups, if even parity is desired? (a) 1100, (b) 1001, (c) 1101, (d) 0010, (e) 000 1100, (f) 1001 0001, (g) 000 1101.

4. Parity is used to check for errors by comparing the received parity bit with a parity bit recalculated from the received group of bits. From the following table, which of the following have parity errors, based on the received parity bit and a recalculated value for odd parity?

Received Character	Received Parity Bit
a. 0011	1
b. 1011	1
c. 1111	0
d. 0000 1100	0
e. 1001 0000	1
f. 1111 0000	1

5. Referring to Fig. 8-9a, a processor is setting up the mode registers (MRs) of the 2651. (a) What pattern should be sent to MR1 to select this condition: one stop bit, odd parity, parity enabled, 8 bit characters, asynchronous 16X rate? (b) What pattern should be sent to MR2 for external clocks, 9600 baud?

The serial I/O of the previous sections is an important method of transferring data from one system to another. However, there are times when serial I/O is not the best choice. Parallel I/O, where a group of bits can be transferred at the same time, is very useful where individual bits must be received (such as those that represent mechanical activity such as switches opening and closing) or where more speed is required (since groups of bits are transferred simultaneously, more bits can be transferred in a given amount of time). Parallel I/O is also used to transfer information from one part of a system to another part of the same system.

There are some practical problems in making parallel I/O work properly and smoothly:

- The system may need more I/O lines than the processor chip is capable of handling directly. Therefore, some method of expanding the number of lines is needed.
- If the data transfer on a line is in both directions, such as when transferring information between computers, the line itself would have to be bidirectional. This means that much more than a simple flip-flop or set of gates is required, because the line must somehow be switchable by the processor and its software to look like an input or an output. This type of "configurable" line takes a large number of gates to implement.
- In many systems, the transfer of data in either direction requires several handshake lines. When data is being transferred to the processor, there must be some way for the processor (or whatever circuit is receiving the data) to know that the signals on the group of parallel lines at that instant is valid data. Otherwise, the system could not distinguish between the actual data and various logic levels on the lines between bursts of data. To make this transfer scheme work, the incoming data is usually accompanied by a *strobe,* a separate signal announcing that this is the data group. The parallel transfer circuit must be able to handle the strobe and look at the data only when a strobe occurs (Fig. 8-10). Handshaking is also required to tell the data source that the previous group of data points has or has not been accepted by the processor. A ready signal is used, similar to the one used for serial I/O.

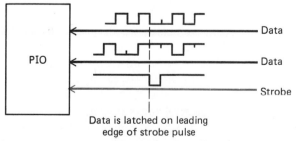

Data is latched on leading
edge of strobe pulse

Fig. 8-10   The strobe function indicates to the IC registers when the data that is coming in should be latched.

- Although the data being transferred is typically in a group of 8 bits, in some cases the processor may wish to ignore one or more of the bits. The I/O system needs some way of masking bits within the group of bits and not reporting changes in the bits that the processor has indicated it wishes to ignore for now. The masking must be flexible, so that the system can change which bits it wishes to mask, as the requirements of the application change.
- Most transfers of data using parallel I/O are of the entire group at one time. The processor may generate a byte that it wants transferred to another part of the system. At some points in the application, the processor may want to read or write single bits only, and leave the rest of the group unchanged.
- Instead of having to poll the input port for new data frequently, the processor would in most cases want to see an interrupt when new data is received. The interrupt might be necessary when a group of bits arrives, or a processor-specified bit changes. The processor would also want an interrupt when data it has written out has been accepted by the receiving end.

A circuit built of individual logic gates, flip-flops, and similar ICs that could perform all these functions would take several hundred chips, occupy a large circuit board and use a large amount of power. Instead, specialized support ICs have been developed that provide all of these functions, and more, in such a way that the IC interfaces conveniently to the signal lines of the processor and is useful for connecting to the outside world or other parts of the same system.

QUESTIONS FOR SECTION 8-6

1. What are four problems in connecting a large number of parallel inputs and outputs to a processor?

2. Give three examples of where parallel input is used. Repeat for parallel output.

3. Why would a processor want to mask input bits?

4. What is a strobe in a parallel I/O situation? Why is it needed?

8-7 THE Z80 PIO

The Z80 PIO (Fig. 8-11), is a 40-pin IC that was developed by Zilog Corporation specifically to act as a parallel I/O port for the Z80 microprocessor. It is used with many other processors as well. The Z80 PIO provides four different modes of operation, and two independent 8-bit ports, yet it only requires one processor address and eight processor data bits (Fig. 8-12). By writing different bits to the various control word registers located within the Z80 PIO, the desired mode can be chosen by the processor and the required bits and bytes can either be input to the processor or output from the processor.

In mode 0, both 8-bit ports (called port A and port B) are output ports. When the processor writes data to either port, a Ready output is activated from the PIO to tell the external device that new data has become available. When the data is read by

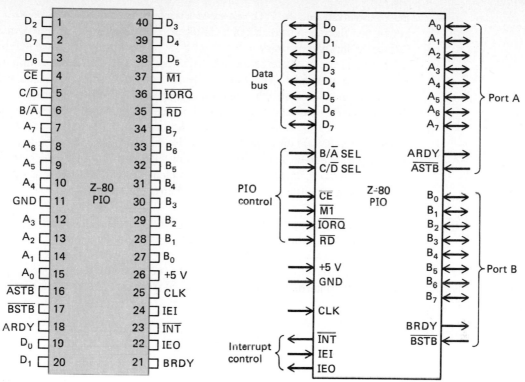

Fig. 8-11 IC pins configuration and their designations for the Z80 PIO. *(Courtesy of Zilog Corp.)*

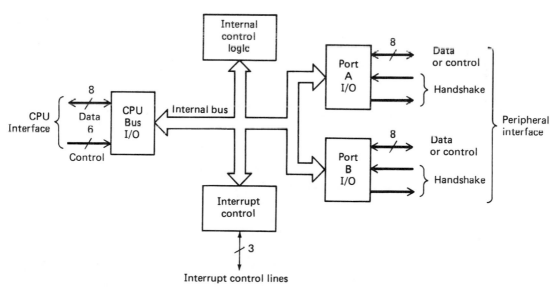

Fig. 8-12 Block diagram of the Z80 PIO. *(Courtesy of Zilog Corp.)*

the external device, the PIO generates an interrupt back to the processor so it knows that the previous data has been taken, and can then write new data to the PIO or go into another task.

Mode 1 is the opposite of mode 0: both port A and port B are configured as inputs to the processor. The external device can strobe a byte of data into the registers of the PIO, which then interrupts the processor. When the processor reads the byte, the PIO indicates to the external device that it is once again ready for new data.

Modes 3 and 4 are more complex; they let the processor use individual bits for either input or output, and let the processor specify in advance which input bits should be masked. In this way, the unwanted bits will not cause a new input interrupt unless the processor wants the interrupt to occur.

Controlling the Z80 PIO

The PIO must be set by the processor to indicate which mode of operation is desired, and the IC needs some other set up information. Four of these control words are shown in Fig. 8-13. The control words reside at the same apparent address, but the PIO is able to distinguish the mode control word, interrupt control word, and the interrupt disable word because the four LSBs of the data are set to

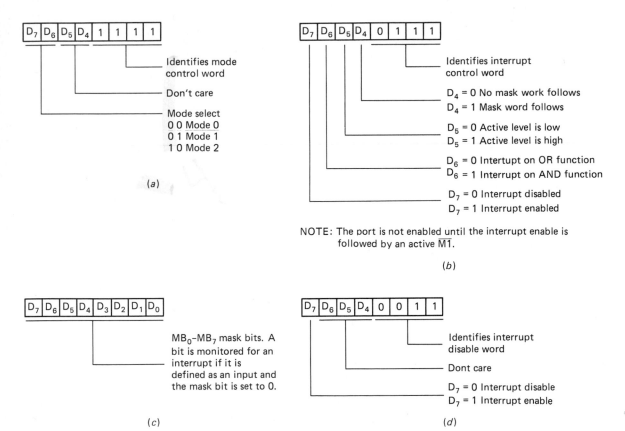

Fig. 8-13 The four setup registers for the Z80 PIO. Each controls a different aspect of PIO operation. (Courtesy of Zilog Corp.)

a different pattern. The mode control word simply tells the PIO what mode is desired. The interrupt control word selects some of the conditions for an interrupt to be generated and tells the PIO if the next byte it receives from the processor is going to be the mask control word or not. The mask control word lets the processor mask out undesired bits. Finally, the interrupt disable word allows the processor to enable or disable the interrupt from the PIO without changing any of the other setup conditions. This lets the processor put the PIO on hold if necessary.

Troubleshooting Parallel I/O ICs

When problems occur in a parallel I/O part of a system, the possible causes are similar to those for serial I/O. Assuming that the processor and its program are working properly, the likely source of the problem is the external circuitry connected to the PIO, the circuit which develops the CE for the PIO, or the PIO itself. Most computer-based systems have special test routines which allow the test technician to send outputs to each port, as directed by keyboard commands. These can be observed with a logic probe or oscilloscope. The test program usually allows the technician to read individual bits as inputs and compare these to the measured logic value on the pins of the PIO to see if there is any difference between the physical level as measured and what the processor thinks is there. Some systems also use a jumper cable which is temporarily installed by the test technician and connects all the data and handshake lines of port A to the lines of port B. The test program then automatically sends data back and forth, and compares what was sent versus what was received. This makes the job of the technician much easier, but requires that the system be designed with connectors for the cable, which is not always available in a system.

QUESTIONS FOR SECTION 8-7

1. How many processor data lines does the Z80 PIO require? How many I/O lines does the PIO provide?

2. What are the four modes of operation of the PIO?

3. What do the control words do in the PIO?

4. How is troubleshooting done for the parallel I/O part of a circuit? What are the likely causes of failures?

PROBLEMS FOR SECTION 8-7

1. Sketch a Z80 PIO used to connect 10 switches and 6 lamps to a system.

2. Referring to Fig. 8-13, a processor wishes to select mode 0 of operation. What should the mode control word be?

3. The processor wishes to be interrupted only if both input bit 1 and bit 0 go low, and so must send the PIO an interrupt control word and a mask control word. The active level for the interrupts is low. What interrupt control word and mask control word should be sent to the PIO?

4. Repeat problem 3 for the case where the processor wishes to be interrupted if either bit 0 or bit 7 go high. What are the interrupt control word and the mask control word?

5. A properly working system suddenly has parallel I/O problems. What are the possible causes for these symptoms?

SUMMARY

Support ICs offload the system processor from many time-consuming routine operations that would prevent the processor from doing other important activities. They also provide the specialized signal interfaces that may be required for the external devices that the processor must handle. The most common applications involve handling bits in parallel, or a serial stream of bits. The processor simply sets up the support IC and loads it with data; or it can read data from the support IC. The support IC is designed to deal with the external signals efficiently. It can also generate a special signal, called an interrupt, to the processor. When this interrupt occurs, the processor suspends its present activities to find out what the interrupt means and service the interrupt event. In this way, the processor is relieved of continuously polling the support IC, which takes time and is inefficient.

REVIEW QUESTIONS

1. What activities would occupy a large part of the system processor time if there were no support ICs?

2. What is the approximate complexity of typical support ICs?

3. What types of information does a processor need to supply to the support IC? What does the support IC send back?

4. How is data transferred between a processor and support IC, and vice versa?

5. Explain why an interrupt is an important feature in a system. What would have to be done if they did not exist? How do interrupts operate?

6. Explain polling as an alternative to interrupts. What is the main drawback of polling?

7. What is the difference between serial I/O and parallel I/O? Which uses more lines? Which is faster at transferring a given number of bits?

8. What unit is often used to measure the rate of serial I/O?

9. Why is a parity bit used? How does it work?

10. What is the function of handshake lines? What would happen if they did not exist?

11. What is a UART? What types of information does the processor have to send to the UART, as part of setup? What does the processor receive back?

12. What are applications which require large numbers of parallel I/O bits? Why can they not be handled directly by the processor?

13. What operating advantage does a mask capability give the system?

14. Why does parallel I/O need a strobe? Where does it come from? What is it used for?

15. What kinds of setup information are needed by the PIO IC? What kind of information does it report back?

REVIEW PROBLEMS

1. The letter C is represented in ASCII code by 0100 0011. The UART is set up for one start bit and eight character bits. What does the serial pattern look like for these additional conditions?
   a. no parity, one stop bit
   b. parity, one stop bit
   c. no parity, two stop bits
   d. parity, two stop bits
   Assume the parity bit is 1 for this character.

2. What is the parity bit required to provide odd parity to these binary groups? (a) 00, (b) 10, (c) 1111, (d) 0111, (e) 1010 1010, (f) 0101 0101.

3. For the binary pattern 0110 1100 with an odd parity bit, give two examples of where the byte would have errors but the parity bit would be unchanged.

4. Sketch the loopback function implemented with a UART.

ADVANCED SUPPORT ICs

As systems have become more complicated, they are also expected to perform more and more functions. At some point, the processor would be unable to meet the requirements of the application, as it tried to count, measure time intervals, put characters on a video terminal, and control a floppy disk drive at the same time.

More complex support ICs offload the processor by handling these tasks efficiently and quickly, leaving the processor to perform calculations and make the decisions. These support ICs are controlled by the processor and are directed by it to handle the tasks in the desired way. The processor sees only the result of the activity of the support IC, since the details are managed by the support IC.

9-1 COUNTERS

The basic function of a counter is to simply count pulses as they come into the system and present the resulting sum to the user and/or processor. Processor ICs themselves could be used for counting, but they are extremely inefficient at counting as soon as the rate of pulses to be counted reaches even a moderate value, about 1 to 10 kHz. In addition, the user often wants to know when the count has reached some specified value. This would require the processor and software to check each value of the count against the specified value, an activity which would absorb a large amount of the processor time. The system may also have a display for the user to see the total count as it occurs, and the processor would have to update the display.

A better choice is to use a counter IC which is specifically designed to count efficiently, be integrated with a processor, and also control a digital display. One such IC is the Intersil ICM7227A, which is a complete four-digit counter in a 28-pin package that can count at up to 2 MHz rates. The 7227A also allows the processor to select if the counting direction should be up or down, to reset the stored count to zero, to specify a count value for comparison, and to read back the count. The 7227 itself does all the counting and comparing and can directly drive a four-digit 7-segment display. This IC acts as the ideal interface between a high-speed signal to be counted, a user display, and a processor.

A typical application for the 7227 is in a unit of test equipment, such as a frequency meter. A frequency meter must count the total number of received pulses in a fixed amount of time in order to determine the number of pulses per second and display this to the user. Many frequency meters also have additional features that can be provided only by a processor—alerting the user when the frequency exceeds a specified value, performing calculations on the frequency values, and storing a sequence of readings for further analysis. The 7227 measures the frequency, displays the results to the user, and also provides these results to the processor.

The 7227 is connected to the processor buses and requires just eight data lines from the processor (Fig. 9-1). Four lines are used to write a control word to the IC, which allows the operating mode to be set. The processor can use these bits to preset the counter to any value, select the counting direction, and cause the value of the count to be stored in a register in the 7227. This register can then be read

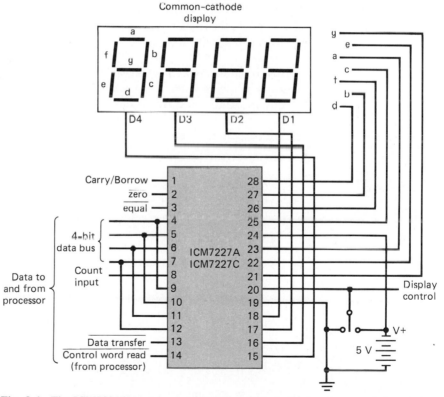

Fig. 9-1 The ICM7227C used in a system. It can count, display the count value on four 7-segment digits, and also report the count to the system processor. (*Courtesy of Intersil, Inc.*)

back by the processor via the other data lines. The IC function normally called CE is called Control Word strobe for this IC. In order to increase the versatility of the 7227, three special outputs are provided. Carry-Borrow is a signal that goes low when the counter goes from 9999 to 0000 (while counting up) or 0000 to 9999 (while counting down). It is used to allow more than one 7227 to be connected together (''cascaded'') to go beyond the four-digit count of a single 7227. The Equal output goes low when the value of the accumulated count and the Preset register are equal. The Zero output provides a low signal when the count is 0000.

To reduce the number of wires between the 7227 and the 7-segment digits, the signals to the display are multiplexed. All the timing for the multiplexing is done by the 7227, so no additional ICs are needed. This means that a practical system can be constructed from a very small number of ICs. A Display Control signal lets the processor or any other signal source (depending on the system design) make all the digits go blank. This is used in applications where the count may be meaningless during part of the system operation, and a meaningless count on the display would confuse or irritate the user.

Troubleshooting Counters

If the counter and system has a display, there are two separate ways to check the operation. A known number of counts is put into the system by the test technician using a pulse generator, and the correct value should appear on the display. The processor program should also read this same value. If the processor has the correct value, the counter is being set up properly by the processor, but the part of the IC that drives the display digits, or the digits themselves, may be faulty. If the processor sees a wrong value, either it is unable to set up the counter properly because of problems with the signal lines between it and the counter, or the signal pulses to be counted are not reaching the counter IC. (Usually they do not go directly to the IC but there is some intermediate ''buffering'' or amplification.)

The presence of both a processor interface and a digital display provides powerful test tools to the technician.

QUESTIONS FOR SECTION 9-1

1. Up to what speeds can processors typically count by themselves? What speed does an IC like the ICM7227A provide?

2. What other count-related activities does a processor have to do besides display a total count value?

3. What features can a processor and counter provide acting together in a piece of test equipment?

PROBLEMS FOR SECTION 9-1

1. A 7227 is counting up. As the count goes from 9999 and wraps around to 0000, what happens to the Carry-Borrow output and the Zero output? Repeat for the count going from 0001 to 0000.

2. A 7227 is connected to a processor, and there are some problems. What might be the cause of the following:
   a. The processor reads correct count from 7227, but the display shows the wrong value.
   b. The processor reads a wrong value of counts, and the display shows the same wrong value.

9-2 COUNTER-TIMERS

Processors do not count well at high speeds, and they do not do a good job at measuring time intervals with great accuracy. This is caused by the limitations in the operating speed of the processor IC and the nature of the software instructions that control the processor.

Most processor-based systems, however, need to keep accurate track of the time and also to measure intervals of time ranging from a few milliseconds to many seconds. The system needs a real-time clock function to show the user the hour, minutes, and seconds, and to identify to itself when certain functions happened. The real-time clock in the system is needed so that the software can schedule certain activities, such as checking for inputs every 0.01 or 0.1 s, for example. It also needs the ability to set different time periods called *time-outs*. These are used when the processor has to communicate with an external device and does not want to be indefinitely waiting for a response. The processor sends a message and wants to wait up to a specific amount of time for the response. If no response arrives, the time-out occurs and the processor goes on to something else.

A typical processor-based system may have several timing functions and intervals in operation at the same time—one for the real time clock, one for a time-out, and another for a different time-out. Because of the difficulty in measuring and generating time periods, specialized support ICs are used.

The Intel 8253 programmable interval timer is a popular counter-timer used with processor ICs to provide the timing functions. It is a 24-pin IC with three fully independent counter-timers (Fig. 9-2, page 232). Each of these counter-timers can count up to a 16-bit value, using either binary or BCD. The count value can be read by the processor if necessary. If the pulses being counted are at a known rate, from a clock, then the time period is directly related to the number of counts. The clocking signal to each counter-timer can be any rate in the range of 0 to 2 MHz, and each counter-timer can have its own clock. Often, the chip is wired to provide three separate ranges of timing: a fast one for short intervals (up to about 10 ms); a medium one for medium intervals (around 0.1 s); and a slow one for longer intervals of several seconds. The flexibility of the 8253 is enhanced by an external gate signal for each of the counter-timers which puts the counter-timer function on hold without losing or changing anything when it goes low. When the gate signal returns to the high state, the counter-timer resumes operation from where it stopped.

The 8253 provides a total of six modes of operation for each counter-timer. The modes allow the processor program to choose the counting or timing function it needs for the application. Discusssions of three of the most important modes follow and they are shown in Fig. 9-3.

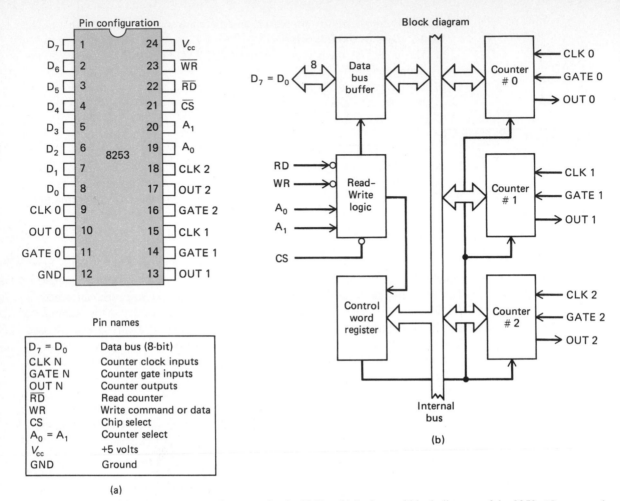

Fig. 9-2 (a) IC pin configuration and pin names for the 8253; (b) An internal block diagram of the 8253. (*Courtesy of Intel Corp.*)

Mode 0: Interrupt on Terminal Count (Fig. 9-3a)

The counter is loaded with a value by the processor and the output is low. The counter begins to count the pulses on its Clock line. When the number of counts reaches the preloaded value, the counter output goes high. This output can be used as an interrupt back to the processor, or as a signal to some other part of the circuit. The mode 0 function therefore lets the processor specify any time interval and be interrupted when that time interval has elapsed.

Mode 2: Rate Generator (Fig. 9-3b)

The 8253 acts as a programmable divide-by-N counter. The processor software sets the value of N into the counter, and then the counter output goes low once every Nth input clock pulse, over and over again. This mode is used to allow the processor to provide signals at varying rates to a system, which may be necessary when the user wants the system to provide different ranges of operation. It may also be used by the processor as an interrupt to provide the real-time clock timing, since the interrupt pulses occur at known time periods.

Mode 0: Interrupt on terminal count

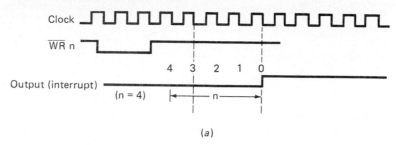

(a)

Mode 2: Rate generator

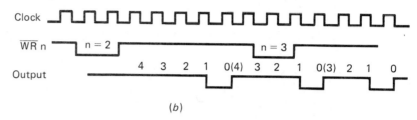

(b)

Mode 3: Square wave generator

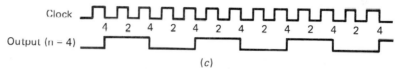

(c)

Fig. 9-3 There are 6 counting-timing modes of the 8253. Three of them are shown in detail: (a) Counting begins when new count = 4 is written to the 8253. The 8253 output goes high when the count decreases to 0. (b) Counting begins when a new value of 4 is written to the 8253. The 8253 output goes low for one clock period once every fourth period. (c) The output of the 8253 goes high for 2 clock pulses, then low for 2 clock pulses. This is similar to mode 2 but the result is a square-wave output. (Courtesy of Intel Corp.)

Mode 3: Square Wave Generator (Fig. 9-3c)

This mode is similar to mode 2, except that the output is a square wave—high for half of the input clock pulses, and low for the other half. Once set up, operation continues indefinitely until the processor changes the mode or sets in a new count value. In many systems, a part of the circuitry may need a square wave to properly sequence events or drive external devices.

The power of these modes is increased by the gate line, which allows either the processor or some other part of the circuit to put the counting or timing operation to hold (depending if the gate is wired to an output of the processor or to a separate line in the circuit). For example, a system may need to determine if a door has been opened more than 1 s. The door switch is connected to the gate line and the counter output is wired as the processor interrupt. The processor sets up the 8253 for mode 0, with a count value corresponding to 1 s. When the door opens, the counter-timer is enabled and begins timing the 1-s period. If the door remains open longer than 1 s, the 8253 output goes high and interrupts the processor. If the door closes before the 1-s period, the 8253 never completes the mode 0 period because the gate line puts the counting on hold and no interrupt occurs.

9-3   CONTROLLING THE 8253

The 8253 is relatively easy to interface to a processor and to control from the processor and its software (Fig. 9-4). The 8253 is designed to look like a 4-byte memory, complete with Read, Write, CS, and address lines, and built in three-state buffers. Using these 4 bytes, the processor can write to any of the three counters or to a Control Word which sets up the operating mode. The processor can also read these same bytes to find out the current value in the counter.

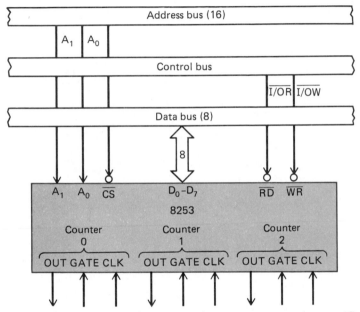

Fig. 9-4 The 8253 is interfaced to the processor buses as a memory element. *(Courtesy of Intel Corp.)*

The operation of these lines is shown in Fig. 9-5. The CS line activates the 8253 when it goes active (low). The 8253 is written to by the processor and read from by the processor just as any other memory IC would be. The 8253 accepts 8 bits of data from the processor when the processor writes, and returns 8 bits of data to the processor during a Read cycle. The two lines of the 8253 that determine which of the four memory bytes are accessed are connected to the two processor LSB address lines, A1 and A0. The CS is decoded from address lines A3 through the most significant address bit of the processor. The figure shows what the state of each line must be to read or write the different locations within the 8253. These states occur automatically as the processor reads or writes the addresses of the 8253.

$\overline{CS}$	$\overline{RD}$	$\overline{WR}$	A_1	A_0	
0	1	0	0	0	Load counter no. 0
0	1	0	0	1	Load counter no. 1
0	1	0	1	0	Load counter no. 2
0	1	0	1	1	Write mode word
0	0	1	0	0	Read counter no. 0
0	0	1	0	1	Read counter no. 1
0	0	1	1	0	Read counter no. 2
0	0	1	1	1	No-operation 3-state
1	X	X	X	X	Disable 3-state
0	1	1	X	X	No-operation 3-state

Fig. 9-5 The function of the registers of the 8253 is determined by the Read, Write, Chip Select, and address lines it receives. *(Courtesy of Intel Corp.)*

The key to controlling the 8253 is the Control Word. The data bits of this word allow the processor to specify to the 8253:

- Which counter is being set up.
- How the data is going to be written to or read from the counter.
- The desired operation mode of that counter.
- Whether the counter should operate in binary (16 bits) or BCD (4 decades).

The Control Word bits are assigned as shown in Fig. 9-6, page 236. The two MSBs of the data, D7 and D6, are for selecting the counter. The next two bits, D5 and D4, tell the 8253 how the 16-bit value of the counter will be loaded or read back. Since the 8253 has a 16-bit counter but only an 8-bit data bus, the count must be handled as 2 bytes. The processor can choose to do 1 byte only, or both bytes in the specified order. Bits D3, D2, and D1 tell the 8253 what mode the counter is to use, and bit D0 sets either binary or BCD operation. After issuing a Control Word to the 8253, the processor then writes or reads the Counter addresses using the format specified by D5 and D4. Any time the processor wants to load a count or read the present count from any of the three counters, it must first send a Control Word telling the 8253 what it is about to do. After the 8253 receives the last byte of data, the specified counter begins running (or holds, waiting for the gate signal to go high).

Troubleshooting the 8253

When counter-timer problems occur that may be related to the 8253, such as no timing signals, incorrect time intervals, or no output, both the processor interconnection and the external connections must be checked:

- The Clock signal for each of the three counters should be present and at the right frequency. Each of the three may have a different clock frequency.
- The gate signal must be high for the corresponding counter to run. If the gate is low, the counter will be stopped, even though it is being set up properly. This gate line may come from the processor, as an output bit, or it may come from

Control Word Format

D_7	D_6	D_5	D_4	D_3	D_2	D_1	D_0
SC1	SC0	RL1	RL0	M2	M1	M0	BCD

(a)

Definition of Control
SC — Select Counter:

SC1	SC0	
0	0	Select counter 0
0	1	Select counter 1
1	0	Select counter 2
1	1	Illegal

RL — Read/Load:

RL1	RL0	
0	0	Counter latching operations.
1	0	Read/Load most significant byte only.
0	1	Read/Load least significant byte only.
1	1	Read/Load least significant byte first, then most significant byte.

M — MODE:

M2	M1	M0	
0	0	0	Mode 0
0	0	1	Mode 1
X	1	0	Mode 2
X	1	1	Mode 3
1	0	0	Mode 4
1	0	1	Mode 5

BCD:

0	Binary counter 16-bits
1	Binary coded decimal (BCD) counter (4 decades)

(b)

Fig. 9-6 The Control Word of the 8253: *(a)* The bits of the control word; *(b)* The meaning of the bits when set by the processor. *(Courtesy of Intel Corp.)*

another part of the circuit or an external device. The processor output may be faulty, or the circuit or external source may be failing to let the signal go high (more likely).

- The output of the counter can go to an external point in the circuit or to an external piece of equipment. It can also go back to the processor as an interrupt. Determine the function of the output in the mode used, and be sure the output signal is reaching the intended destination. For example, if the processor is looking for the output interrupt to tell it that a certain interval has passed, and the interrupt line between the 8253 and the processor is faulty, the software may never update the user real time clock.

If the system allows the test technician to write and read memory locations as part of the test procedure, the Control Word and counter registers can be set and the operation of the 8253 tested in slower, easier to observe modes, using an oscilloscope or voltmeter.

QUESTIONS FOR SECTION 9-3

1. For what reasons do most processor-based systems need timer functions?

2. What is the relationship between a counter and a timer?

3. What is a time-out?

4. What are the three most important modes of operation of an 8253? What would each be used for?

5. What does the external gate line do? What are the two possible functions of the 8253 counter-timer output?

6. How is the 8253 connected to a processor? How is it addressed?

7. What does the processor tell the 8253 via the Control Word?

8. Why does the processor have to load count information into the 8253 one byte at a time?

9. What signal lines should be checked first when troubleshooting an 8253?

PROBLEMS FOR SECTION 9-3

1. An 8253 is located in the processor memory beginning at memory location 1010. To what address does the processor write when it is writing the Control Word?

2. An 8253 is being used to time an interval of 250 ms. The clock to the 8253 is running at 1 kHz. What binary value should be written to the counter? What BCD value?

3. Repeat the previous problem for an interval of 10 s using a 10-Hz clock.

4. Using a 1-MHz clock, what is the maximum interval that can be timed with the binary mode? With the BCD mode?

5. A test technician has access to the 8253 and wants to check it out by causing it to generate a continuous stream of 1-kHz pulses, which are easily seen on the oscilloscope. The clock of counter 1 is running at 1 MHz, and the counter output is available at a test point. (a) What Control Word should be used? (b) Assuming the 8253 is at addresses 1000 and up, what data should be written to what address(es), to load the count value into counter 1?

6. The technician wants to read the count of counter 2 in binary mode. (a) What Control Word should be sent? (b) What address should be read for the count bytes?

9-4 CRT CONTROLLERS

The most common interface between a computer system user and the computer itself is a terminal with a keyboard and video screen. The screen is referred to as a *cathode-ray tube*, or CRT, which is the technical name for the large glass bulb which is the heart of the terminal. The large flat end of the CRT is the screen on which characters are displayed to the user. (The CRT is used in a regular TV, too.)

Although it may seem that putting the characters in the screen is a relatively simple task, in reality it is quite complicated. In order to understand the difficulties, it is useful to understand how the CRT and its circuitry actually put characters on the screen.

At the back of the CRT is a small group of electrical elements (Fig. 9-7). One of these is a cathode, which gives off electrons when heated by electrical current. These electrons are directed to a spot on the face of the CRT by controlling electric fields or magnetic fields, produced in metal plates (for electric fields) or coils of wire (for magnetic fields) located between the cathode and the screen. The screen is coated on the inside with special chemicals called *phosphors,* which glow when struck by the electrons of the cathode. It is this glowing phosphor coating that is actually seen. Also near the cathode is a wire mesh, called a *grid,* which can be controlled by a voltage to repel the electrons and keep them from reaching the screen. This allows the intensity of the electron beam reaching the screen to be controlled, and thus the brightness of the characters on the screen is adjustable. For most computer applications, the brightness is fixed at two levels (on and dark) or three levels (bright, regular on, and dark).

For a character to appear on the screen, the steering fields and the grid have to

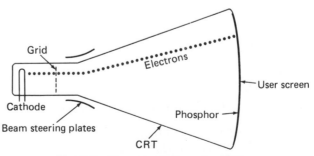

Fig. 9-7 The construction of a CRT.

be controlled precisely to put the electron beam in the right place and make sure it is on at the right time. One way of doing this is to have the steering fields direct the electron beam to "paint" the character on the screen just as it would be written by hand. This is called *vector graphics* and is used in some advanced systems which draw complex shapes on the CRT. Unfortunately, vector graphics requires a large amount of complex circuitry and substantial processor activity. As a result, vector graphics is not used for CRTs which display characters and simple lines. Instead, a method called *raster scan* is used. The controlling fields are used to generate a series of lines (rasters) across the face of the CRT (Fig. 9-8). The electron beam

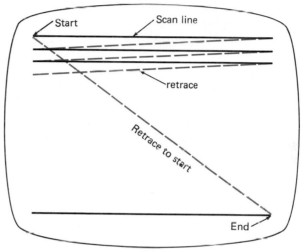

Fig. 9-8 The raster scan on the face of the CRT.

starts in the upper left-hand corner of the screen and moves left to right across the screen, and top to bottom. The result is a series of lines across the screen. The vertical and horizontal movement of the beam is controlled by two circuits operating independently to control the fields in both directions. When the electron beam reaches the end of a line, it is brought back quickly to the beginning of the next line. This is called *retrace,* and during retrace the electron beam is shut off by the grid so the user does not see the retrace lines. At the bottom right corner, a frame has been completed, and the beam goes back to the upper left. (Raster scan is also used on home TVs.) The entire screen must be redrawn, or refreshed, at least 30 times per second, even if there are no changes. The phosphor only glows for a fraction of a second, and the refresh is needed to prevent the image from flickering. The actual character is formed by a group of dots in a rectangular bar called a *matrix*. A typical group size used in a CRT display is seven dots in the horizontal direction by nine dots in the vertical. As the electron beam sweeps across, it is turned on and off to form the first row of dots for all the characters on the first line of the screen. It then retraces and forms the second row of dots, and so on (Fig. 9-9, page 240.). This is easier for the CRT to do than to form each character completely, one at a time. However, the raster scan method still requires considerable support circuitry.

The processor software does not want to be involved in directly controlling the

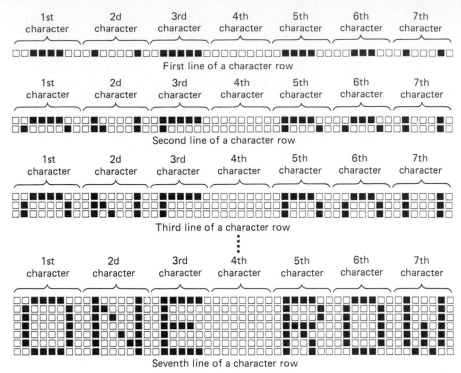

Fig. 9-9 How a row of characters is formed by lines of dots as the raster scans across the face of the CRT. *(Courtesy of Intel Corp.)*

beam, which would require keeping track of where the raster is and what dots must be presented to form each character. A practical system might also need some additional features, such as a cursor (a highlighted area on the screen), underlining, reverse video (where the dark and light areas are reversed like a photographic negative), and blinking characters. The processor would be unable to provide these features. The processor software instead would prefer to write the characters to be displayed to an area in its memory and have them taken from this memory and put on the screen, with any special features added.

Making the proper character dots appear for a desired character involves one more key piece, a ROM called a *character generator*. For each possible letter, number, and symbol that might be displayed, the character generator ROM has stored the pattern of dots that form the character, in a row by column format.

A line counter tells the character generator what line (of the nine lines that form the character) is presently being traced by the raster. The code corresponding to the character itself (usually ASCII code) is also sent to the character generator. The character generator takes these two codes and outputs a group of 1's and 0's, corresponding to the dots that must be on and off to form that line of that character. These 1's and 0's are used to control the electron beam.

The scheme would be inefficient if the circuitry for the CRT had to access the processor memory continuously for each raster line to find out what character the processor wishes to display. Instead, the CRT control circuitry reads from the processor memory the entire row of characters that will make up the next group of raster scans. This is usually done during the retrace time, so the viewer sees no delay in setting the next row onto the screen. The CRT control circuitry must therefore have Read-Write memory (a RAM buffer) of its own to store the characters that it has fetched from the processor memory.

The method actually used to control the electron beam of the CRT to produce the correct dots involves a master clock, several counters, a parallel-to-serial shift register, and some circuitry. The master clock frequency is the rate at which dots are to be generated. The counters are:

- A divide-by-N row counter, which indicates which row of dots within the line is being illuminated by the raster, where N is equal to the number of rows per character.
- A divide-by-N character counter, which shows the number of the character within a line. N is the number of characters per line.
- A divide-by-N line counter, which shows which line of the screen is being displayed.

Each of the counters is connected to the master clock, which also drives the circuitry of the CRT, which in turn causes the raster itself to occur (Fig. 9-10).

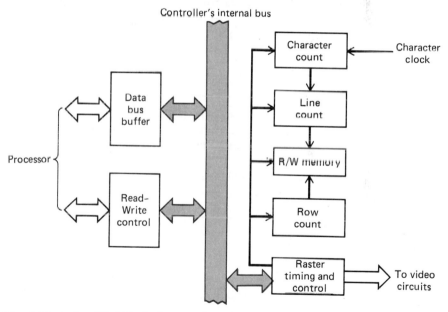

Fig. 9-10 A simplified block diagram of a typical CRT controller. Note the three counters required.

Careful coordination between the counters and other circuitry makes the display work properly. For the sake of discussion, assume a display of 24 lines with 80 characters on each line. Each character is formed by the dots of a 7 × 9 matrix. The raster starts at the upper left corner of the CRT screen. Assume the raster is about to return to the upper left corner from the lower right.

The line counter shows that the raster is on line 1. The line, row, and character counters show that a new line is about to begin. During this retrace period, the CRT controller circuitry accesses the processor memory and reads the entire next line of characters to be displayed. (The characters are usually stored by the processor with 8 bits representing each character, often in ASCII format.) The entire line is stored in Read-Write memory of the CRT control circuitry. In this way, the processor memory does not have to do an access operation while the line is being displayed, since these accesses take time and could cause flickering on the screen. Next, the three counters show that the first row of the first line is about to begin. The stored character bits for this first position are passed on as input bits to the character generator, along with the row count. The character generator has been programmed to contain the dot sequence for each row of all characters (letters, numbers, symbols) as a byte of 1's and 0's, with 1 representing a dot and 0 meaning no dot. The 8 bits go to the parallel-to-serial shift register. As the master clock pulses cause the raster to move one dot position to the right with each pulse, the shift register outputs a serial stream of 1's and 0's, corresponding to whether each dot should be on or off. In this way, the first row of the first character is put on the screen. At the end of the dots of the first character, the character counter shows that the raster has advanced to the position of the next character. The process of getting the stored character begins again. The second character is taken from the RAM, and sent to the character generator along with the row count, and the proper dots for that character go to the shift register.

At the end of the first row, the row counter advances by 1, while the character counter goes back to the beginning. The entire process begins again, except that the character generator provides the sequence of dots for the second row of the first line. This continues for each row of the first line. After the last row, the line counter advances, and during the retrace time the next line of characters is read from the processor memory.

One feature of the character generator IC is that it can be programmed at the factory to provide different character shapes, or type styles, on the screen. Depending on the dot sequences it provides, it can make a letter look square or curved, large or small, as long as it fits into the 7×9 matrix (Fig. 9-11). Special symbols for a particular application can also be provided this way.

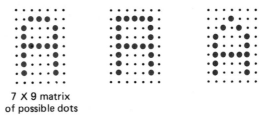

7 X 9 matrix
of possible dots

Fig. 9-11 The letter A formed by dots within the 7×9 dot matrix. It can be in different styles or fonts.

Controlling the CRT and providing the right dot sequence at the right time is a complex job requiring many ICs, typically about 150 of the 7400 series. For this reason, CRT controller ICs are used.

In order to make the CRT control function more reliable and less expensive, single-chip controllers such as the Intel 8275 were developed. This 40-pin IC provides nearly all the functions needed to drive a CRT and put the proper characters on the screen. The few functions it does not provide are the circuitry required to be able to read the processor memory for the character row, the character generator, and some relatively simple circuitry which combines the horizontal raster signal, the vertical raster signal, and dot control signal into the specific type of signal needed by the CRT (Fig. 9-12).

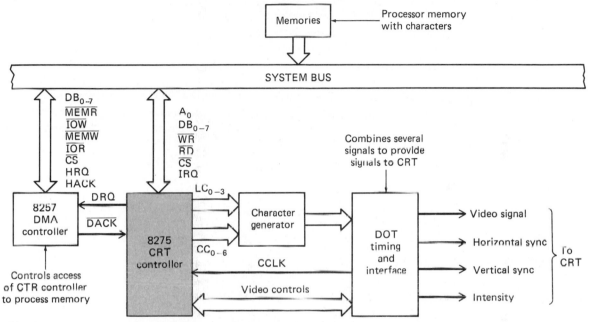

Fig. 9-12 Block diagram of a CRT controller as used in a circuit. *(Courtesy of Intel Corp.)*

The features of the 8275 that make it attractive for many applications are what it can do beside controlling the CRT in a single, inflexible format. The 8275 can be programmed by the processor software to allow from 1 to 80 characters per line, from 1 to 64 lines per screen, as well as 1 to 16 lines per character. This means that different-size characters can be provided on a screen by putting fewer characters on the display area, or that smaller CRTs can be used with fewer characters and rows than a large one. Since the number of lines per character can be changed, a system can use a special character generator for non-English languages such as Japanese which require more detail, using more lines per character.

The 8275 is also capable of providing "visual attributes" once the processor specifies which ones it wants for various characters. These attributes are blinking characters, highlighting characters (brighter), reverse video, and underlining.

Many computer systems have a cursor (indicator) on the screen which shows the user where the next typed character will go. The 8275 can draw the cursor at any position that the processor specifies to it by row and column numbers. It also lets the processor select the type of cursor: blinking underline, blinking reverse video block, nonblinking underline, or a nonblinking reverse video block.

Because of the powerful features and flexibility of a CRT controller like the 8275, the setup sequence is quite complex. Like the 8253 IC, the 8275 looks to the processor like a Read-Write memory, complete with three-state buffers. The 8275 occupies only one memory location, and through this location the processor accesses the IC to give it setup information. The setup is only done once, when the system is turned on. After that, the 8275 functions continuously and automatically to fetch characters from the processor memory and put them on the screen. The setup can be changed, if desired, by the processor while the screen is on. The first setup byte indicates the type of information that is going to be sent (number of rows, columns, visual attributes, cursor type, etc.) and then that byte is followed by up to four additional bytes with the actual setup numbers.

Troubleshooting CRT Controllers

A CRT controller circuit, even with a powerful IC like the 8275, is complex and difficult to troubleshoot with test equipment such as oscilloscopes. Signals are continuously going high and low and data is constantly being moved as the CRT circuit develops the raster scan, counts, fetches characters, and puts them on the screen. Therefore, the best way for finding faults in these circuits is to logically look at what is not working properly and relate that to possible causes. For example:

- If some characters always come out perfectly but others are always incorrect, the character generator is most likely faulty.
- If characters are sometimes good and the same character is sometimes bad, the memory access circuit or controller RAM may be bad.
- If the same dot position of every character is bad, then the parallel-to-serial shift register in the controller may be faulty.
- If the entire screen is unstable and rolling, the master clock may be bad, the counters in the CRT controller could be faulty, or the actual horizontal and vertical drive circuitry in the CRT assembly could be bad.
- No display or all dots on means a bad CRT or the dot control line is stuck.

Try to clearly observe what is faulty: the whole display, certain characters, certain rows, some columns, and so on, and determine the link between the observed problem and the CRT control circuitry.

Many systems have special screen test routines that the service technician can call up from the keyboard. These routines fill every character position on the screen with the letter A, then B, and so on, so that performance can be checked and a detailed list of what is working and what is not can be developed. The test patterns also put repeated lines of A through Z, 0 through 9, and symbols on the screen for further visual checks.

QUESTIONS FOR SECTION 9-5

1. What is a CRT? How does it illuminate an area of the screen?

2. What is raster scan? Retrace?

3. What is refresh? Why is it done even if the screen is unchanged?

4. What forms the actual characters on the screen? How does raster scan form a line of characters?

5. What is a character generator? What does it do? What additional feature does it provide?

6. What are the line, row, and character counters? The master clock?

7. How do the counters, character generator, Read-Write memory, and parallel-to-serial shift register combine to put characters on the screen?

8. Give examples of visual attributes.

9. What is the cursor?

10. Why is setting up a CRT controller complicated? What does the 8275 look like to the processor?

11. What makes troubleshooting CRT control circuits difficult? What is the recommended procedure?

PROBLEMS FOR SECTION 9-5

1. A CRT is designed to show 12 rows of 24 characters. How many bytes of Read-Write memory does the controller need?

2. A 12 line by 24 character screen is refreshed 30 times per second. Each character consists of 7 dots horizontally by 9 dots vertically.
   a. How many dots per line are there?
   b. How many lines per characters are there?
   c. How many characters on the screen?
   d. How many dots per character matrix?
   e. How many dots per screen?
   f. What is the master clock rate?

3. For the previous problem, what is the divide-by-N value for (a) the line counter; (b) the character counter; (c) the row counter?

4. A CRT screen has the problems listed. What are the likely causes?
   a. Only half the rows appear on the screen.
   b. Left-hand part of every character is missing.
   c. All letter A's are missing a part.
   d. Display rolls left to right.
   e. Display is stable, but many characters are wrong.

9-6 IEEE-488 CONTROLLERS

For many years, a piece of test equipment was used as an isolated instrument. A signal generator put out signals, a frequency meter measured their frequency, and a voltmeter measured voltage. As microprocessors became common, designers started adding them to instruments to make the instrument more capable and versatile. The user of the equipment did not have to perform detailed calculations based on the numbers from the instrument, because the "intelligence" of the instrument could do this. Many of the manual adjustments that were so time-consuming and error-prone, such as selecting the correct range of operation, could be done automatically under the control of the microprocessor. However, each instrument still acted independently from all the others that might be used in a test setup.

In many test and instrumentation applications, several different instruments are used with each performing a specific specialized function. Consider the testing of a radio (Fig. 9-13). The radio needs a signal source to generate signals into the antenna connection. It also requires a voltmeter to measure the voltages at certain points within the radio, a spectrum analyzer to check frequencies at some key internal points, and a distortion analyzer to monitor the audio quality at the output. Each of these instruments has to be set and varied during the test, and the results have to be recorded manually.

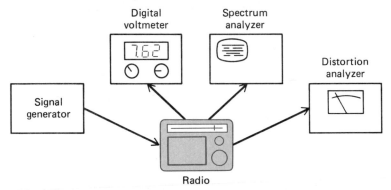

Fig. 9-13 Testing a radio with individual instruments.

It would be much better if there were a way to link all of these instruments together, along with a computer acting as a controller. This controller would send commands to the various instruments, which would take the place of manually setting ranges, output levels, and so on. In addition, the controller could take readings from each instrument automatically, which would minimize the need for someone to copy down all the values. Results and summaries could be printed or drawn on a plotter. The ability to link instruments and have them communicate would allow powerful testing systems to be built, using instruments from various manufacturers.

The key to this link is a standard that defines the rules for the link. This standard would specify the signals on the link, the type of connector to be used, and the way that messages are passed back and forth. Such a standard was first put into place by the Institute of Electrical and Electronic Engineers (IEEE) in 1975, and it was given the number 488. This IEEE-488 is often referred to as the "488" standard, or the GPIB (for general-purposes interface bus). Over 300 manufacturers now make equipment of all types that can communicate on the IEEE-488 bus.

Some of the rules specified by the IEEE-488 bus standard are:

- Up to 15 devices may be on the bus.
- Signals on the bus must have defined voltage levels and shapes.

- A total cable length of 20 meters (m) (60 ft) and specific cable and connectors must be used. The cable uses 24 wires: 8 for data, 8 for control, and 8 for ground.
- Each instrument on the bus has a unique address, which is set by the user with small switches built into the instrument. These addresses are used to direct messages to an instrument or to indicate to the receiving instrument who the sender is. It is possible to have two identical units on the bus at different addresses.
- Devices on the bus are categorized as listeners, talkers, or controllers. The *controller* would send messages and receive them. There can be only one controller in a system. The controller is usually a small computer, with an interface to the IEEE-488 bus. *Listeners* are devices which can only receive messages, and *talkers* can only send them. An instrument can be a talker or listener, depending on its function.

Examples of talkers, listeners, and talker-listeners:

Talker: A voltmeter which makes and reports readings but has no settings that can be made by the controller.
Listener: A power supply which can be set to any output voltage by the controller, but can report nothing back.
Talker-listener: A voltmeter that not only makes readings but also can receive settings and ranges from the controller.

Many less expensive IEEE-488 instruments do not allow the front panel switches and controls to be set via the bus, only by the user. This is the difference between a talker only and a talker-listener for this application.

Not every IEEE-488 system needs a controller. A very simple test setup might just have a talker and listener. A voltmeter might talk, and a printer to print the readings might listen. However, most systems do have a controller because of the additional system flexibility that a controller provides.

Using IEEE-488–based equipment to test radios is shown in Fig. 9-14. The controller would be programmed to run the specific tests required. It would put prompting messages on the screen for the test technician, such as where to put the

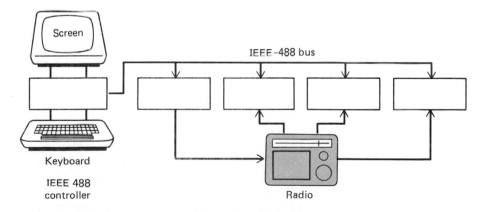

Fig. 9-14 Testing a radio with IEEE-488 instruments.

test probes, what settings to use on the radio itself (not an IEEE-488 device, of course), and what results to look for. The controller would then set the output of the signal source to the proper level, and instruct the voltmeter, spectrum analyzer, and distortion meter to go to certain ranges and make the proper number and type of readings. It would receive these readings and compare them against a list of what was acceptable and what was not, and determine if the radio under test had passed or failed. The controller might also print summaries of the test results on a printer and draw graphs of the results on a plotter, also connected via IEEE-488.

The standard rules for intercommunication between instruments defined by IEEE-488 is quite complicated. There are approximately 100 groups of tables, charts, and diagrams which spell out what happens under certain conditions. This is because the various instruments may be in different states of operation, and each instrument needs to be able to handle all applicable states. It is similar to driving a car—there are a few basic rules, but there are also rules and guidelines to follow under special circumstances such as intersections, when the car stalls, when two cars meet at stop signs, and so on. The IEEE-488 standard covers every possible circumstance, in order that everything be clear to all instruments, but the rulebook is thick. Any attempt to write a computer program to perform according to these rules would be complicated, probably filled with errors, and take a lot of the available processor memory and time to execute. Soon after the IEEE standard was adopted, manufacturers of the ICs began to develop support ICs that actually implement the charts and diagrams of the standard. The processor of the system only has to indicate the message it wants sent, and to which instrument, and the support IC takes care of the actual communication and message handling in both directions. The same support IC can be used in a controller, talker, listener, or talker-listener.

9-7   AN IEEE-488 SUPPORT IC

The Texas Instruments TMS9914 is an IC designed to manage the IEEE-488 bus and greatly offload a system processor. It is a 40-pin IC that connects to the processor and to the IEEE-488 bus. In operation, the processor first set up the TMS9914 to the desired operating conditions and specific IEEE-488 mode desired for the instrument itself.

When the processor of the instrument wants to send messages, it indicates the way it wants the message sent (the IEEE-488 specifies several different types of message) and the message. If the instrument is also a listener, the processor instructs the 9914 as to the kinds of message formats to receive. The 9914 actually handles the bus handshaking and signalling on the eight control lines and eight data lines. All messages on the data lines are sent as a series of bytes, called *bit parallel–byte serial format*.

The Physical Interface to the Bus

The IEEE-488 specification requires that the signals put onto the bus have specific voltage levels, along with other defined electrical characteristics and the ability to withstand short circuits. This last feature is needed so that a failure of the interface of one instrument or a power shutoff will not cause failure in another attached piece of equipment or a bus shutdown. Support ICs such as the 9914 have the capability of providing both the IEEE-488 management function but not the proper

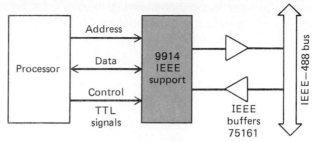

Fig. 9-15 The 9914 IEEE support IC and buffers for the IEEE bus.

physical interface. Instead, bus buffers are available to take the signals directly from the 9914 and interconnect them to the bus (Fig. 9-15). One such buffer is the SN75161 from Texas Instruments. It is an IC with eight bidirectional buffers, designed to connect to the 9914 and provide the proper signal levels to the bus, receive the signals from the bus, and provide TTL signals to the 9914.

Troubleshooting IEEE-488 Systems

Although there are very few components in the IEEE-488 interface using support ICs, problems can be difficult to solve. Most of the problems can be isolated to one of these causes:

- A failure of the support IC or bus buffers. The apparent symptom is that the IEEE-488 bus does not work at all or works very poorly—messages do not get through, messages are sent but not received.
- Some problem with the system software. This is the most likely cause of problems.

Troubleshooting an IEEE-488–based system requires logical thinking and often some specialized test equipment, such as a bus analyzer designed for monitoring and indicating what is happening on the IEEE-488 bus. The power and flexibility of the bus means that there are many possible configurations and messages that may be sent, and often the problem is not equipment failure, but a bug or mistake in setup or in the program used to control the processor and IEEE-488 support IC. Check these items:

1. Are all instruments on? Are the addresses of the instruments properly set by their switches? The physically set address must agree with the address number that the program uses for that instrument.

2. Are any of the system instruments working and communicating with the controller? If so, the circuitry of the controller is probably good, and the problem may lie in the hardware or software of the nonworking instrument, or the software of the controller. If nothing is working, the IEEE-488 support IC in the controller or its buffers to the bus may be bad, or the controller is incorrectly set up.

3. Even if nothing seems to be working properly, the setup instructions issued by the controller for the bus may be wrong. Remember that the IEEE-488 bus allows many types of instruments to be connected. Perhaps the configuration has been changed since the program was last run, and the controller has not issued new statements reflecting this.

4. If only one instrument seems to be bad, try substituting another identical one for the apparently bad one. This would show if the problem was a single bad unit.

In general, the complexity of IEEE-488 software and configurations is much greater than the complexity of the circuitry, and problems are usually due to software or actual operational steps. The IEEE-488 specification defines a great deal, but it does not define the actual contents of the messages between instruments. The same function may be called out differently on equipment from different manufacturers. For example, a voltmeter from manufacturer A might require this message to set the range: RANGE 10 VOLTS. A similar voltmeter from manufacturer B might use VOLTS RANGE 10. Both are allowed, but a program that is designed for one voltmeter would not work with the other. This is why starting up an IEEE-488 system can be difficult and requires the designer and programmer to read the manuals of each piece of equipment connected carefully.

Even systems that have been running properly for weeks can develop problems that are due to programming bugs. Most pieces of IEEE-488 equipment have a normal operating mode and also some special routines that they use when something abnormal happens, such as an input value that is out of the range allowed. Unless the controller has been programmed to handle the rare special message (which itself may use a different group of IEEE-488 commands and responses than routine messages) the controller will stop working correctly. This situation may occur infrequently and so be very hard to repeat and fix.

Setting up a system based on IEEE-488 calls for careful planning and study. Fixing one that no longer works also requires careful study and much of the logical thinking that went into the initial setup.

QUESTIONS FOR SECTION 9-7

1. What is the IEEE-488 standard? What does GPIB stand for?

2. Give two examples of how interconnected instrumentation might be used to test a product.

3. How many devices are allowed on the bus? How are they uniquely identified?

4. What is a talker? A listener? A talker-listener? Give two examples of each. When would an instrument that is only a listener become a talker-listener?

5. Why are support ICs essential for an effective IEEE-488 interface on an instrument?

6. Why are buffers needed between the support IC such as the TMS9914 and the bus itself? What happens on the bus if an instrument is turned off?

7. What makes troubleshooting new or existing IEEE-488 systems difficult? What kind of special test equipment is needed?

8. What are the first and obvious things to check?

9. What should be checked next? How can a system which has been working for months suddenly malfunction and yet have no circuit problems?

10. What would you check in a system with a controller, a frequency meter, and two identical voltmeters on the IEEE-488 bus?

11. If a system is working properly most of the time, but under certain conditions it malfunctions and these conditions and the malfunction could be consistently repeated, what would you suspect? How would you verify this?

9-8 PARITY AND ERROR DETECTION AND CORRECTION

The use of parity to detect errors in a serial communications link was discussed in a previous chapter. Recall that a parity bit is generated at the transmitting end and sent along with the other data bits. The receiver recalculates this parity bit based on the received bits and compares it to the received parity bit value. If they are the same, there was no parity error; if they differ, an error has occurred between the point where the data was transmitted and the point it was received. For odd parity, a 1 is sent if the number of 1's in the data group is even, while in even parity a 1 is sent if the number of 1's in the group is odd. Another way to say this is that odd parity makes the total number of 1's odd, while even parity makes the total number of 1's even.

There are many other situations where parity can be used. In a computer system with hundreds of thousands or millions of bits of memory, there is a small but definite chance that a single bit will be in error (change its logic state) between the time the memory location is written to by the processor and read back later. The causes of these errors are varied: a memory bit cell that was not perfectly formed in production, a noise pulse on the power-supply lines that is too small in value to affect most bits except a few more sensitive ones, or even natural alpha particle radiation that occurs in the material used to package the IC. The bit failure may be hard (permanent) or soft (temporary). For these reasons, many memory systems include parity checking in their design.

Consider a memory that consists of 1K bytes (Fig. 9-16). Each byte requires 8

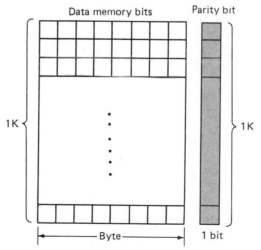

Fig. 9-16 A section of a system memory showing the memory bytes and the parity bit for each byte.

bits, of course, and a ninth bit is added for parity. As the processor writes new data to memory a special parity circuit determines the parity of these 8 bits and stores that bit in the ninth location. When the processor reads the byte, the parity circuit determines the parity again and compares it to the parity bit already stored. If they are different, the parity circuit signals the processor with an interrupt. The processor may then do several things:

- It may display a message to the system operator if there is a screen or printer attached.
- It may turn on an error light on the front panel.
- It may keep a log of the number and time of these errors, so the operator can examine the health of the system.
- It may cause the processor to go to a special routine which assumes that something is failing and therefore regular processor operation should stop.

Most systems use odd parity instead of even. This is because the odd parity for a group of all 0's is 1 (zero is considered to be an even number). When examining bits with an analyzer or oscilloscope, signals are easier to study and interpret if there is some change in the signals as they are observed. A pattern of all 0's could be confused with a circuit line that was stuck at 0. Odd parity guarantees that at least one bit will be a 1. Similarly, it ensures that a byte of all 1's will have a 1 parity bit.

9-9 PARITY GENERATOR AND CHECKER ICs

An IC to develop the parity bit is relatively simple compared to the support ICs studied in previous sections. One such IC is the 74180 of the TTL family, available from many manufacturers of TTL. This IC uses a pattern of XOR and AND gates to provide both even and odd parity at the selection of the circuit designer. It can handle up to nine input bits, which allows for a full byte of data plus an expansion from another 74180 so that more than a byte of data can be operated on. The processor is not connected directly to the parity bit, but to this IC (Fig. 9-17). In the system operation, the 74180 is connected to the parity bit memory location. When the processor writes data to memory, the 74180 calculates the parity value and writes it in the parity location. When the processor does a Read cycle, the 74180 reads the data bits and generates a parity output which is directed to a comparator where the already stored parity bit and this recalculated parity bit can

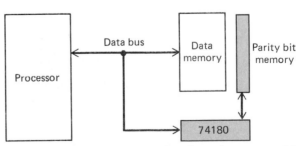

Fig. 9-17 The parity bit memory is connected to the parity generator/checker IC.

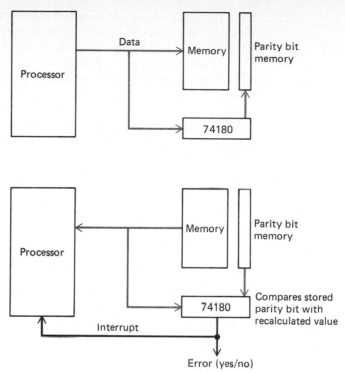

Fig. 9-18 Use of the parity generator/checker IC during write and read cycles *(a)* During write, the 74180 determines the value of the parity bit and writes it to parity memory. *(b)* During read, the 74180 recalculates the value of the parity bit from the stored data and compares it to the stored bit.

be compared (Fig. 9-18). The output of the comparison can go to the processor as an interrupt and to an indicator lamp on the user panel.

There are two drawbacks to parity in terms of improving the reliability of a circuit or system:

1. The parity bit indicates an error has occurred but cannot show which bit was in error.

2. A parity bit, whether odd or even, can only detect an odd number of errors. In a single byte, for example, the parity bit will show if 1, 3, 5, or 7 bits have changed. If 2, 4, 6, or 8 bits have changed the parity bit will be "fooled" and show no parity error:

		Parity Bit (Odd Parity)	
Original byte	1101 0100	1	
1 bit in error	1111 0100	0	Error detected
2 bits in error	1110 0100	1	Error missed
All bits in error	0010 1011	1	Error missed

While parity provides a level of increased confidence in the reliability of a memory system, it really does not go far enough to provide the error-free performance that a system with large amounts of memory needs.

The solution to this problem involves applying advanced mathematical theories to develop circuits for error detection and correction functions.

QUESTIONS FOR SECTION 9-9

1. Where is parity used besides serial communications? Why?

2. What is the rule for even parity? Odd parity?

3. What are the causes of bit errors? What is a hard error versus a soft error?

4. How does parity work in an 8-bit-wide memory system? When is it generated? When and how is the error detected?

5. Identify three things a system might do if a parity error is detected.

6. Why is odd parity usually used?

PROBLEMS FOR SECTION 9-9

1. For a 1K × 8 bit memory, how many additional memory bits are needed to store the parity bits?

2. Calculate even parity for: 0000 0000, 1111 1111. Repeat for odd parity.

3. Two 74180 ICs are cascaded so that the parity output from one is the ninth bit input to the other. Both are set for odd parity. A 16-bit data word is sent to the 74180, 1 byte to each. Show that the combination of ICs produces the correct overall value of parity for the inputs a, b, and c, by calculating odd parity for all 16 bits and also by taking the parity of the first byte and using it along with the second byte.
   a. 0100 1100 1111 0010
   b. 1111 0000 1111 1111
   c. 0101 1010 0101 1101

9-10   ERROR DETECTION AND CORRECTION

In 1950 the mathematician R.W. Hamming used advanced algebra theory to show that it is possible to provide groups of check bits along with a group of data bits, and that these check bits could be used for detecting and correcting errors. From this analysis came practical error detection and correction (EDC) circuits. Many EDC circuits are called Hamming circuits as a result.

In an EDC system more than a single parity bit is used. The theory shows that a group of 16 data bits (2 bytes) requires 6 check bits to provide the following EDC characteristics:

- Detection and correction of *any* single-bit error.

- Detection of *all* 2-bit errors and *some* more-than-2-bit errors.
- Detection and correction of *any* single-bit error in the group of check bits themselves.

Many computers with large amounts of memory, in the range of half to one megabyte (Mbyte) and larger, use EDC circuitry. The cost of the additional memory ICs for the six check bits is relatively low, and the reliability and time between failures is increased by 100 times.

Hamming theory shows that 6 bits are required to perform EDC on a 16-bit word, for a total of 22 bits. It also shows that a group of 4 bytes (32 bits) requires a check group of 7 bits. A check group of 8 bits is required for a word of 8 bytes (64 bits).

The actual circuitry to perform EDC uses several hundred gates and would occupy a typical circuit board. Several IC manufacturers offer EDC circuitry in a single IC, such as the Am2960 from Advanced Micro Devices (Fig. 9-19).

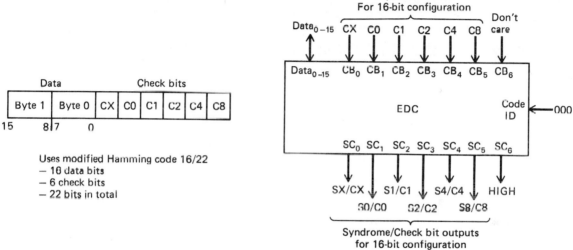

Fig. 9-19 The bit format and IC configuration for the AM2960. *(Courtesy of Advanced Micro Devices, Inc.)*

This 48-pin IC provides all the circuitry needed to perform the check bit generation, error detection, and error correction for 16 bits. By cascading additional Am2960s, more than 16 bits can be handled because the appropriate interconnecting signals are provided.

In operation, the EDC circuitry monitors the data lines as the simple parity circuit does. When the processor writes data to memory, the data flows through the EDC circuit where the six check bits are calculated. These bits are determined by the XOR (and XNOR, or XOR followed by inversion) function operating on different groups of 8 bits within the data bytes (Fig. 9-20), which is the same as generating an even parity bit for the group. (The Hamming theory shows which bits must be combined in what way to provide the proper check bits.) When the processor reads data, the data and the check bits pass through the EDC. The EDC

Fig. 9-20 How check bits are formed from the 16 data bits using XOR (4 check bits) and XNOR (2 check bits).

Generated check bits	Parity	Participating data bits															
		0	1	2	3	4	5	6	7	8	9	10	11	12	13	14	15
CX	Even (XOR)		X	X	X		X			X	X		X			X	
C0	Even (XOR)	X	X	X		X		X		X		X		X			
C1	Odd (XNOR)	X			X	X			X	X	X				X		X
C2	Odd (XNOR)	X	X				X	X	X				X	X	X		
C4	Even (XOR)			X	X	X	X	X	X							X	X
C8	Even (XOR)									X	X	X	X	X	X	X	X

The check bit is generated as either an XOR or XNOR of the eight data bits noted by an "X" in the table.

recalculates the check bits and compares them with the stored values that have just been read out of memory. If they are the same, there is no error. If they differ, however, the EDC can look at the pattern of old and new check bits, determine where the error is, and correct it before allowing the data to continue back to the processor. The processor receives the data as if it had been read directly from memory, and it may also receive and interrupt indicating that an error had occurred. If the EDC comparison and analysis of the two groups of check bits shows that a 2-bit error occurred, it uses a different interrupt so the processor can distinguish between an error that was corrected and one that could not be corrected.

Troubleshooting Parity and EDC Circuits

Parity circuits are relatively easy to troubleshoot with the proper equipment such as a logic analyzer which shows the binary logic state of all bits and the parity bit simultaneously. (Be sure that the memory itself has been tested first, of course.) The analyzer is connected to all the bits, and then the parity bit generated by the parity circuit is compared by the test technician against the data bits. The correct value of the parity bit can be calculated manually from the data bits and verified versus the value that the parity circuit provides. The other part to check is the comparison circuit, which should compare the parity bit read from memory with the recalculated parity bit. To see if this is working, a single-bit error must occur, and if the memory is working properly, this may not happen when needed. The best approach is to cause an error in the stored parity by forcing it to the incorrect state. (This forcing usually is difficult to do since there is no provision for it in most designs, and wires usually have to be disconnected and clip leads used, which is time-consuming and may actually short circuit something else.) If the parity bit is in the incorrect state, the parity generator-checker IC should show that a parity error has occurred.

Verifying the operation of an EDC IC uses the same general principles as the parity generator-checker, but in practice is a little more complex. The logic analyzer must be able to show the state of all 22 lines at once (for a 16-bit memory system with six check bits). The correct values of the check bits can be calculated manually from the chart in the data sheet of the EDC IC which shows which data bits are combined to form which check bits (Fig. 9-20). This is more time-consuming than manually figuring out a single parity bit. The method is useful, however, since it can be done by hand.

Just as with the parity checker-generator, it is more difficult to force an error to occur in a working system. One of the check bits must be forced to the wrong logic state, but generally the design of EDC ICs and their interconnection in the circuit is such that doing this will cause major malfunctions. Therefore, this is not recommended without especially designed test fixtures and equipment.

QUESTIONS FOR SECTION 9-10

1. What are the two weaknesses of parity for error detection? Will parity detect four errors in a byte? Five errors?

2. What is EDC? How is it different from parity checking?

3. What three features does an EDC circuit with 6 check bits on 16 data bits provide?

4. What reliability improvement does an EDC typically offer?

5. How many check bits are required to perform an EDC on 4 bytes? 8 bytes?

6. How are the check bits calculated in an EDC IC?

7. What does an EDC IC do to the data bits if it detects a single-bit error? 2 bits in error?

8. Why should memory be tested first, before checking the parity and EDC circuits and ICs?

9. What instrument is used for troubleshooting parity and EDC ICs? How?

10. What is the problem from a test standpoint when testing parity and EDC ICs?

PROBLEMS FOR SECTION 9-10

1. A 16-bit data word is used with EDC circuitry. Using Fig. 9-20, calculate the check bits CX, C0, C1, C2, C4 and C8 for these data bits:
   a. 0000 0000 0000 0000
   b. 1111 1111 1111 1111
   c. 0110 1000 1011 0011

2. A memory system of 1K × 8 bits is built from eight individual 1K × 1 bit memory ICs.
   a. What happens when a single memory IC is removed and parity is used?
   b. What happens and why if the same IC is removed where EDC is used?

3. a. Use truth tables to show that an even parity bit is equivalent to the XOR function for two inputs A and B.

   b. Repeat for three inputs A, B, C. (The XOR of A, B, and C is the XOR of A and B, and that result XORed with C.)

   c. Repeat a to show an odd parity bit is the same as the XOR and NOT function, called XNOR.

9-11 FLOPPY DISK CONTROLLERS AND INTERFACES

Users of computer systems need a way to store their programs and data so they can be used later. Even a computer with megabytes of memory will not have room to store all the user information in the system RAM. Users also may want to have a way of removing their program and data for security reasons, and a memory storage method that is nonvolatile, meaning it will not lose the bits even if power fails. Of course, low cost is important too, so that the method of data storage can be used by systems in many applications.

A method of data storage which meets all these requirements for removable mass storage is magnetic tape, similar to cassette tape. The 1's and 0's are written on the tape by a Write head, and read back by the same head. A logic 1 is stored by magnetizing a tiny section of the tape one way, and a 0 is recorded by magnetizing it another way. Magnetic tapes can store several thousand bits per inch, and a tape can be several hundred or even thousand feet long.

There are two drawbacks to tape as a storage medium, however. Access to anything on the tape is serial—to locate data at the end of the tape means searching through the entire length of the tape. Even in a high-speed search mode, this can take minutes and is frustrating for many users. The tape also does not readily allow the data or program to be expanded in the middle, because all the subsequent 1's and 0's would have to be moved further along. As a result, tape is used primarily when long, continuous blocks of data must be stored, such as insurance records or historical data.

A more convenient method of data storage uses the same magnetic principle as tape but gets around the limitations of tape by using a flat disk of magnetic material. This disk is made of coated flexible plastic and is called a *floppy disk*. It rotates like a phonograph record, but at a much higher speed. Data is written to the disk as it rotates. The disk itself is divided into tracks, and each track is further divided into sectors (Fig. 9-21). A typical sector holds 128 bytes. Each sector must also have an ID tag showing the number of that section along with some other headings information. To read a track sector, the magnetic head of the floppy disk drive must be positioned by a motor over the correct track, lowered to the track, and then it must read all the sectors until it finds the ID of the one it is looking for.

The floppy disk overcomes the drawbacks of tape. The data on any part of the disk is available as soon as the head is positioned (a few milliseconds), so there is no serial search as there is through the tape. For this reason, the disk is sometimes referred to as a random access storage device. If the amount of data stored has to be changed, new sectors can be used as needed and linked through the IDs of the sectors by keeping track of which sectors are being used for the data.

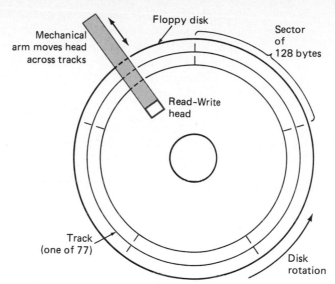

Fig. 9-21 The organization of a floppy disk.

The attraction of the floppy disk is the low cost, quick access, and flexibility. Controlling a floppy disk is technically complicated. Some of the issues that must be handled include:

- The motor which positions the head must be told which track and sector to find.
- The motor which moves the Read-Write head has to be precisely controlled to a few thousandths of an inch of position.
- The actual writing and reading operation has to be controlled precisely so that the data is written or read when the correct sector is under the head. Since the disk is spinning at 1800 revolutions per second, highly accurate timing is required.
- The logic levels used for the 1's and 0's of the data must be converted to levels and formats that are compatible with the magnetization process. It is not possible to write TTL signals directly to the disk. They must be reconverted when the disk is read back.
- There must be some way of determining if any of the data read back from the disk has an error as a result of problems in the writing, a defect in the disk, or a fault when reading.
- The track and sector IDs and other information must be written to the beginning of each sector.
- The processor must be informed when a disk-related operation is completed, so it can specify the next disk activity.
- There must be some way of formatting a blank disk out of the box, to put a basic ''road map'' of tracks and sectors on the disk, even when no data is present.

All of the above factors mean that a processor would have to devote a large amount of its resources to directly managing a disk. Even then, the timings are so critical during disk activity that many errors and false starts would be common. Special disk controller ICs make the task practical for nearly any processor.

One such IC is the FD1771-01 from Western Digital Corp. This floppy disk controller (FDC) has 40 pins and is the interface between the processor and the disk mechanism (Fig. 9-22). The disk controller responds to general instructions

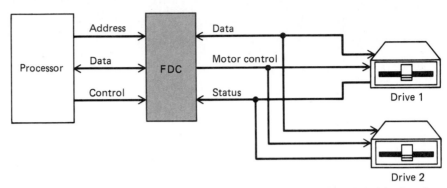

Fig. 9-22 The role of a floppy disk controller as a processor to floppy disk drive interface.

from the processor and implements the detailed steps that make the instruction happen. In operation, a disk contains a few special tracks called the *directory*. This directory is a table of contents for the disk. It lists the name of every program or block of data on the disk, and the numbers of the tracks and sectors that contain the actual bits of the program or data. When the processor wants to read a program, it instructs the disk controller to find the directory and report the contents. The controller moves the head, puts the head down, and reads the directory back to the processor. If the processor sees the desired item in the directory, it tells the controller IC which tracks or sectors to read back. The controller once again finds the specified tracks or sectors, reads the data on the disk, and passes this data back to the processor. As the data is read by the controller, it checks the bits for errors by examining check bits (similar to parity and EDC) that it put on the disk when it wrote the data there previously. If there is an error, the controller may try to read again, since the error may be caused by a speck of dirt or poor contact between the head and the disk. If the bits cannot be read back without error, the controller IC reports this to the processor. As new data and programs are stored on the disk, the controller updates the directory on the special tracks of the disk, under direction of the processor.

A FDC such as the FD1771-01 looks to the processor like a few registers or memory bytes. The processor writes control words and data to the FDC, and receives back data and status information. There is also an interrupt line from the FDC to the processor, so the FDC can notify the processor when an activity is complete or an error has occurred.

Troubleshooting FDCs

The interaction between the processor, FDC, and floppy disk drive is complex and requires special equipment to troubleshoot completely. Special test equipment designed for this purpose is usually required. More common equipment like oscilloscopes and logic analyzers can also be used, but only in a few special cases.

The most practical method of localizing a problem is to substitute known good units for the suspect ones. Common problems and some means of localizing the specific problems before other means are used are as follows:

- The floppy disk itself may be defective, since they are not rugged and can be damaged by fingers, dirt, and scratches. Try writing or reading another disk on the same drive unit and computer system.
- There are several different track and sector and ID formats commonly used, and a disk written with one format cannot be read by a drive set up for a different format. Be sure that the disk to be read is compatible with the drive in use.
- Next, use another floppy disk drive. This is not as difficult as it sounds, since most computer systems have two floppy disk drives to allow convenient copying of data from one disk to another. Try to read and write the second drive. The drives can also be interchanged, since they are identical except for the cable to which they are connected. Simply swapping the cable will also swap drives. (Sometimes there is a small jumper in the drive which also tells the system which drive is number 1 and which is number 2, and these have to be interchanged also.) The presence of two drives is a very powerful troubleshooting tool. An FDC such as the FD1771-01 can manage both drives, so writing to both drives or exchanging drive mechanisms can point to a bad drive or bad FDC. If one drive is OK but the other does not work, suspect the drive, or cabling. If neither drive works, the FDC may be bad.
- The problem may be with the mechanical alignment of the drive. A drive has many tight tolerances and sometimes these are exceeded, especially after excessive shock and vibration. The symptom here is that a drive can read disks that it has written but not disks which were written by another drive unit.
- Finally, most computers have verification programs which allow a user to test drives with a special command typed at the keyboard. These verification programs are similar to the Read-Write memory test studied in previous sections. They write various patterns to every sector on the disk, and then read them back and compare.

QUESTIONS FOR SECTION 9-11

1. Why is removable mass storage needed in a computer system?

2. How is tape used? What are the two drawbacks?

3. What is a floppy disk? What is its structure? How is data accessed? Why is it faster than tape?

4. What is a random access device? How are longer programs put on disk in place of shorter ones?

5. What are four problems in using and controlling a floppy disk and its mechanism?

6. What is the disk directory? What does it contain? How does the processor use the disk directory?

7. What does the FDC do? What does the FDC look like to the processor?

8. How is a floppy disk system checked out if there are problems?

9. Why would a mechanical tolerance problem sometimes allow a floppy disk system to read back disks it has written, but not disks written by other drives?

10. What is a verification program?

SUMMARY

When a system has to perform more complicated functions, and interface to more sophisticated devices, the support ICs become more important to the success of the system. Advanced support ICs are available which count and check for specific values and which provide precisely measured time intervals. The video display CRT requires large amounts of data transfer to make the proper characters appear on the screen, and the CRT controller is used for this. Test instruments may be interconnected via a standard party line which is supported by a controller IC. Errors may occur in one or more of the thousands of bits in the system memory, but there are ICs which can detect these errors and even correct them without interfering with the regular sequence of events. Finally, the floppy disk unit used for storing bits on a removable disk requires control of the mechanical elements as well as the reading and writing of the bits.

For each of these functions, support ICs are used to provide a more efficient operation. Without these ICs, the system processor would probably not have enough time to perform calculations and make decisions. These ICs are normally set up by the system processor to a desired set of initial conditions, and then the support IC runs independent of the processor. They send either data or status information back to the processor as required.

REVIEW QUESTIONS

1. What are the two main reasons that counter ICs for processor systems are needed?

2. How does a counter IC give a boost to system performance?

3. Give three examples of the need for timers in systems.

4. Explain the similarity and difference between a counter and a timer.

5. What are three modes of operation of a timer IC that are often required? For what purposes?

6. Why does a gate line on the 8253 increase the versatility of the 8253?

7. How does the processor set the 8253 to the conditions it wants? What types of information must be initialized?

8. How many memory locations of the processor does the 8253 occupy?

9. What causes an image to appear on the screen of any CRT?

10. How long does the image remain if the screen is not refreshed?

11. How are the individual dots which form a character turned on and off so that the desired character is seen?

12. Is each character formed one at a time, or is another technique used? Explain.

13. What counters are needed in a CRT controller? What does each one do?

14. What additional flexibility does a CRT controller provide to a system that a controller built of fixed counters and registers could not provide?

15. Why was the IEEE-488 standard developed? How has it allowed computers to be used to control other equipment?

16. What are the three modes of IEEE-488 operation? Explain the differences.

17. Why is a support IC almost essential to provide an IEEE-488 interface in a device?

18. What type of buffer is required on IEEE-488 signal lines? What two features must the buffers provide?

19. Explain how parity is used in memory.

20. How is memory parity similar to parity in serial communications?

21. What is a soft error versus a hard error? Which is more difficult to trouble-shoot and deal with when fixing equipment? Why?

22. Explain when the parity bit is generated and when it is checked when writing to and reading from memory with parity.

23. How many bits does it take to provide parity for a 1K memory with 4-bit words? 8-bit words? 16-bit words?

24. Why is parity usually not set to even?

25. What does a system usually do when the parity circuit detects a parity error?

26. Discuss the pros and cons of parity versus EDC. Which provides more features, is easier to implement, and costs more to provide?

27. What is the concept behind EDC? How is it a more complicated concept than parity?

28. How many check bits does EDC require for a 16-bit memory location?

29. What error types will EDC detect and correct? What type will it detect only?

30. Why is it hard to test memory circuits with parity and even harder to test those with EDC?

REVIEW PROBLEMS

1. A counter has a 16-bit capacity. It is combined with a clock that is running at 1 kHz. What is the maximum period of time that it can measure?

2. An 8253 is set to measure periods of 1 s. The gate line, however, is connected to an external switch that is on (high) for 50 percent of the time. What is the actual time period that the 8253 measures? Repeat for a gate that is high 20 percent of the time.

3. An 8253 is being used to time an interval of 100 ms. The clock is running at 5 kHz. What binary value should be written to the counter? What BCD value?

4. Most CRT screens used with computers provide 25 rows of 80 characters each. For such a screen, how many bytes of RAM does the controller need?

5. The screen on a typical CRT is refreshed 30 times per second. Using the CRT data of the previous problem, and a 7×9 character matrix, answer the following:
   a. How many dots per line are there?
   b. How many lines per character?
   c. How many total characters on the screen?
   d. How many dots per character matrix?
   e. How many dots per screen?
   f. How many dots per second does the controller provide to the CRT?
   g. What is the divide-by-N value for the line counter, character counter, and row counter?

6. A 64 Kbyte $\times$ 16 bit memory uses 64K $\times$ 1 RAMs. How many ICs are required for the memory? If parity is added, how many additional ICs are required?

7. Repeat the previous problem for EDC being added.

8. If the same total memory size is made up from 8K $\times$ 8 RAMs, how many additional ICs of the same size would be needed for EDC?

9. Calculate the odd parity bit for these bytes: (a) 1010 1010; (b) 0000 0001; (c) 1000 0000; (d) 1111 0111; (e) 1110 0111.

THE INTERFACE TO THE ANALOG WORLD

10

This chapter examines the ICs which are the interface between the digital system and the world of real, continuous signals. Analog-to-digital converters (ADCs) provide a digital number corresponding to the analog value of a signal, while digital-to-analog converters (DACs) allow a computer system to provide analog output.

The principles of operation of DACs and ADCs are complicated, especially for the different varieties of ADCs. However, the IC takes care of most of the operation, so the user and troubleshooter of these components do not have to understand every subtle point of operation. Some ADC circuits are made more versatile and cost-effective by using sample and hold circuits or multiplexers for the incoming analog signals.

10-1 ANALOG SIGNALS AND BINARY REPRESENTATION

The difference between analog signals and digital signals was discussed in Chap. 2. An analog signal is continuous and can be read to any resolution within the overall range, limited only by how closely the user wants to interpret it. A digital signal, however, presents only specific, discrete values to the user, and there can be no values other than these.

Most signals that exist in the world outside the computer system are analog signals. Any system that needs to measure temperature, pressure, voltage, energy, distance, or similar physical variables, or needs to control things like power, heat, lights, or position, must be able to interface with these analog signals. Real-world interfacing involves analog signals for two reasons: The physical variable being measured is itself analog, and the device used to convert the physical variable into an electrical signal (called a *transducer*) is also an analog device. Some examples are:

- Thermocouples, which convert temperature to voltage.
- Strain gauges, which provide a varying voltage related to how much pressure is applied to them.
- Energy meters, which develop a voltage proportional to the amount of electrical energy being consumed.
- Valves which respond to an electrical signal and open or close partly or completely depending on the value of the signal, and so control the flow of a liquid or gas.
- Electric heaters, which provide more or less heat depending on the value of a control voltage.

In order to build a system that is useful for real-world measurement and control, it is necessary to use special ICs to interface analog signals to the computer digital signals. These ICs are ADCs and DACs.

Binary Representation of Analog Signals

When binary numbers are used to represent analog signals for use with an ADC or DAC, there must be a clearly defined relationship between the true analog value and the way it is represented by the binary number. This means that for every value of the analog signal, the corresponding digital number that represents it can be calculated. Of course, the reverse is also true. Each digital value must yield a predictable analog value. In addition, there must be a way to understand the limitations of using a group of bits to represent an analog signal value.

Consider an analog signal that can vary between 0 and +1 V. This might represent the weight of packages being checked on a scale, with 0 meaning the scale is empty and 1 V meaning the scale is at its full load of 1000 pounds (lb). The computer system needs to read the 0 to 1 V signal so it can keep track of the individual weights for recordkeeping and shipment information. Therefore, the analog output is converted to digital format by an ADC and this digital format is used by the computer system.

The number of bits used to represent the analog value determines what is known as the resolution of the converted value. If the converter uses 4 bits, it can generate any one of 16 digital values. (Why?) This means that it has divided the 1-V analog signal into any 1 of 15 groups, or steps (Fig. 10-1), and each step has a value of $1/15 = 0.06667$ V, or $1000/15 = 66.67$ lb. The 4-bit converter provides resolution of 1 part in $2^4 - 1 = 15$. It also means that *any* voltages *within* the step size will be converted to the same digital value. A package weighing 70 lb will produce the same digital value as one weighing 75 lb.

For some applications the resolution of the 4-bit converter is enough, but for most, the uncertainty of the relatively large steps is too great. These applications require a converter with more bits of digital output (Fig. 10-2). Converters are available with 8, 10, 12, 14, or 16 bits of resolution. The system designer decides how many bits are needed based on the required performance of the system. Generally, converters with more bits cost more, so the system uses as few bits as needed.

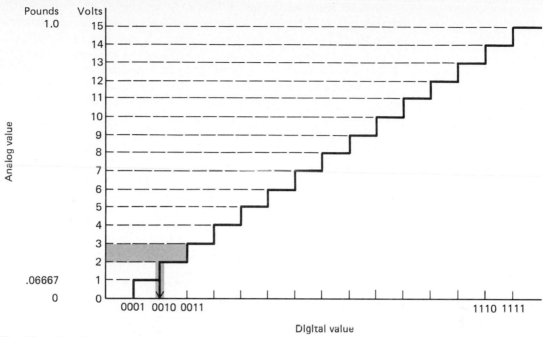

Fig. 10-1 The digital representation of an analog number divides that number into "steps." Any analog value within the step size translates to a single digital number. Example: from 2 to 3 is 0010.

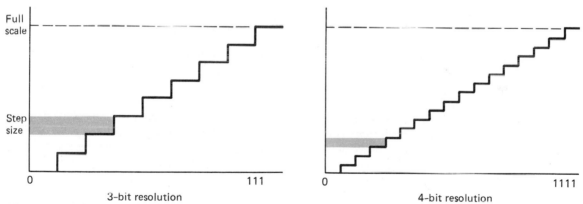

Fig. 10-2 Additional resolution decreases step size. Four-bit resolution has steps that are one-half those of three bit resolution.

In the previous application with the scale, it was decided that the resolution needed was 1 lb. The converter has to take the 0 to 1 V signal and divide it into at least 1000 steps so that 1 lb out of 1000 will be seen. A 10-bit converter provides $2^{10} - 1 = 1023$ steps, so that would be used. With this converter, each count of the binary number will represent $1/1023 = 0.0007775$ V, or $1000/1024 = 0.9775$ lb. The uncertainty therefore is within the requirement.

The same reasoning applies to DACs. In some applications the DAC analog output is used to control the position of a mechanical slide that can go over a range of 0 to 10 in. The analog voltage actually controls the motor which moves the slide to the proper location. A 6-bit converter would accept any number from 00 0000 to 11 1111 from the computer and generate an analog output with 63 steps. The 10-in range would be divided into $10/63 = 0.15873$ in, and the control could not be to any finer position. If the system required the ability to position the slide to 0.04 in, then a resolution of $10/0.04 = 250$ is needed. This could be provided by an 8-bit converter with 0000 0000 corresponding to the 0-in position, and 255 being the 10-in position.

Remember that resolution is not related to accuracy. Resolution is the number of divisions of the full range. Accuracy is how correct the reading is when compared to some absolute, external standard. Resolution is sometimes expressed in the percent of the value of 1 bit to the whole range. A 10-bit converter has $1/1024 = 0.1$ percent resolution; a 12-bit converter has $1/4096 = 0.025$ percent resolution.

QUESTIONS FOR SECTION 10-1

1. What are some sources of analog signals? What types of devices need analog signals?

2. When is an ADC needed? When is a DAC needed?

3. What is resolution in a converter, either ADC or DAC? What is the drawback of too little resolution? How would this problem be overcome?

4. What is the relationship between the number of bits of resolution and the number of steps the range is divided into?

5. What are common commercially available resolutions?

6. What is the difference between resolution and accuracy?

7. A thermometer is used to measure the temperature in a vat of liquid. The user is not sure if the thermometer readings are good, so the readings are compared at several points with a laboratory thermometer. The two thermometers agree to as fine as can be seen at all points. Is this accuracy or resolution that has been checked?

8. A metal ruler has graduations to 0.1 in. It is used at low temperatures, and so it has contracted. It is then used at higher temperatures and has expanded. What is affected: the accuracy or the resolution? Explain.

PROBLEMS FOR SECTION 10-1

1. How many steps are provided by a 6-bit converter? an 8-bit unit? a 12-bit unit? a 14 bit unit?

2. Repeat problem 1, and express the resolution as a percentage of full scale.

3. A 6-bit ADC is used to read a signal from a scale. The scale has a maximum range of 250 lb. Can this converter provide readings to within 2 lb? 4 lb? 6 lb?

4. An 8-bit DAC is used to position a robot arm to a point along a 100-in path. Can this DAC allow positioning to 0.025 in? to 0.035 in? to 0.5 in?

5. A 12-bit ADC is used to measure a temperature in the range from 0 to 700°C. What will the resolution of the reading be, in °C?

10-2 CONVERTER SCALING AND BINARY CODES

Analog signals in the real world can have many different values and ranges. They can be 0 to 1 V, 0 to 10 V, 2 to 6 V, and so on. The range of the signal is determined by the transducer that is being used. Converters, however, are designed to operate only on a specific range, usually 0 to 1 V or 0 to 10 V. Most ADCs expect to see a signal in either one of those ranges, regardless of the number of bits of conversion they provide. Similarly, DACs provide a 0 to 1 or 0 to 10 V output (some DACs are designed to provide current output, usually 0 to 1 or 0 to 20 mA). There must be a way to scale the actual signal to match the fixed range of the ADC or DAC.

The solution is to use an external amplifier circuit that amplifies or reduces the analog signal from the transducer into another analog signal with a range that matches what the ADC uses. (For DACs, the amplifier takes the analog output of the DAC and scales it to match the range needed by whatever the DAC is really controlling.) For example, if the signal that needs to be digitized is from a thermocouple measuring temperature, it is usually very small and may be up to 100 mV only, while the converter may be a 0 to 10 V unit. A 10-bit converter would have a resolution of approximately 10 mV, and therefore the 100-mV signal would be divided into 10 very coarse steps of 10 mV each. The external amplifier would be used to boost the 100-mV signal up to 10 V, and let the system make use of the full range of the converter. In this way, the converter is used to its full resolution potential (Fig. 10-3). The situation is analogous to measuring fractions of an inch with a yardstick; the yardstick has enough range, but only a small part of it is used and therefore the effective resolution is poor.

The external scaling can be used to match the DAC output to the application

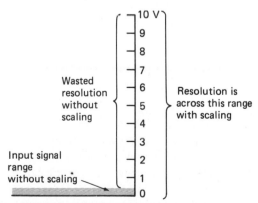

Fig. 10-3 The loss of resolution occurs if the input signal has a different range than the converter.

requirement. A DAC that is providing 0 to 10 V output may be driving a valve that would like to see only 0 to 1 V. Certainly, the DAC can provide that output range, but then the resolution of the DAC from 1 to 10 V is wasted. An amplifier that actually scales the DAC output down would be used, so as to distribute the DAC resolution over the range actually needed. In the opposite case, a valve might need 0 to 20 V from a DAC that provides 0 to 1 V. An amplifier that multiplies the DAC output by 20 would be used.

The understanding of scaling lets the relationship between the digital value of the ADC or DAC be related to the actual analog voltage, and to the value of the physical variable that the transducer, or signal, is measuring. First, consider the 10-V range ADC or DAC, with 8, 10, 12, 14, or 16 bits of resolution. Figure 10-4 shows the maximum ADC or DAC value, the resolution for each count of the binary count, and the binary value for a signal of half the full scale (a 5-V signal).

Number bits resolution Maximum ADC or DAC count	8 1111 1111	10 11 1111 1111	12 1111 1111 1111	14 11 1111 1111 1111	16 1111 1111 1111 1111
Binary value for ½ scale	0111 1111	01 1111 1111	0111 1111 1111	01 1111 1111 1111	0111 1111 1111 1111
Maximum converter count	255	1023	4095	16383	65535
Resolution of each count, volts	0.0392	0.00977	.00244	.00061038	.0001526

Fig. 10-4 This table summarizes converter counts and volts/count for 8- through 16-bit resolution.

Note that a single bit represents a different amount of signal voltage, depending on the number of bits (and therefore the resolution) of the converter, since the maximum count is divided into the full-scale range. The computer system that is dealing with the ADC and DAC must be programmed to know the number of volts that a single bit represents, so that it can convert the binary value received from the ADC to the actual correct input value or determine what binary value should be sent to the DAC to generate the correct output value. For example, for a 10-bit ADC over a range of 0 to +10 V, each count of the ADC represents 10/1023 or approximately 10 mV. If the 10-V full-scale signal represented 500 lb, then each count of the ADC would be $0.01 \times 500/10 = 0.5$ lb. If the 10-V full-scale signal represented 1000 lb, then each count would be $0.01 \times 1000/10 = 1.0$ lb. The conversion would normally be made by the program in the processor that reads the ADC. This program would know what the full-scale signal was and the value of each count of the ADC. It would then perform the necessary multiplication and division to scale results properly.

Other Binary Codes

Up to this point the binary value of all 0's has represented 0 value, while all 1's has meant full-scale value. This simple binary system is often used, but other binary codes are sometimes needed to handle different applications.

One such situation is where the ADC or DAC must handle negative analog signals, such as −10 to +10 V. The most common way of doing this is to use the MSB of the binary group as a sign bit. If this MSB is 0, then the value is positive.

If it is a 1, then the binary number is negative. Thus, 1011 would be −3, while 0011 would be +3. It is important to note that if a converter is used over a minus to plus range, then for a fixed number of bits of resolution the step size is doubled (Fig. 10-5). For example, a 10-bit converter used over a 0 to +10 V range yields a

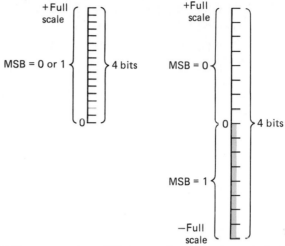

Fig. 10-5 Using the same number of bits to span from   to   results in doubling the step size as compared to 0 to + only. The resolution is therefore only one-half as good.

step size of 10/1024, about .01 V, while used over a −10 to +10 V range the step size is 20/1024, about .02 V. The difference is a factor of 2, since the total range to be covered has doubled, but the same number of bits is available. Other codes are sometimes used where it is easier to build or use the DAC or ADC with these special codes. Some of these codes allow conversion from one format to another more easily. These include BCD, Gray code, and two's complement (used for mathematical operations). Regardless of the code, the basic operation of the ADC and DAC remains the same.

QUESTIONS FOR SECTION 10-2

1. Why is scaling required for signals that are too large for the ADC input of DAC output?

2. Why is scaling done for signals that are much smaller than the converter range?

3. How many times better is the resolution of a 10-bit converter than an 8-bit unit? Repeat for 12 bits versus 8, and 16 bits versus 8.

4. What must the processor of the system do with each ADC reading to properly determine the weight in pounds, temperature in degrees, and so on?

5. When is simple binary ADC coding insufficient? What is done to represent minus values?

6. What is the effect on step size of taking an 8-bit converter and using from −10 to +10 V instead of from 0 to +10 V?

PROBLEMS FOR SECTION 10-2

1. A 10-bit ADC is designed to be used with a 0 to $+10$ V signal. However, the signal coming into the ADC is only 0 to 5 V. What is the number of bits of effective resolution that the system will achieve without any scaling? What will the step size be in volts? Repeat for a signal of 0 to 2.5 V.

2. For each of the signal input and ADC input ranges listed, indicate what scaling factor would be needed and if the scaling circuit should multiply or divide the input signal.

V Signal range	ADC input range, V
0–5	0–1
0–1	0–10
0–10	0–10
0–0.01	0–10
0–100	0–1

3. A 4-bit converter is used in regular binary format. What value is represented by a count of 0111? 1111? 0101? 1101? Repeat for a 4-bit converter where the MSB is used for the sign.

10-3 DIGITAL-TO-ANALOG CONVERTERS

An IC used for digital-to-analog conversion must have a way of converting the bits of the digital word into an analog signal with a value proportional to the number represented by the digital word. This is done using digitally controlled switches connected to resistors of specific values.

Figure 10-6 shows the internal circuitry of a 4-bit DAC. An external voltage reference, typically 10 V, is used by this DAC, in addition to the power supply for the DAC itself. Each bit is connected to an internal "switch" which controls the flow of current from the reference. When the digital signal on the bit line is a 1, the switch allows reference current to flow through the resistor associated with that bit. When the digital signal is a 0, the switch diverts the current to ground, and so no current flows through the resistor of that bit. All the currents that have not been steered to ground are then added together and scaled by the operational amplifier (op amp).

The key to making this scheme work is the values of the resistors used for each bit. The resistor of the MSB bit 3, has a value which can be called R Ω. The resistor of bit 2 then has a value of $2R$, and so on to the LSB, bit 0, which has a resistor of value $8R$. The weighting of the resistors is binary, with the smallest-value resistor used for the MSB, and the largest resistor connected to the LSB. Since the current that flows through the resistor is, by Ohm's law:

$$I = \frac{V}{R}$$

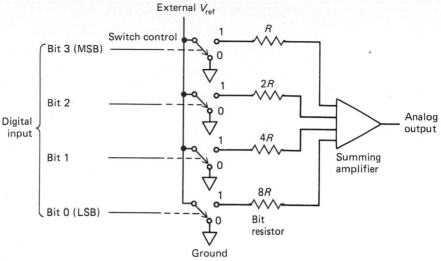

Fig. 10-6 The internal circuitry of a 4-bit DAC. The reference voltage and amplifier may be outside of the IC. The resistor values span from R to 8R.

where I is the current, V is the reference voltage, and R is the resistance, the effect of the binary weighting is that the current flow through each resistor is twice the flow of the adjacent LSB. The current flow is proportional to the relative binary positions of the bits. If the LSB has 1 mA of current, then the next bit has 2 mA, then 4 mA, then 8 mA, and so on. It is these currents that are added together to create an analog value from a digital value. The bits of the digital group control the switches that allow the current to be added or ignored. For example, assume a 4-bit DAC with a 1-V reference. The resistors are ⅛, ¼, ½ and 1 kΩ for the MSB through LSB. The current flows therefore are 8, 4, 2 and 1 mA through the resistors. If the binary value is 1001, the switches for the MSB and LSB will direct the resistor currents to the summing points, and the total current will be 8 + 1 = 9 mA. When the binary value is 1111, all the switches will steer the current to the summing point, for a value of 8 + 4 + 2 + 1 = 15 mA.

The concept of R, $2R$, $4R$, and so on is a good one in theory, but there are some practical problems. As the number of digital bits increases, the range of resistor values needed becomes very large. Mathematically, if the smallest resistor is R, then the largest one is 2^{N-1} times R for an N-bit DAC. A 10-bit DAC would need a range of 2^9 or 512:1 in resistor values. This wide range is very hard to fabricate on a IC and the resistors are very prone to drift in value as the temperature changes or the IC is used at different voltages. The wide range is not tolerant of production process variations or in circuit operational changes. To overcome this problem, most DACs use a variation on this resistor method, called the R-$2R$ ladder (Fig. 10-7).

In the R-$2R$ ladder, the current flow is still controlled by switches connected to the digital bit lines. Instead of a single resistor for each bit, resistors of value R and $2R$ are connected at a node as shown. The current that flows divides at the node, with exactly half going to the left and half going to the right. This gives the effect

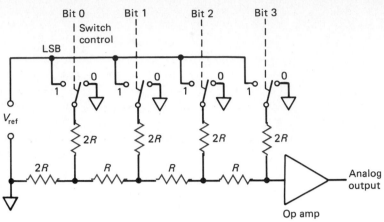

Fig. 10-7 The R-2R method used for a DAC reduces the wide range of resistors needed.

of the actual current going to the summing point being exactly one-half the amount of the adjacent more significant bit. The practical importance is that all the resistors are two very close values, R and $2R$. This is much easier to fabricate on the IC, and the drift and other problems associated with a wide span of values are much less.

The DACs discussed to this point are really incomplete without additional circuitry which allows them to interface easily to the processor and to the outside analog world. Some additional features are needed for a practical DAC.

A Practical DAC

In order for a DAC to be interfaced to a processor, the data from the processor must be latched so that it remains unchanged even after the processor goes off to other program steps. Otherwise, the processor might write the digital value, but the DAC might also respond to data from the processor that is intended for other parts of the system.

There are two common ways for a DAC to be interfaced to a processor:

- As a memory element, similar to the support ICs of the previous chapter
- As an external parallel device, connected via a PIO.

Some DACs contain built-in registers, as shown in Fig. 10-8a, which can be written to by the processor. These registers have a CE or strobe line, which causes the data to be latched into the registers. Other DACs do not have these built-in registers, but instead require external registers such as made up of 7400 series ICs (Fig. 10-8b). From the processor perspective, either interface operates the same. The processor writes the desired DAC data to an address which has been defined as the address of the DAC. The address is decoded and a CE is generated which causes the data to be stored in the latch. As soon as the digital input bits change, the switches inside the DAC change to provide the appropriate output.

If the DAC is connected to the processor via a PIO, then latches are not needed since the PIO contains these. When the processor wishes to change the analog

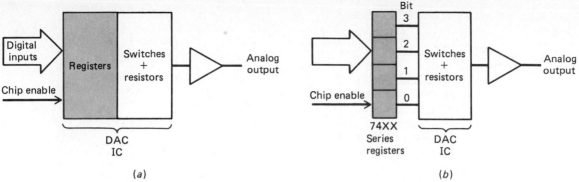

Fig. 10-8 (*a*) IC with built in bit register; (*b*) IC with bit register external to DAC IC.

output value, it must first set up the PIO. It then writes the digital value to the PIO, which acts as the latch for the DAC.

The DAC output follows changes in the digital input bits very quickly. Typical values are from 100 ms to 10 ms for the analog output to reach the new value specified by the digital bits.

A DAC also needs a voltage reference to provide the precise current values controlled by the digital switches. This reference must be accurate if the DAC is to be accurate. Even more critical, the reference must be stable over time and temperature so that the DAC output does not drift. Some DACs require an external reference, while others have the reference built into the DAC IC. The internal reference requires fewer components and is usually preferred in a circuit design. However, it also restricts the flexibility of the circuit. Some designs require unusual reference values for the particular application. For these situations, an external reference would be used.

The output of a DAC is the current summed from all the internal switches. This current must be buffered to isolate it from malfunctions in the external signal, or to change the current range that is supplied from the DAC switches. Many systems require a voltage output rather than current. For all three of these needs, an op amp is used to isolate, scale, or convert current to voltage. The op amp may be part of the DAC IC, or it may be an external/internal circuitry of the DAC and the external world. In a typical application, the DAC may produce 0 mA for all 0's input, and 1 mA for all 1's input. The application may require 0 to 20 mA, however, to control a valve which in turn controls fuel flow. The op amp would isolate the DAC from the valve, and at the same time multiply the DAC current by 20.

A typical DAC for 8-bit applications is the AD558, manufactured by Analog Devices (Fig. 10-9, page 276). This DAC contains the latches, a CE, and separate CS line for strobing data into these latches and controlling the operation. The use of two lines provides flexibility in operation, since the DAC can be set up via these lines for latching of data, or for allowing the data to flow directly to the DAC switches (called transparent mode). The AD558 also has a built-in op amp to provide voltage output. The user can select either of two output ranges (0 to 7.26 V or 0 to 10 V) by connecting designated pins of the IC together. This DAC is extremely easy to use in applications where 8 bits of resolution is enough.

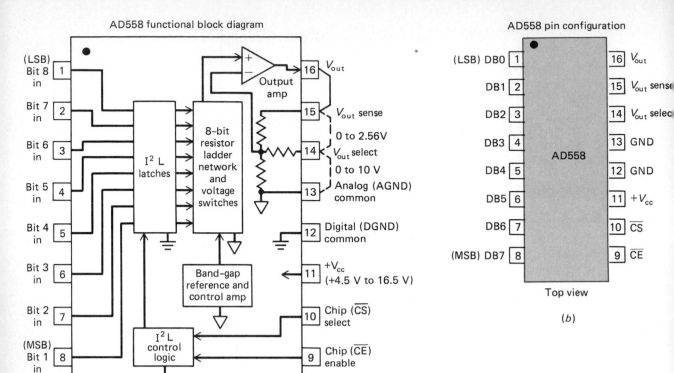

Fig. 10-9 The block diagram and IC pin configuration of the AD558 DAC, an 8-bit unit with built-in registers (latches), reference, and output amplifier. *(Courtesy of Analog Devices, Inc.)*

QUESTIONS FOR SECTION 10-3

1. What is the central element of a DAC? What is the principle of operation of a DAC?

2. What relative values of resistors are used in a DAC?

3. Which bit has the largest resistance value? Why?

4. What is the practical problem with the method of using a wide range of resistor values in a DAC? What is the alternative method used?

5. In what two ways can a DAC be interfaced to a processor?

6. What elements does a DAC need besides the internal switches? What is the role of these additional elements?

7. How is a DAC loaded with the new digital value? How fast does the DAC output respond with a new analog output value?

8. What are the two important performance characteristics of the DAC reference? When is it used and what are the benefits of an internal reference? An external reference?

9. What are the three reasons for the op amp in the DAC function?

PROBLEMS FOR SECTION 10-3

1. An 8-bit DAC uses a 1 kΩ resistor for the LSB. What are the values of the other resistors for the other bits?

2. A 6-bit DAC uses 10 kΩ for the MSB. What are the resistor values for the other bits?

3. A 6-bit DAC has 1 kΩ, 2 kΩ, etc., resistor values. The reference is 10 V. What is the output current for these binary inputs: 01 1011, 11 0100, 00 1100, and 11 1111?

4. For the DAC of problem 3, what scaling factor is needed to provide 100-mA output current at full scale (all bits = 1)? What scale factor is needed so that each binary count (step) is 1 mA?

10-4   DAC APPLICATIONS AND TROUBLESHOOTING

The most common application of a DAC is to provide a desired analog output current or voltage under control of the system processor. This analog output can then go to a mechanical valve, robot arm motor, or heating system.

DACs with external references can be used in some additional applications. For example, if a processor needed to control the level of a music signal in a sound system, an external reference DAC would be used. The music signal would be connected as the reference, and then the digital word written to the DAC would control how much of this signal reaches the output (Fig. 10-10). This application is called a *digitally controlled multiplier,* or *gain control,* and allows the processor to control the amplitude of a signal without having to actually handle that signal.

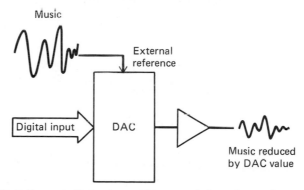

Fig. 10-10   Digitally controlled gain, used to control the volume of a music signal.

There are even some DAC applications that do not require a processor at all. One such common application is called a *waveform generator*. In some situations, it is necessary to generate a very complicated signal over and over, for test purposes or as a signal to be used by another system. This complex waveform might be a radar or sonar signal, or a simulation of the electrical signal produced by motion detectors when an earthquake occurs.

A DAC is used to do this along with a ROM, counters, and a clock oscillator

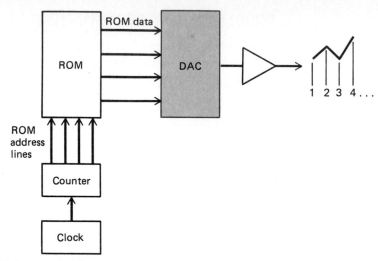

Fig. 10-11 Block diagram of a DAC used in a waveform generator.

(Fig. 10-11). Each address of the ROM contains a binary value equal to the desired analog value of the waveform at one point (Fig. 10-12). The first address of the ROM contains the binary number equal to the desired analog value of the waveform at time 0. The next address contains the binary value for the next time period. The waveform generator works as follows: The clock oscillator puts out pulses, which are counted by the counter. The counter outputs act as address lines to the ROM. As the ROM sees each address, it puts the binary data at that address on the data bus. The data bus sends the binary data to a DAC, which converts it to analog. The waveform generated in this way can have any shape, so this type of generator is sometimes called an arbitrary waveform generator. The resolution of the output waveform is determined by the number of bits in each address of the ROM, which must be the same as the number of bits of the DAC. The rate at which the waveform is generated depends on the clock oscillator frequency. A clock operating at 10 Hz will generate 10 addresses per second and thus 10 points of the waveform per second. The total number of points of the waveform is

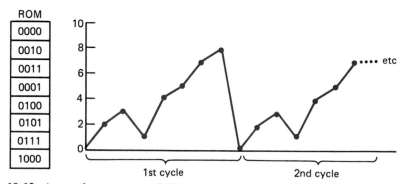

Fig. 10-12 A waveform generator ROM: 8 locations, 4 bits each, and the resulting waveform.

determined by the size of the counter, which also is the number of address locations in the ROM. Typically, these waveform generators can provide 1024 or 2048 such points, as rates that vary between very slow (only a few new points per second) to very fast (several hundred thousand points per second). The faster rate would be used to recreate a radar signal, while the slowest rate might simulate an earthquake signal.

Troubleshooting DACs

Problems with DACs fall into two categories:

- Very subtle failures, such as drift or small changes in the absolute accuracy of the output, or problems with the less significant bits (which represent very small parts of the total analog output).
- Major failures, which are easily noticeable.

The subtle failures are hard to detect and require special test instrumentation to measure and verify. Fortunately, these types of failures are rare in a system that was working properly. The major failures are more common. These major failures can be any of the following:

- The voltage reference, whether internal or external, has failed. The result would be that the analog output would be way off from where is should be. For DACs with external references, the voltage of the reference can be checked with a digital voltmeter. Internal references cannot be checked, except on some special DACs which bring the reference value out to a pin of the IC.
- The latches may not be working properly. They may not be accepting new data, or may be stuck at a 1 or 0. In some cases, only a few of the latches have failed, and in others all the latches or the strobe signal may not be working properly.

Most systems have a special test program which the test technician can use to write binary values to the DAC from the system keyboard. The latches, if external, can be checked by writing known binary values to the DAC and checking if they are present at the output of the latches and the input of the DAC. The strobe or CS line should be present (this requires an oscilloscope). If the latches are internal to the DAC, they cannot be checked directly, but writing different binary patterns to the DAC and examining the DAC output with the voltmeter will show if any of the more significant bit latches are bad. (The less significant bits are often so small that it is hard to measure them and detect any changes accurately, especially for 12-, 14-, and 16-bit DACs where a single bit represents 1 part in 4096, 16,384, or 65,536 of the output value.)

If the latches are good, the internal circuitry and switches in the DAC may be the problem. These can be checked only by writing different values to the DAC, just as for DACs with internal latches. For example, if the MSB latch is stuck at 0, then the DAC will provide outputs from 0 to one-half full scale (for regular binary coding) or from 0 to + full scale but not 0 to − full scale (for DACs where the MSB represents the sign of the output).

Finally, the op amp of the DAC may have failed. The usual failure is that the output is stuck at 0 or at the full-scale value and cannot be changed. This same symptom can also be caused by latches which are all stuck at 0 or 1, so caution is needed. If the op amp is internal, it cannot be checked separately from the rest of the DAC IC. An external op amp can be checked in some circuit designs by measuring the output of the DAC IC before it goes to the op amp and comparing it to the op amp output. Many designs do not have provision for this, however.

DAC Circuit Adjustments

Digital IC circuits do not have points for the test technician or service person to adjust, except for some special cases. This is because the digital world is absolute, with logic 1 and logic 0 levels as the only values that can exist in the circuit.

The digital-to-analog interface, however, often has several adjustments or "trims." The analog voltage may have to be slightly tuned to give a more accurate result. These trims are usually done at the factory, but sometimes they are incorrectly done, or inadvertently changed. If they are set wrong, the DAC may appear to be faulty when it really is not.

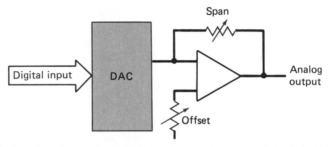

Fig. 10-13 The location, schematically, of two adjustments (trims) in DAC circuits.

Most DAC circuits have two adjustments to compensate for the two sources of error that typically occur whenever a digital value is converted to analog (Fig. 10-13). *Offset errors* occur because the DAC may not be putting out exactly zero when the digital input is all 0's (Fig. 10-14). The entire output of the DAC is offset by a small amount: 0 may come out as 0.01, 5 may come out as 5.01, and so on. It is as if a perfect ruler was used, but the 0 point was not lined up with the edge to be measured. This adjustment is sometimes called the *zero adjust*. The adjustment is made by loading all 0's into the DAC and observing the analog output with a

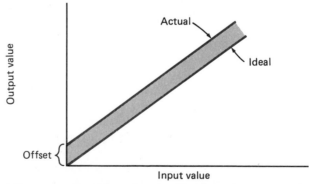

Fig. 10-14 Offset error: the entire output is offset from the correct value by the same amount.

voltmeter. The adjustment is made by turning a small variable resistor called a *potentiometer*, or *pot*, and watching the output value. The test procedure or specification for the circuit says how close to 0 the reading must be for the overall system to meet its accuracy specifications. Typically, the pot is turned until the output reads 0.0001 V or as close to 0 as required.

The second type of error is called a *span* or *gain error*. Even if 0 output is exact, the output of the DAC at full scale may be off by a small amount (Fig. 10-15).

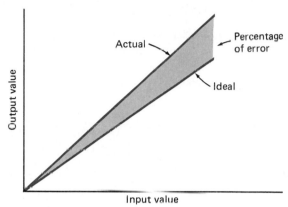

Fig. 10-15 Span error: the actual value is some percentage greater (or less) than the correct value.

This may be because the reference is not putting out exactly the desired value or the op amp is not perfect when it multiplies the DAC output. This span error is revealed when each step of the DAC output is off by a constant percentage: 1 V is 1.1, 2 V is 2.2, and so on up to 10 V reading as 11 (10% high). This is analogous to a rubber ruler, set at 0 exactly but then stretched. (Note that for offset error, the entire output is off by a constant value, while for span error the output is off by larger amounts as the output increases). The trim for span errors is done by setting the DAC to full-scale output while monitoring the output value with a voltmeter. The trim pot is then adjusted to make the full-scale output read within the specified range. Typically, the 10-V output would be adjusted to be within 0.001 or 0.0001 V of 10 V (called a + and − specification).

The two trims can correct the inaccuracy of the DAC. They can improve the accuracy to almost any desired value, if done carefully with proper equipment. They can do nothing to improve resolution, of course, which is determined only by the number of bits.

Some circuits have no trims or only one of the two trims. This is because highly accurate DACs are used (which are more costly), or the untrimmed performance will be adequate for the application, or the errors can be calibrated out by a special routine in the system processor program which compensates for these errors by adding correction factors. This software compensation is more common in newer equipment, where the power of the processor is used with less expensive (and so less accurate) ADCs and DACs.

QUESTIONS FOR SECTION 10-4

1. What is a special application of a DAC without a fixed reference value? What is this function called?

2. What is a waveform generator? How does it use a DAC? How does a waveform generator work?

3. What determines the resolution of the waveform values in a waveform generator? What determines the number of waveform values per second? The total number of points that make up the waveform?

4. What are the two types of DAC failure symptoms?

5. What are the sources of a major failure in a DAC circuit?

6. When can the reference be checked? How?

7. When can the latches be checked? How?

8. What is the visible effect of a bad MSB in a DAC?

9. What are the two types of DAC circuit errors? How are they seen? How are they corrected? What is the correction procedure?

10. Why don't all DAC circuits have trims?

11. Explain how an untrimmed DAC circuit could still be useful in a system.

PROBLEMS FOR SECTION 10-4

1. A multiplying 4-bit DAC is used to control the volume on a video recorder. The volume is set by the internal processor, responding to signals from the remote control. The volume is presently set by 0111. What digital value is needed to cut the volume in half? What digital value is needed to double the volume?

2. A waveform generator is being studied. The ROM is very small, with four locations of 2 bits each.
   a. How many different analog output values are possible?
   b. How many points will the waveform have?
   c. Assume the contents of the ROM are 00, 11, 01, and 01. Sketch the waveform that results.
   d. The counter operates at a rate of 20 pulses per second. How many times per second is the waveform repeated?

3. For each of the observed symptoms of a DAC, what is a possible cause of failure?
   a. All bits work and all steps are present, but the steps are one-half the size they should be.
   b. All steps are twice the size they should be, and no output difference is seen when 100 and 101, for example, are sent to the DAC. Prove the answer by sketching this for a 3-bit DAC.

c. The step sizes are OK, but there are some large jumps in analog output for certain digital input values.

4. A 3-bit DAC should put out 0, 1, 2, . . . , 7 V exactly. What errors (offset, scale, and amount) would cause these observed analog output values:
   a. All outputs are 0.1 V higher than what they should be.
   b. All outputs are 10 percent than what they should be.
   c. The outputs are 0.1, 1.2, 2.3, 4.5, 5.6, 6.7, 7.8 V.

5. A specification calls for adjusting the full-scale output to within 0.1 percent of the nominal output. What is the allowable error, in volts, for a nominal 10 V output?

6. A DAC specification says that the output is guaranteed to be within 5 mV of the full-scale output of 10 V. Does this meet a need for an output that is good to 0.01 percent?

7. A system must be able to control an analog output to within 0.025 percent of full scale. How many bits of resolution are needed?

10-5 ANALOG-TO-DIGITAL CONVERTERS

ADCs perform the opposite function of DACs. They take an analog signal and convert it to a digital number corresponding to the value of the analog signal. ADCs are more complicated in operation than DACs. All DACs use the same principle of operation and differ only in the number of bits and whether the latches, reference, and op amp are internal or external to the IC. For ADCs, however, there are different types of operation and new specifications and interface lines required.

Like DACs, ADCs are available with different numbers of bits of resolution. The most common values are 8, 10, 12, 14, and 16 bits. There is another specification that is critical to ADCs and that is much less of an issue with DACs: the speed of conversion. DACs present the new analog output almost instantaneously after receiving a new digital value. With ADCs, the situation is quite different. Depending on the type of ADC and the number of bits of resolution, this conversion time can take anywhere from several microseconds to many milliseconds. The faster ADCs are generally more expensive than the slower ones, and they may be affected by unwanted noise that is present along with the analog signal. The conversion time specification is sometimes called the *sampling rate,* which is the number of conversions per second, that the ADC could perform if it began the next one as soon as the previous one was completed.

For DACs, the processor writes the digital value and can go on to do other things, since the DAC will automatically provide the analog output. For ADCs, the processor must request a conversion via a start line and does not get the resulting digitized value immediately. The digital number is available only after the conversion process is complete. ADCs usually have a signal line (or lines) that provide status information on the ADC back to the processor (Fig. 10-16). These lines can be read by the processor to tell it that the ADC is busy doing a conversion, and therefore a new one should not be requested, or that a conversion has just

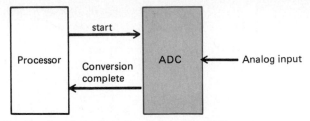

Fig. 10-16 The interface from the processor to the ADC.

been completed and that new digitized information is available. This conversion-complete or end-of-conversion line may be used in the circuit as an interrupt to the processor. The processor requests a conversion, then goes off to other activities. The conversion-complete interrupt occurs some time later and causes the processor to go to a special interrupt-handling routine which then gets the new value from the ADC.

The digital number provided by the ADC can be in several formats. The most common are regular binary, regular binary with the MSB used for the sign bit, and a special format called two's complement, which is useful in some arithmetic operations.

QUESTIONS FOR SECTION 10-5

1. What specification is critical with ADC that is much less a factor with DACs? Why is this spec not a problem for most DACs?

2. What is the range for conversion times for ADCs?

3. What is the difference in the interaction between a processor and a DAC versus a processor and an ADC?

4. What are the interface lines between a processor and an ADC? How are they used in the conversion sequence?

10-6 THE SUCCESSIVE APPROXIMATION ADC

The successive approximation (SA) type of ADC works by using a DAC along with some additional support circuitry. The performance of the SA type of ADC is determined almost entirely by this DAC. The support circuitry consists of two items: a special circuit that loads unique digital values into the DAC SA register, and a comparator circuit that compares the value of the analog output of the DAC to the analog signal to be digitized (Fig. 10-17). If the DAC output is smaller than the signal to be digitized, the comparator output is a logic 1. If it is larger, the output is a logic 0. The DAC, comparator, and special circuit are all part of one IC.

The SA type of DAC works by comparing the analog signal to be digitized against the values loaded into the DAC. Consider a 4-bit ADC which has a 4-bit DAC in it. First, the SA register of the DAC is loaded with 1000 and the DAC output is compared to the external analog signal. If the DAC output is larger, the 1

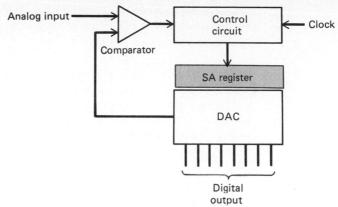

Fig. 10-17 Block diagram of the SA type of ADC.

in the MSB position is changed to a 0 by the comparator. If it is smaller, the MSB position value is left at 1. Next, the process is repeated with the second from MSB position of the DAC loaded with a 1. The comparison is done again, and the same rule of changing the bit or leaving it unchanged is applied. The procedure is then repeated for the third from MSB bit, and the last bit, the LSB. At the end of this divide-and-compare scheme, the DAC contains the digital value corresponding to the analog input. What this SA method does, in effect, is to compare the input to half-scale (the MSB), then compare the remainder of the analog input to half of half-scale (the second MSB), and so on. (Recall that each binary position represents one-half the value of the position to its left.)

For example, consider a 3-bit SA type of ADC. To make the example simpler, the range of the inputs that this ADC can handle is 0 to 7 V. Thus, 000 is 0 V, 001 is 1 V, and so on up to 111 being 7-V input. For an analog input of 5.2 V, the following would happen:

- The DAC would be loaded with 100, which would generate an output of 4 V to the comparator.
- The comparator would see that the analog input is greater than the DAC value and would therefore leave the 1 in the MSB position.
- The DAC would next have a 1 in the second bit position. This would cause an output of 6 V.
- The comparator would see that the analog signal is less than the DAC output and so change the value in the second bit position to 0.
- The DAC would finally have a 1 loaded in the LSB position. Now the DAC has the value 101, and this produces an output of 5 V. The comparator sees that this is less than the analog input value, and therefore leaves the 1.
- The conversion is now complete. The 101 value is the closest digital value that can be achieved with this relatively low resolution ADC. The ADC end-of-conversion line indicates to the processor that the conversion sequence is complete.

One advantage of the SA method is that the conversion is complete as soon as all the comparisons are made. The number of comparisons required is equal to the number of bits of the DAC. Therefore, the total conversion time is very short. The DAC used within the ADC must be very accurate and stable for the ADC to give accurate results.

A Practical SA ADC

An ADC IC must be interfaced into the system in a way that is convenient for the processor. The ADC also requires a clock to step the internal logic of the ADC through the comparison sequence. Most ADCs are designed to convert signals that are 0 to +10 V. If the analog input is not in this range, or does not fit the range properly, an external amplifier is used between the analog input and the ADC to scale the signal properly. The DAC in the SA type of ADC, like all DACs, also needs a reference voltage.

A typical commercially available ADC of this type is the AD7574 from Analog Devices (Fig. 10-18). This 18-pin DIP IC can interface to the buses of a processor like any memory element. It provides an 8-bit conversion in 15 μs. Built into the IC are all the pieces needed for a complete ADC except the reference voltage. The control logic, DAC, SA register, and comparator are all included. The AD7574 itself uses a single power supply of +5 V.

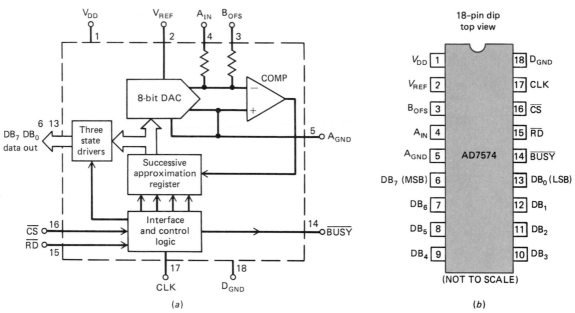

Fig. 10-18 Block diagram of AD7574 SA ADC and pin configuration. *(Courtesy of Analog Devices, Inc.)*

In order to initiate a conversion, the processor writes to the AD7574 at the address specified by the circuit design. This address is decoded by a decoder, which then makes the CS line go low, while the Read-Write line is also low since the processor is performing a Write cycle. (For this ADC, this sequence eliminates the need for a start line.) The ADC begins the conversion, making its busy line go low while the conversion is in progress. The processor can repeatedly poll this line, or it can use it to cause an interrupt when busy returns to the high state, indicating conversion complete. The processor then performs a memory Read cycle at the address of the AD7574, which turns on the three-state buffers and puts the converted value on the data bus so the processor can read it. This sequence is shown in Fig. 10-19.

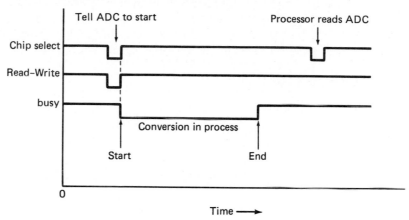

Fig. 10-19 The sequence of signals on the lines between the processor and the AD7574 ADC.

The entire conversion process is relatively transparent from the processor viewpoint, since it does not need to get involved in the details of the conversion.

QUESTIONS FOR SECTION 10-6

1. What is the heart of the SA type of ADC?

2. What other elements are needed?

3. How does a SA type of ADC work?

4. What is the role of the comparator? What determines the conversion time?

5. What is the sequence of events from the processor perspective when doing a conversion with an AD7574?

6. Does the AD7574 have a start line? What makes it begin converting?

PROBLEMS FOR SECTION 10-6

1. A SA type of ADC has 4 bits of resolution. The ADC is designed to handle signals in the 0 to 15 V range.
   a. What is the value, in volts, of each count that the ADC provides?
   b. An analog input of 11 V is to be digitized. Go through the steps of operation for this ADC. What are the internal values, at each step, in the SA register?
   c. If each step of the conversion takes 20 μs, how long does the entire conversion take?

2. The original 4-bit ADC of a system is being replaced with an ADC with one more bit, to provide extra resolution. If nothing else is changed (the scaling circuit, or the conversion factor that the processor uses):
   a. What was the full-scale digital value for the old ADC?
   b. What is the full-scale digital reading for the new ADC?
   c. For a full-scale voltage of 10 V, what was the value of a single bit for the old ADC? What is it for the new ADC?
   d. What was the reading for a 5-V signal with the old ADC? What is the reading for the same signal with the new ADC?

10-7 THE INTEGRATING ADC

Another common type of ADC is called the *integrating ADC*. This type of converter requires no DAC at all. Instead, it uses an internal voltage signal called a *ramp*. This ramp voltage begins at 0 and increases with time in an extremely straight line to a maximum voltage. Since the rate of increase is constant and precise, each value of this voltage can be related to the time when it occurred.

The integrating converter also requires a clock signal (usually provided externally), a binary counter, and a comparator (Fig. 10-20). In operation, conversion

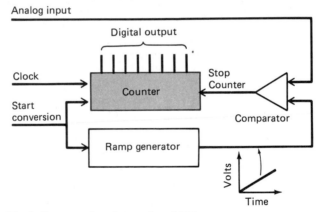

Fig. 10-20 Block diagram of an integrating ADC.

begins with the counter being reset to all zeroes, while the ramp voltage begins to increase from 0 and the counter receives pulses from the clock. The analog input is compared continuously to the ramp. When the ramp voltage exceeds the analog input voltage, as determined by the comparator, the counter is stopped. The value in the counter therefore is directly related to the analog input, since the period of counting was determined by the value of the analog input. The counter has added, or integrated, the clock pulses that occurred during the ramp.

The detailed operation of an integrating ADC is seen by the examples of Fig. 10-21. The analog input to be converted is at 5.2 V. The ramp waveform of the ADC increases from 0, in this particular ADC, at a rate of 1 V per 0.1 s. A 3-bit counter is used, and the clock is putting out pulses at a 10-pulse-per-second rate.

The ADC receives a start signal from the processor. It then begins counting clock pulses. After 0.1 s, the ramp is at 1 V, still below the value of the analog

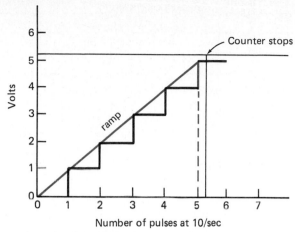

Fig. 10-21 Operation of a three-bit integrating ADC for an input of 5.2 volts.

input. The counter reads 001. After 0.2 s, the counter reads 010, and the ramp value is 5 V. As the time passes the 0.52 s point, the ramp is at 5.2 V and the comparator senses that the ramp is now equal to the analog input. The comparator disables the counter from accepting any further Clock pulses. The value in the counter is 101, which is the closest digital value.

Integrating converters are capable of giving high resolution (12, 14, and 16 bits) at low cost. When a faster clock and more counter stages are used, however, the conversion time is relatively slow since the counting cannot be done at a high rate in a practical implementation of this scheme. This slow conversion is not a drawback in many applications where the signal to be converted does not change quickly, such as temperature, pressure, or power-supply voltages.

Practical integrating ADCs use interface lines to the processor just like the SA type of ADC does. Since the conversion time is slow (typically 10 to 20 ms) the processor almost always uses the busy line as an interrupt so it can do other program activities while the conversion is in process. The integrating ADC is a good example of how digital IC functions such as counters can become the building blocks for new functions.

A Simple ADC Application

Most ADC applications use a processor in combination with the ADC. The processor initiates the analog-to-digital conversion, takes the digitized reading that results, and then does some further numerical processing on this reading.

There is one very common application that does not require a processor, however. This is the digital voltmeter (DVM), used to provide a digital readout of a voltage in a circuit. The DVM is the single most important piece of electronic test equipment used by engineers, service technicians, and anyone who is dealing with the design, debugging, and repair of electrical and electronic equipment.

The DVM consists of an input scaling circuit, an ADC, and a multiple-digit 7-segment readout, usually three or four digits, as shown in Fig. 10-22. Operation is straightforward. The user sets the range switch to select the desired input signal range, or maximum value. This is usually in steps of 1, 2, 5, 10, and 20 V, or in

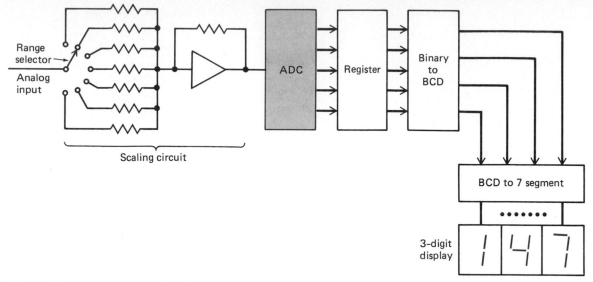

Fig. 10-22 Block diagram of a digital voltmeter.

steps of 0.2, 2, 20, 200 V. This range switch internally picks the necessary scaling resistors to make the voltage being measured fit into the input range of the ADC. The ADC is usually the integrating type, and it is driven by an oscillator which puts out pulses at a fixed rate, typically 30 per second. These pulses are used as the signal to start a conversion. At the end of a conversion, the ADC latches its digital value into a register. The value stored in the register goes through a binary-to-BCD conversion circuit, and then to a BCD-to-7-segment decoder-driver and the display digits. The user sees the digital value on these digits.

QUESTIONS FOR SECTION 10-7

1. How does an integrating ADC work? Why is it called integrating?

2. What are two major benefits of integrating ADCs? What is a possible drawback in some applications?

3. What is the typical conversion time of an integrating ADC?

4. What determines the resolution of the integrating ADC? What determines the conversion speed?

5. How is an ADC used in a DVM? What kind of ADC is used? Why?

6. What role does the processor have in a DVM?

7. What does the DVM user select? Why?

PROBLEMS FOR SECTION 10-7

1. An integrating ADC has a 100-Hz clock. If the ADC counter is 10 bits, how long does it take to convert a signal that is full-scale? Half-scale?

2. An integrating ADC uses a ramp that increases at the rate of 10 V/s. The ADC is designed to handle signals that are 0 to 10 V. How long does it take for the conversion to be complete if the analog signal is 5 V?

3. Show why increasing the clock frequency in an integrating ADC shortens the conversion time. Show why increasing the counter size increases the resolution, but lengthens the conversion time. If the counter is increased by 1 bit in length, what must be done to the clock value in order to maintain the same conversion time?

10-8 MULTIPLEXERS AND SAMPLE AND HOLD CIRCUITS

The systems in which ADCs are used sometimes require special circuitry between the analog signals to be digitized and the analog input of the ADC. Multiplexers allow many analog signals to share a single ADC. *Sample and Hold* (S/H) circuits are needed when the analog signal may be changing in value while the conversion process is taking place.

If there are many analog signals coming into the computer system, it may be very expensive (due to board space, power, etc.) to use an ADC for each such channel. An example of this would be a system that monitors energy usage in a building. There may be hundreds of temperature points to check, but they are checked very infrequently, perhaps once every 5 min. For such an application, a single ADC would be used. Between the ADC and the hundreds of analog signals would be a multiplexer, or mux. This mux allows only one of the many inputs to reach the ADC (Fig. 10-23). (This is similar to the mux of Chap. 4, where one of many digital signals was selected. Here, the selected signal is an analog signal.) The computer using the ADC also controls the mux and so specifies which channel it wishes to have as the analog input to the ADC. The mux is basically a computer-

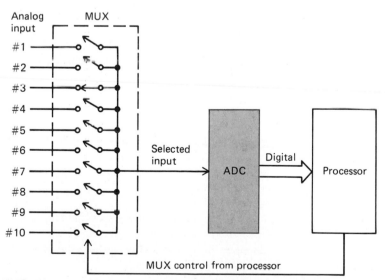

Fig. 10-23 The multiplexer for many analog inputs. Only one switch is on at a time; here, number 3.

controllable array of switches. The switches are used to select the desired signal. The mux does not change the analog signal or do any processing on it—it acts only as a path-selection system.

The sequence of operation when a mux is used is as follows: The computer processor sends out a digital code indicating which channel of analog input is desired, usually via a PIO. The mux then sets the on and off states of its switches so that this signal is the analog input to the ADC. The processor next requests a conversion, and since it specified the channel, it is able to associate the digital value received with the analog point it came from. The processor then specifies another channel to the mux, and the processor begins again. Note that while analog signals into an ADC can be muxed to share this single ADC, the same is usually not practical for DACs. This is because DAC analog outputs usually have to be maintained at the desired value, and muxing a DAC output would lose the output.

The S/H is used in applications where the analog signal to be converted may be changing during the conversion process. Depending on the amount and the direction of change, and the timing of the conversion steps, the ADC could produce results that are completely different from the analog input value during the conversion time. These false digital values would be useless in any calculations or decisions that the processor had to make but would not be recognized by the processor as incorrect, since the processor has no way of seeing what is going on at the analog side of the ADC.

The S/H is a special circuit that is connected between the analog signal and the input of the ADC. It has the ability to capture the analog value of that signal and hold it for anywhere from several hundred microseconds to hundreds of milliseconds (the S/H designer uses components that provide the necessary hold time for the ADC). When the processor sends the start conversion signal to the ADC, the S/H first samples the value of the signal and holds it, and then the conversion process begins (Fig. 10-24). As a result, the value of the analog signal as seen by

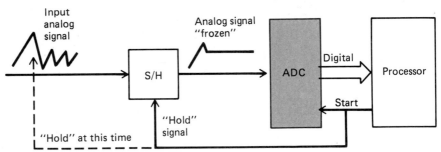

Fig. 10-24 Operation of a Sample and Hold. The input waveform value is held fixed during conversion.

the ADC during conversion is unchanging. In this way, the digital value that results from the conversion is accurate and not misleading.

In some cases, the S/H is used to allow a slow but less costly ADC to be used with faster-changing signals.

QUESTIONS FOR SECTION 10-8

1. What is a mux? What are two reasons why it might be used?

2. What does the mux do with the ADC?

3. How is the mux controlled? Why can or cannot DACs use a mux?

4. What is S/H circuit? Where is it needed?

5. What happens if there is no S/H where one is needed?

6. How does the S/H work with the ADC?

PROBLEMS FOR SECTION 10-8

1. A mux is used to select one of eight signals to go to an 8-bit ADC. How many bits does the mux require from the processor for selecting the desired input signal?

2. A system has 100 analog input channels. The conversion time of the system ADC is 0.1 s. The application needs to read each analog input once per second.
   a. Can a mux and one ADC be used?
   b. If not, how many ADCs are needed?

3. A system is converting an analog signal into a 10-bit value, with a converter that takes 0.01 s for the conversion. The input signal may change at a rate of up to 100 V/s. The application requires that the conversion be accurate to within 0.5 V. Show why a S/H is or is not needed. Repeat for a specification that requires conversion accuracy to within 2 percent of the true signal value.

10-9 TROUBLESHOOTING ADCs

Problems with ADCs can occur either with the analog side or the digital side of the circuitry. A known value of analog signal is required as a test input. This signal can come from a power supply or battery, and the exact value is not critical as long as the value is first measured with a voltmeter. Failures of ADCs usually fall into two categories:

- No conversion activity takes place, and so no digitized value is reported to the processor.
- The digitized value that does come back is grossly incorrect.

To find the cause of a digital reading that is way off, it is necessary to make sure that the analog signal from the external point is actually reaching the ADC analog input. If there are no muxes or S/Hs involved, this is usually straightforward. The known voltage is attached to the system input in place of the actual voltage, and this signal is followed through the input circuitry and scaling amplifier to the analog input point of the ADC IC, using a voltmeter or oscilloscope. The value of the known signal should be set to about one-half the maximum signal that the system transducer normally provides. The value as measured by the voltmeter or scope may change, of course, if the signal goes through any scaling amplifiers. This is usually marked on the system schematic or in the test procedure. For example, a 0 to 100 mV signal from a thermocouple may be amplified by 100 times to use the full 10-V range of the ADC. A 50-mV signal would be used for test. It should read 50 mV before the scaling amplifier, and 5 V after the amplifier. Most failures in this front end cause the input to be either 0 V at the analog input, or full scale, so a half-scale voltage will be easily distinguished from these.

If the system uses a mux or S/H, the test procedure must have a way to control these so that the analog signal can be directed to the ADC. For systems with muxes, the mux should be set to a specific channel, and the known voltage put on this channel. The voltage can then be followed through the mux and into the scaling amplifier, and then to the analog input of the ADC.

For systems with S/H circuitry, the known input voltage can still be used. However, some failures of S/H circuits that involve the hold circuitry are such that a steady voltage from a power supply will be passed through properly, but the real varying signal will not be held during the conversion. This shows up as different-size errors in the digitized value, and the size of the error is random, depending on when the digitizing sampling occurred relative to the changing of the analog input value.

For cases when no conversion activity takes place, the problem may be that the start signal does not reach the ADC (if the ADC uses a separate start signal). The ADC may of course be faulty and not recognize this signal even if it does reach the ADC. The other causes of the problems may be that the ADC does not generate the busy signal (or conversion-complete signal) or the signal does not reach the processor, so the processor does not know that a conversion is complete. These faults have no relationship to the actual analog signal value being converted.

A complete test of the ADC in a system involves a test routine that is usually provided by the system for the service technician. This routine lets the technician request a conversion and see the result on the system screen or display. A known value of input would be connected to the system analog input, and a conversion takes place. The digitized value is then displayed and can be compared to the known value. If a multiple-channel system and mux is involved, the test routine asks the technician to type in the desired channel number, and then the routine sets the mux up properly. This test routine can be used to check the signals to the mux (if any) and between the ADC and processor. The signal path from the analog input to be measured, through the scaling amplifiers, and to the ADC input may have some offset and scale errors, which would cause the ADC to provide incorrect digital values. These errors are usually small and can be taken care of via two adjustments. A 0-V signal is first used to set the offset calibration pot so that the ADC reads all 0's. The input voltage is then set to full scale, and the scale pot adjusted so that the ADC provides the full-scale digital value.

QUESTIONS FOR SECTION 10-9

1. What are the two main groups of ADC problems?

2. How is an ADC input checked? What value is often used? Why?

3. Is the value of the analog input at the ADC necessarily the same as it is where it comes into the system? Why or why not?

4. If there is no start signal, what does the ADC do? If there is no busy or conversion-complete signal reaching the processor, what happens?

PROBLEM FOR SECTION 10-9

1. An ADC shows the following symptoms. What could be causing the failure?
   a. No new converted value reported.
   b. All values are one-half what they should be.
   c. The conversions for values from 0 to half-scale are good; above half-scale the MSB stays at 0.
   d. The conversion takes twice as long as it should.
   e. The results of the conversion are different and wrong for each conversion done—they are random and unrelated to the signal.

SUMMARY

In order to allow digital systems to interface to real-world signals— temperatures, speed, pressure, and motors, for example—there must be a way of interfacing the analog signals of the real world with the digital signals of the circuit and system. ADCs and DACs are the solution. They provide the conversion needed. ADCs and DACs can have differing values of resolution, depending on how many bits they use to represent a voltage or current. Higher resolution generally costs more than less resolution. Another factor in the performance of ADC is the speed of conversion, which can be anywhere from milliseconds to microseconds, depending on the converter design.

Most practical DACs use a ladder of resistors internally, where the resistors have one of two possible values. For ADCs, there are two predominant ways to build them: the successive approximation method and the integrating type. Both kinds have technical advantages and disadvantages. ADCs and DACs often have scaling circuits between them and the outside world to match the range of the ADC or DAC to the range of the analog signal, so as to make maximum use of the resolution that the ADC or DAC can provide.

In some ADC applications, the ADC is preceded by a multiplexer which lets many analog inputs share one ADC. A S/H circuit may also be included to hold the value of the analog signal while the conversion process takes place.

REVIEW QUESTIONS

1. Why are ADCs and DACs needed?

2. Is a DAC built by simply running signals into the output of an ADC?

3. What is the relation between DAC or ADC bits and the resolution of the digitized value?

4. Does high resolution mean high accuracy? How can there be one without the other?

5. What is the role of scaling? Why can't the ADC or DAC simply be made with the required range?

6. For every extra bit in a DAC or ADC, by what factor does the resolution increase?

7. When a processor reads a value from an ADC, is the digitized value in correct units (pounds, inches etc.)? Explain.

8. How are negative values often represented in digital format?

9. How does an R-$2R$ DAC function? Why is the R-$2R$ scheme used?

10. What are the main elements of a DAC?

11. How is DAC performance calibrated? What is the meaning of offset error and gain error? Can there be one without the other?

12. Most DACs require a very accurate and stable reference. Which kind does not? Why?

13. What is conversion time for ADCs? Why is it much more complicated an issue for ADCs than for DACs?

14. Explain why the ADC-to-processor interface is more complex than the processor-to-DAC interface.

15. What are the two major categories of ADCs? How do they differ in operation?

16. Discuss two pros and cons of integrating ADCs versus SA type of ADCs.

17. What functional elements studied so far are used in a DVM?

18. For an integrating DAC, what determines the resolution? The speed of conversion?

19. Repeat question 18 for a SA ADC.

20. Why is a multiplexer sometimes used? What feature does it offer in a system? What is the drawback?

21. Why is a S/H sometimes needed? Where would it not be needed?

22. Discuss three problems that can be observed in an ADC circuit and their possible causes.

23. How is a mux used with an ADC different from the mux of Chap. 4? How is it smaller?

1. For a 16-bit ADC, how many steps are there? What is the resolution in percent of full scale?

2. For the 16-bit converter, what does each bit represent if full scale is 10 V?

3. A 10-bit converter is used to read a signal that represents 50 lb. How many pounds does each count of the converter represent?

4. A 10-bit ADC is scaled to handle signals of 0 to 10 V. The incoming signal, however, is never larger than 1.25 V. How many bits are effectively being used? What is the actual resolution as a percent of the actual full-scale input value?

5. What value would be loaded into a 6-bit DAC to represent the following values: 28, 50, 47? Repeat for a 6-bit DAC where the MSB is the sign bit. How would -20, -28, and -10 be represented?

6. A 4-bit DAC uses a resistor of 1 kΩ for the MSB. A DAC with R-$2R$ resistors uses 1-kΩ resistors for the R value. Compare the value of the largest resistor in each of the DACs.

7. A waveform generator needs to generate a waveform with eight points. Each point can be 1 of 16 values. The entire waveform must be repeated once every second.
   a. What size memory is needed?
   b. What clock rate is needed?
   c. How many bits must the DAC have?
   d. What contents should the memory have to make the waveform look like a square wave? Like a pulse that is full-scale 25 percent of the time and at 0 the rest of the time?

8. A DAC should put out 0, 1, 2, 3, . . . , 7 V exactly. Instead, it puts out the following values:
   a. -0.1, 0.9, 1.9, 2.9, 3.9, . . . , 6.9
   b. 0, 0.9, 1.8, 2.7, 3.6, . . . , 6.3
   What is the problem for these cases?

9. A specification says that a 10-bit DAC should be accurate to within 1 bit. The DAC is supposed to put out 10.00 V, but puts out 9.999 V. Does it meet the spec? What about an output of 10.2 V?

10. An integrating ADC is often designed to eliminate noise that is picked up in the ac line. This noise has a frequency of 60 Hz, so the integration period for full scale should be one 60-Hz cycle. For a 10-bit conversion, what should be the clock rate?

11. A signal is changing at 100 V/s. The S/H circuit is used to allow a slow converter to be used. If the converter is capable of 10 conversions per second, how much can the input change between conversions?

11

MICROPROCESSORS

The microprocessor is a very complex IC which provides many of the functions of a full-scale computer. The internal circuitry of the microprocessor consists of many of the elements studied to this point—registers, memory, buffers—connected in a way to allow the microprocessor to follow a program (a sequence of instructions). This program allows the microprocessor to perform the desired operation and yet use different sets of instructions depending on the results of previous operations. This flexibility in using a set of instructions, rather than fixed wiring, and exchanging exactly which instructions should be used, are what give microprocessor-based systems their power and versatility.

The processor implements (executes) the instructions in a precise way, with a sequence of individual substeps determined by its design. Different programming languages are used to develop the sequence of instructions required for the microprocessor application, and all the languages result is a series of 1's and 0's, the only language the IC itself can understand.

The microprocessor must also have a way to receive inputs from external devices and generate outputs to these devices, such as a printer or keyboard. Working with microprocessor systems requires both the wiring schematic and a memory map, a diagram of the contents of the memory associated with the microprocessor system.

Some applications use single-chip microcomputers, which are low to moderate performance microprocessors with many of the additional pieces for a complete system built into the IC. These single-chip microcomputers are often used for intelligent, low-cost control of electromechanical devices.

The microprocessor is often called a "computer on a chip," and that is an accurate description. The microprocessor is a single IC that contains many, and in some cases nearly all, of the individual functions that are needed to provide the features of a computer. The term central processing unit (CPU) is also used sometimes. The microprocessor became a practical IC device as the technology of designing and fabricating large ICs reached a point in the 1970s that tens of thousands of gates and boolean functions could be built on a single IC. To do this required that the companies that manufacture ICs develop new techniques of design, planning, and testing so that the ICs would work, could be made at low cost, and would not consume so much power that they would burn themselves up. Microprocessors are now available with differing capabilities to meet the specific needs of the many applications they are used in.

The features that are needed in a computer include:

- The ability to read instructions from memory, understand these instructions, and implement them.
- Ability to handle a wide range of instructions, including comparing numbers, performing arithmetic, and manipulating bits of a digital word.
- A decision-making capability, based on what the instruction says to do and the specific data that the instruction is examining.
- The ability to handle data (numbers) and store and retrieve these numbers under the control of the program of instructions.
- A way to input and output numbers, letters, and information to and from the system.

The microprocessor as a computer is in many ways nearly identical to larger computers which handle more data at higher speeds. The difference is that the microprocessor incorporates many of the separate elements needed to build such a computer onto the single IC. In fact, the exact functions of a microprocessor can be duplicated by wiring together hundreds or thousands of small- and medium-scale ICs. The parts and assembly cost of such a computer, along with the space and power consumption, would make such a computer impractical for most real projects. The applications of microprocessors are extremely wide ranging. In simple applications, the microprocessor is used to replace individual transistors, relays, and so on, at lower cost. In more complicated situations, the microprocessor allows functions to be introduced into products that could not have been before because too much circuitry would have been required.

An example of how a microprocessor allows new features to be added to an old function is the household thermostat. A standard thermostat is a mechanical device. Heat-sensitive metal strips cause electrical contacts to open and close as the temperature of the room goes above or below the desired temperature value. It is difficult to have this kind of unit allow for heating and cooling patterns to vary with the time of day (cooler at night, warmer at wake-up time, and moderate in midday when the house is often not occupied), or day of the week. A small microprocessor system can be used to provide these features and more. The microprocessor, via an ADC, measures the temperature. It decides whether to turn the boiler on or off by following instructions to compare the measured temperature with the temperature that the user has entered via a keyboard. But the microprocessor system can also have a clock to keep track of the time of day and can automatically lower the selected temperature by a few degrees at different hours. It can also be programmed to have different time and temperature patterns for different days of the week. As a result, the Saturday and Sunday temperature pattern can be different from the Monday through Friday pattern. This sort of flexibility and functionability is simply not possible with a mechanical or simple electrical thermostat.

Hard-Wired versus Programmed Logic

These terms are often used to describe the way a product function is implemented. In hard-wired logic, the function needed is obtained by wiring together the required logic gates. An example would be a counter, such as was built from flip-flops in Chap. 3. Another way to build a counter would be to have a microprocessor system programmed to add 1 to a number each time the microprocessor system saw that another pulse has passed (Fig. 11-1). This would be called *programmed logic*. The advantage of programmed logic is that many features can be added or changed, such as comparing the count against some maximum limit, averaging the count over several runs of pulses, and so on. It is then reasonable to ask why hard-wired logic is used at all. The reason is that for a given function, hard-wired logic is much faster in operation. For a microprocessor to do anything, there are dozens and often hundreds of small steps that have to be executed. This takes time. The hard-wired operation with ICs is nearly instantaneous. In applications where speed is not a problem, such as thermostats or traffic light control, the microprocessor would be used. However, many applications require much faster operation and so hard-wired logic may be needed. Often, systems have a combination of both—hard-wired logic where highest speed is necessary, and a microprocessor where less speed but more flexibility and features are desired. A hard-wired counter might be combined with a microprocessor; the counter does the actual counting, and the microprocessor reads this count and makes decisions and performs calculations based on the count value.

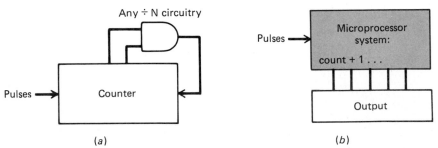

Fig. 11-1 Hard-wired versus programmed logic: *(a)* Shows a hard-wired collection of gates to count to a specific number; *(b)* Shows a microprocessor system which is programmed to add 1 to a count for each new pulse.

QUESTIONS FOR SECTION 11-1

1. What technical developments have made the microprocessor practical?

2. What three characteristics does a computer need?

3. How is a microprocessor similar to a larger computer?

4. How does a microprocessor-based system allow new features to be implemented in existing functions?

5. What is hard-wired logic? Give an example, and its pros and cons.

6. Give an example of programmed logic, and its disadvantages and advantages.

11-2 PARTS OF A PROCESSOR

A processor has many internal sections that all function in unison to enable the computer operations to occur. Each of these sections plays a critical role in providing the needed signals at the right time. They form the CPU of any microprocessor-based computer.

Most of these internal sections are not connected to the pins of the microprocessor IC and therefore do not connect to the external circuit. However, it is very important to understand how the microprocessor operates in order to effectively deal with processor based circuits. The entire approach to debugging and troubleshooting must take into account that these systems are structured very differently from those without processors. In a traditional system, such as a radio, the signal goes through clearly defined blocks or stages (Fig. 11-2). At each point it is

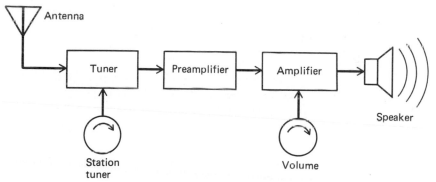

Fig. 11-2 Block diagram of a radio system. The signals flow from left to right, into and out of each block.

amplified, modified, or changed, but the signal can be examined going into and out of each stage. If the signal goes up to a stage but does not come out as expected, the problem is usually within that stage. Generally speaking, these stages are independent of each other.

In a microprocessor-based system, this independence is not the case. The functional blocks of the system—microprocessor, RAM, ROM, and support ICs—do

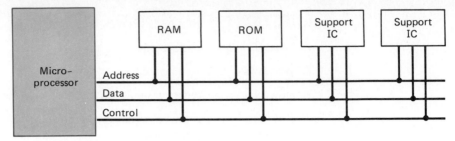

Fig. 11-3 A typical microprocessor-based system. Each section interacts with all the others, so correct operation of the system requires that all the others work properly.

not function on their own (Fig. 11-3). Each block requires the proper operation of the others to work. The processor must read instructions from ROM, implement them, initialize support ICs, receive inputs from I/O ICs, and so on. There is continuous handshaking between blocks, too. In many cases, it all works or nothing, or only a little works. This can make working with a nonfunctioning system very frustrating without the following three things:

- The proper test equipment.
- A test plan and instructions appropriate for the system.
- An understanding of the principles of operation and interaction of the various system components.

A block diagram of a microprocessor is shown in Fig. 11-4. The key elements are:

- The program counter, which develops the addresses which go onto the system address bus.
- The instruction decoder, which interprets the instructions that are fetched from memory and causes them to be implemented within the IC.
- Internal registers, which are used for storage of intermediate data when the microprocessor is implementing more complex instructions.
- An accumulator or ALU where calculations or operations on data (Add, OR, etc.) can be performed.
- Clocking circuitry, which paces the operation at precise intervals. Typical clock rates are between 1 and 10 MHz.
- Control and sequencing logic, which is the overall manager and coordinator within the IC and for external control signals.
- Buffers for the address, data, and control buses.

These elements can be seen in a photograph of an actual IC die (Fig. 11-5).

The simple operation of addition illustrates the many internal steps that are performed by the processor IC. The processor must successfully perform each of these just to add two numbers and produce a result. A memory IC contains the instruction that an addition is to be performed and also the data (the actual numbers to be added) for this example.

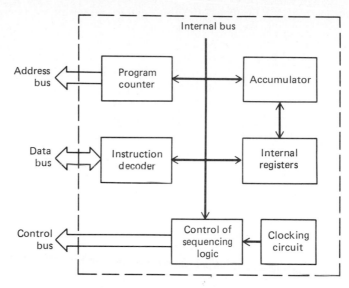

Note: Buffers for 3 buses are not shown.

Fig. 11-4 Block diagram of the key parts of a microprocessor IC.

Fig. 11-5 Photo of an actual microprocessor die (MC68020 type). The key function blocks in the silicon die are marked. *(Courtesy of Motorola, Inc.)*

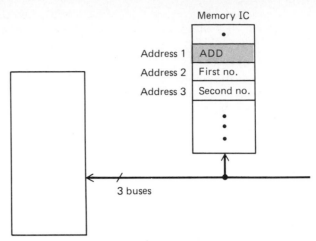

Fig. 11-6 The memory IC contains the instructions and the data. The program counter develops the addresses of the Add and the two numbers to be added.

Operation begins when the processor does a memory Read cycle at the address specified by the program counter (Fig. 11-6). The data brought into the processor on the data bus from this memory address is the binary code for the Add instruction. The instruction decoder recognizes this pattern, and it knows that the Add instruction needs the numbers to be added. The instruction decoder indicates this to the control logic. The control logic advances the program counter value by one, and causes another memory Read cycle to occur. The next address in memory is then read, which contains the first number to be added. This number is read by the processor and put into the accumulator.

The control logic advances the program counter again, so that the next memory address can be read. The second number to be added is read into the processor, and this second number is actually added to the first number in the accumulator. At this point, the sum of the two numbers actually has been computed, but it is located inside the processor IC, in the accumulator. It is not accessible to the computer user yet.

Another instruction is needed to get the sum out of the processor and to a location that is external to the IC. This might be a Move instruction, which specifies what data is to be moved, and to where. After the Add instruction is complete, the control logic once again advances the program counter, and the processor reads the next instruction from memory. A Move instruction would be used at the next location of memory to tell the processor that the number in the accumulator is to be taken out. The memory location after the Move instruction would specify the memory address to which the accumulator value should be moved.

The operational pattern repeats itself as the microprocessor performs the steps required for the application. The program counter advances, the next instruction is obtained by the processor from memory, additional data is also fetched from memory, and the instruction is completed. The program counter then advances to allow the next instruction from memory to be read.

If all the processor could do is read addresses in fixed, repeated order based on the value of the program counter, the processor would not be very powerful. It could only repeat the same steps over and over. But there is a very powerful alternative. Based on the results of a calculation, or comparison of numbers, the processor itself may decide to change the program counter value to a very different number so that a new section of instructions is read from memory. For example, an instruction could be executed which has this effect: ''Compare the next two numbers. If the first is larger than the second, continue fetching instructions from the next address. However, if the second number is larger than the first, begin fetching instructions from an entirely new block of addresses.''

This ability to make decisions and modify the flow of instructions is what makes processors and computers so powerful. They are not fixed in the steps they take— they can vary the path depending on the circumstances. The program is a map to follow—and the direction to be taken at the various forks in the road depends on the situation when the fork is reached. That decision on which fork to take is made by the processor itself, based on instructions it has been given.

QUESTIONS FOR SECTION 11-2

1. What are the key functional blocks in a microprocessor?

2. Why is obtaining proper system operation sometimes difficult? What three things are required?

3. Explain the role of each of these functional blocks in adding two numbers: the program counter, the instruction decoder, the internal registers, ALU, clock, and the control and sequencing logic.

4. What makes the concept of an IC which implements instructions so powerful? What can the program be set up to do to itself under specific circumstances?

11-3 PROCESSOR INSTRUCTIONS

There are many instructions that processor ICs are designed to follow. Some are very simple, while others involve many steps and memory cycles. The many types are needed in order for the processor to have the efficiency and flexibility needed in the various applications. The overall group of instructions available is called the *instruction set*.

In the previous section, the Add instruction was studied. The memory locations immediately after the Add instruction contained the two numbers to be added. This would be a limitation for most systems. If the system is a vending machine that is counting coins, then the instructions might be in a ROM, but the numbers to be actually added would be in RAM (having gotten to the RAM from a coin-identifying mechanism which determines the coin type and value). If the numbers are stored in PROM then the same pair of numbers will be added each time the Add instruction is fetched, and so nothing new is calculated. However, a system with the instruction in ROM and the numbers in RAM would be able to add whatever pairs of numbers came along. Somehow, the Add instruction has to be able to tell the control logic that the number in the next memory location is not the

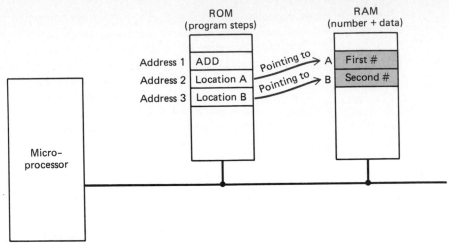

Fig. 11-7 The Add instruction where the data after Add points to other locations, which contain the numbers to be added.

actual number to be added, but the memory *address* of where the number can be found (Fig. 11-7). This simple extension to the Add instruction completely opens up the addition function. It takes more processor cycles to perform the Add, but the effect is that Add is now the key to a system that can add whatever numbers are required. The sequence would be as follows:

- Get Add instruction, of the type that indicates addresses of the numbers rather than the numbers themselves.
- Advance the program counter, get the address of the first number.

Note that the *datum* retrieved is a number that is an address value, and it gets used as such.

- Use that address as the new program counter value, and read in the first number to be added (which may be in a RAM location in another IC).
- Return the program counter to the address of the instruction, advance it again, and get the address of the second number.
- Use that second address as the new program counter value, and get the second number.
- Perform the addition, leaving the sum in the accumulator.

The instruction set for most microprocessors includes several types of Add instruction, which allow the address of the numbers to be added to be specified in various ways. The person writing the program for the application would use the Add instruction type that is most suitable. The processor instruction decoder would recognize each Add type as it occurred and perform the proper steps for that particular Add instruction. The types of instructions that the instruction set provides and that a processor can execute (in the various address modes) include:

- Arithmetic instructions, such as Add. The instructions can either leave the result in the internal accumulator, to be used by the next instruction, or they can put them in a specified location in external RAM.
- Branch and Jump instructions. These cause the processor to examine various conditions and cause the program counter to go to a new value if the condition is set. For example, there is an instruction to Jump if the number in the accumulator is 0. This would allow the program to perform some check on bits. If these bits represented the position of alarm switches, the intent of the program might be to Branch to a routine that says everything is OK if all bits are 0. If all bits are not 0, there is an open door or window, and the processor would try to find out which one it is.
- Binary operation instructions. These can cause all the bits in the accumulator to be shifted one position to the right or left, and also allow the processor to perform AND, OR, XOR, etc. If the processor was trying to determine which door or window was open, the instructions of the program would tell it to AND various masks with the bits from the door switches. It would then see if the result of this operation was a 0 or not.
- Move instructions. These allow data to be moved from external memory locations into the processor, and vice versa, and also allow the processor to be set up for the next desired operation on the data. A Move instruction then allows the result of the operation to be stored in external memory for later use.

All functions of a system with a processor are built up using these and similar very simple instructions. Fortunately, the final user of the system does not see each of these detailed instructions and all the steps that are needed to make even simple operations such as putting messages on a screen occur. However, a person troubleshooting a system or helping to develop the first mode of the system would see these instructions. The exact sequence might have to be checked with proper test equipment to make sure that each part was occurring at the right time and that the right data was being transferred.

Interrupts form a special group of instructions. An interrupt is the result of a special signal line to the processor working with a special set of instructions that the processor has been programmed to provide when the signal line goes active.

The manufacturer of the microprocessor IC defines a special location for the interrupt instructions. This location, which is a memory address or group of addresses, contains the interrupt vector, or location of the instructions to be executed when the signal line for the interrupt goes active. The sequence involves a sudden change in the program counter value, without losing the old value (Fig. 11-8):

- The processor is executing the regular, noninterrupt sequence. The program counter is advanced to bring in new instructions and data. If a Branch or Jump occurs, the program counter changes to reflect this.

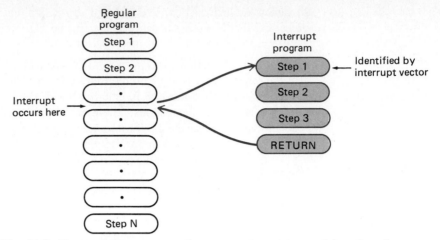

Fig. 11-8 How an interrupt causes the program to go to a special routine, then resume where it left off.

- At some point (which can be anywhere), the interrupt occurs. The processor sees this and saves the present value of the program counter in an internal register with the IC. It then goes immediately to the memory address associated with the interrupt (the interrupt vector) by substituting that address for the one the program counter was using just previous to the interrupt.
- The processor executes the instructions at the interrupt program addresses. The last instruction is a Return, which tells the processor that the block of interrupt instructions is completed.
- The processor then retrieves the program counter value that it saved when the interrupt occurred and resumes the execution of instructions from that address, as if the interrupt had never occurred.

The effect of all this is that the overall program can continue through the desired sequence of instructions, yet be ready and able to handle the special interrupt event when it occurs.

QUESTIONS FOR SECTION 11-3

1. What is an instruction set? What role does it serve?

2. What are the main groups (types) of instructions? What do they do?

3. What do Branch and Jump instructions do? Why are they needed?

4. What instructions are used for an interrupt? How does the interrupt scheme work? Why is it needed in many applications?

5. What is the interrupt vector? What does it do?

The complicated steps that are executed by a processor for each program instruction are not seen by the user of the system or even the person who writes the program. However, these steps cause the address, data, and control bus of the system to perform in precise sequences. It is important to understand a typical sequence because troubleshooting often requires examining this sequence with special test instruments.

For this example, consider a processor that has a 4-bit address bus, a 4-bit data bus, and two control lines, one indicating that a memory cycle is taking place and the other showing if it is a Read or Write cycle. The Add instruction will be studied, and this particular Add is one which specifies to the processor to Add the contents of one memory location to the contents of another and place the result in a third. For convenience, call these locations A, B, and C.

The four address bits allow the processor to address 16 memory locations. The first eight of these, 0000 through 0111, are in a ROM IC which contains the program. The second eight, 1000 through 1111, are in a RAM where the actual numbers to be added are located and where the result will be placed. The system clock provides a pacing signal to the processor. As the steps of the Add instruction are implemented, note in Table 11-1, page 310, how the necessary information (data) is taken from memory, and then used for subsequent steps. This data can be either the number to be added, or it can be the address location of the numbers.

Another common operation required for a microprocessor instruction is to move data from the internal accumulator to a memory location. This data could be the result of a calculation, or it could be setup data for a support IC which is connected so that its internal registers look like memory locations. The Move sequence, for example, might be instructed to move data from the accumulator to location 1011. The Move itself might be instruction 0011, and in this example it is located at address 1000. The operation sequence begins when the processor program counter generates address 1000, and the processor does a read at that address as is shown in Table 11-2 on page 311.

Note that each instruction really requires two major activities by the processor. First, the processor must fetch the actual instruction and any data associated with it. Then the processor must actually execute, or implement, the instruction. This process of fetch and execute continues with each step of the program of instructions (Fig. 11-9). As soon as the execution of the instruction is completed, the

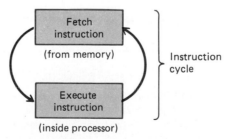

Fig. 11-9 A complete instruction cycle consists of 2 parts: *(a)* The instruction and associated data is fetched from memory. *(b)* The instruction is executed inside the processor.

TABLE 11-1
PROCESSOR STEPS TO IMPLEMENT ADD INSTRUCTIONS

Clock Pulse #	Program Counter (Address Bus)	Data Bus	Memory Cycle	R or W	Description
1	0000	1011	Yes	R	1011 is code for this type of Add. Instruction decoder recognizes this.
2	0001	1000	Yes	R	Address of location A is obtained from next ROM location.
3	1000	0011	Yes	R	The number at A is read. Note that the location was data in the previous step, and now is the address value.
4	0010	1001	Yes	R	Address of location B is obtained.
5	1001	0001	Yes	R	The number at B is read. As in step 3, the location was the data in the previous step.
6	Not relevant	Not relevant	No	Don't care	The actual addition is performed. The address and data buses have no meaning at this cycle.
7	1011	1010	Yes	R	Read address of location C.
8	1010	0100	Yes	W	The result is written out to location C.

processor goes and fetches the next instruction. The location of the next instruction may change depending on the results of the execute just performed, of course.

These particular sequences are developed by the instruction decoder working with the control logic within the microprocessor. Microprocessors from different manufacturers have similar instructions but they differ in specific details. Therefore, this sequence is typical and shows the types of steps, but when working with a particular IC the data sheet or manual for that IC must be used.

TABLE 11-2
PROCESSOR STEPS TO IMPLEMENT MOVE INSTRUCTIONS

Clock Pulse #	Program Counter (Address Bus)	Data Bus	Memory Cycle	R or W	Description
1	1000	0011	Yes	R	0011 is code for Move instruction. Decoder recognizes this.
2	1001	1011	Yes	R	Address of destination is obtained from next location.
3	1011	Accumulator contents	Yes	W	Accumulator contents are written to 1011.

It may seem that there is no point to examining the many steps that make up a single instruction. In general, when everything is working properly, this is true. However, when parts of the system malfunction, there are cases where instruments will be connected to all the address, data, and control lines, and the actual execution of each instruction will be monitored. This type of troubleshooting is used when the malfunction is subtle—it may be that a single bit in memory is bad. For example, the effect could be that the processor should have fetched the Add instruction 1011, but instead got a Move instruction 0011. The program would then perform a Move rather than an Add, and would soon crash, or be unable to function in the way it was supposed to.

In the detailed example studied, the address was 4 bits, and the data was 4 bits. Microprocessors are available today in 4-, 8-, 16-, and 32-bit versions of address and data. More address bits allow more memory locations to be accessed. This would be useful for applications with very long programs or that had to store and retrieve large amounts of data. Additional data bits provide two features. Larger numbers can be stored and handled, which allows for more precision in a single operation. Also, the number of instructions available is 2^N where N is the number of data bits. This is because the instruction itself is a data word stored in memory. More instructions allow the microprocessor designer to design some very powerful single instruction into the overall instruction set. Some advanced instructions allow a number to be shifted, added, and compared with a single instruction statement, rather than three individual ones. This means that the programmer's job is easier, and the number of program steps required is shorter (and less program memory is used).

QUESTIONS FOR SECTION 11-4

1. Do all microprocessors use the same instruction set? The same steps within the instruction? Why is this normally not a concern to the user?

2. What are the steps that the microprocessor must go through in order to perform a simple Add of two numbers?

3. Why do we study these instruction steps, since they are not normally seen by the user?

4. Explain how and why the information on the data bus may be an address value.

5. Microprocessors are available with what numbers of data bits? What is the advantage of more data bits? More address bits?

6. What are the two steps that a processor implements as it processes an instruction?

PROBLEMS FOR SECTION 11-4

1. A program is designed to begin at address 1000 and end at 1100. The interrupt routine begins at 1110 and ends at 1111. The interrupt vector is stored at location 0010.
   a. What is the value stored at location 0010?
   b. Where is the Return instruction stored?
   c. The program counter is at 1010. The processor gets an interrupt. What two addresses does the program counter next indicate?

2. An Add instruction is used to add 3 to a number in RAM. This particular Add instruction, coded as 1000, says: Add the next number to the number in memory at the location specified (in this case 1100). What does the instruction sequence look like?

3. A Move instruction moves the contents of the processor ALU to a memory location specified. If the Move is 0111, and the location desired is 1000, what does the sequence look like?

4. How many memory locations can a 4-bit (address) microprocessor directly access? Repeat for 8 and 16 bits. What is the largest integer number that can be directly represented by a 4-bit (data) microprocessor? Repeat for 8 and 16 bits.

5. An Add instruction is designed to add the contents of location A to the contents of location B and place the result in location C. If A = 1000, B = 1010, and C = 0111, which of these addresses will appear as data on the data bus? In what order?

11-5 PROGRAMMING LANGUAGES

The microprocessor IC receives its instructions in the form of patterns of 1's and 0's. This digital pattern is the only format it can understand. However, a person trying to develop a program for an application, or trying to troubleshoot a system, would have a difficult time writing the program using 1's and 0's. There is too much room for error, confusion, and misinterpretation when using long lists of binary numbers.

A program to add two numbers in this machine language and then put the result in another location for later use might look like this:

Program Steps in Memory	Comment
1011	Perform addition numbers whose locations are given next.
0010	Location of first number.
0011	Location of second number.
1010	Move result to location given next.
1100	Location where result is to be stored.

This method of programming is clearly impractical for anything but a very short and simple program.

It is important to understand the concept of machine language programming because the contents of the memory IC used with the processor IC are just these binary patterns, and they may have to be examined as part of the debugging and troubleshooting routine. But practical systems need a better way of writing the program instructions, and so the binary machine language is rarely used to write programs. This new method would allow the programmer to refer to instructions by their names rather than the binary pattern that they present to the processor IC.

Assembly language is the programming language that is one step removed from the actual bits. In assembly language, the programmer can refer to memory locations by names and can also call numbers by name rather than the memory location of the number. For example, the location where the result of an addition is to be stored can be called ''sum'' and then each time this value is needed, the programmer simply refers to it by this name. A typical assembly language program might look like this:

Assembly Language Statements		Comments
ADD	A,B	Add numbers called A and B
MOVE	SUM	Move the results to sum

The programmer types in the assembly language statements, but they cannot be run as such. They have to be converted to the machine language format. A special program called an *assembler*, often provided by the manufacturer of the microprocessor IC, is used to translate the assembly language statements into the machine code of 1's and 0's that the processor actually will see (Fig. 11-10a).

Assembly language is easier to write than machine code. It is also a little more readable and can produce programming statements that run efficiently and effec-

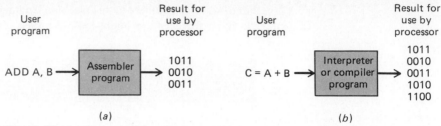

Fig. 11-10 *(a)* The assembler takes assembly language statements and converts them to binary processor code. *(b)* The interpreter or compiler does the same for high level languages.

tively for the application. However, it still can be awkward and error-prone since the programmer has to use a very un-English-like language and keep track of many small details of the program. The assembly language program translates on a line-by-line basis to machine code, so the format of the assembly language must follow all the characteristics of the instructions that the processor can handle. This can be a drawback in that a program written for one processor could not be used for another processor type, since the instruction sets of the processors are different.

The languages that most systems use are a step above assembly language. These are referred to as *high-level languages*. The most common such languages are BASIC, FORTRAN, and Pascal. In these languages, the statements are much closer to English and are much more convenient for writing and checking, especially as the program gets longer. A program in BASIC to add two numbers would be simply:

C = A + B   Which takes the numbers A and B, adds them, and calls the result C.

The high-level language format also requires a special program, provided by the manufacturer of the system or an independent company, to translate the BASIC, FORTRAN, or Pascal statements to machine code. This translator is called a *compiler* or *interpreter*, depending on how it functions (Fig. 11-10*b*). A single line of high-level language program translates to anywhere from 5 to 20 lines of machine code. The advantage of the high-level language is that it is easier to write and produces the same result as many lines of much harder to write assembly language program. It is also "transportable" from one type of processor to another, since the compiler or interpreter takes care of translating the statements into the precise machine statements for the specific processor IC. One disadvantage of high-level languages is that they result in code that is not as efficient as assembly language. The resulting machine code has more lines than assembly language (and thus requires more memory) and also is slower to execute the overall application since there are more steps to go through. However, in many applications the cost of the additional memory is small compared to the ease of using the higher-level language, and the speed is not a problem since faster processor ICs are available.

All languages, whether high level, assembly, or machine, must eventually produce a sequence of binary groups. These groups are the actual instruction codes and information that the processor retrieves and executes.

QUESTIONS FOR SECTION 11-5

1. What do the instructions that the processor sees actually look like?

2. Why is programming in the form that the microprocessor actually sees impractical? Why is it a problem for someone testing a system?

3. What is assembly language? What are its advantages and disadvantages? Give an example of an assembly language instruction. How is assembly language converted to the machine code format of the processor?

4. What is a high-level language? Give an example of a statement in a high-level language. What are the three advantages? The disadvantages? How is the high-level language translated into the machine code?

11-6 MICROPROCESSOR I/O

The capabilities of a processor IC are useless unless there is a way to get external information into the IC and get results out. The most convenient way is to use the appropriate support ICs, discussed in Chap. 6. The method used to electrically interconnect the microprocessor with the support IC is selected by the performance, cost, and technical requirements of the product. Each of the various methods requires a slightly different approach in debug and troubleshooting activities.

Most processor–support IC interfaces use one of these designs:

- Memory mapped, where the support IC appears as a small block of memory to the processor.
- Programmed, where the support IC is a special device addressed in a unique way by the processor.
- Direct memory access (DMA), in which the flow of data between the support IC and the system memory bypasses the processor itself.

In concept, the simplest interconnection is the memory-mapped I/O. An address decoder identifies the specific address or addresses that are assigned to the support IC. The processor software reads from and writes to the support IC as it would read or write any data in a memory location using the address, data, and control lines. For example, to read data from a counter IC, the program would simply execute a Read cycle, at the address of the counter IC. Troubleshooting of memory-mapped input and output uses many of the same tools and techniques as checking memory ICs: Processor memory Read and Write cycles are initiated, and the data sent and received is checked. Memory-mapped I/O requires a processor Read or Write operation for each data word, and so it is relatively slow for transferring large amounts of data.

Programmed I/O is used with processors that can indicate to the rest of the system if the current processor cycle is a memory cycle or an I/O cycle, via special control bus lines. Programmed I/O does not use any of the memory addresses of the system. The processor does I/O by executing special I/O instructions. These cause data to be put onto the data bus (for output) or read from the data bus (for input), but the control bus lines do not indicate a memory Read or Write, so the memory ICs ignore the data. In this way, programmed I/O can transfer data to or from support ICs under the control of the user program.

The programmed I/O can identify different ICs by the address that is placed on the address bus by the processor. The support ICs can be uniquely addressed and not overlap with the memory ICs of the system (Fig. 11-11).

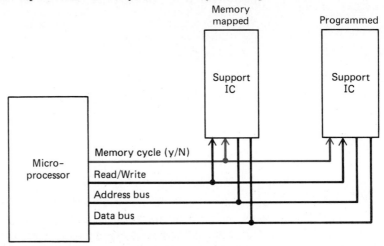

Fig. 11-11 Memory-mapped and programmed I/O. For memory mapped, the processor indicates a memory cycle. For programmed I/O, it indicates a non-memory cycle.

Direct Memory Access I/O is a technique for achieving very fast data transfer between a processor and an external device. A typical application would be the transfer of large groups, or blocks, of data between a floppy disk and a computer system. The goal is to get all the data from the disk and into the memory of the computer as quickly as possible. The processor does not need to look at the data at the time of transfer. Therefore, if there were a way to transfer the data with minimum processor interaction, high total transfer speeds could result. This is in contrast to memory-mapped or programmed I/O, where each data word is handled by the processor itself and must pass through the processor (which acts as a bottleneck).

DMA I/O operates by using special lines of the microprocessor IC and a support IC called a DMA controller (see Fig. 11-12). These ICs work together to identify

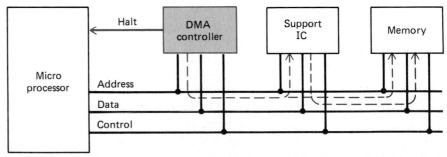

Fig. 11-12 Direct memory access. Once DMA is set up, the DMA controller halts the microprocessor and controls the address and control lines to send data directly to-from memory. The addresses are supplied by the DMA controller, and the data is never seen by the microprocessor.

where the data is to come from (or go to) and the number of data words to be transferred. Once the data transfer is set up, the processor is put on hold and is effectively out of the transfer path while data flows between system memory and the IC that is supplying or receiving data (Fig. 11-12).

DMA operation details are complicated, and following a DMA transfer requires studying the operating cycles of the processor, DMA controller, and other circuitry in the system.

QUESTIONS FOR SECTION 11-6

1. Why does a system need I/O capability?

2. What are the three types of I/O that a microprocessor can have? What are the characteristics of each? The advantages and disadvantages?

3. How is I/O checked when testing a system?

4. What is the signal line difference between memory-mapped and programmed I/O? What is the instruction difference from a software perspective?

5. What is DMA? Why is it used? Why may it not be used?

6. Why is DMA very different from the two other types of I/O?

11-7 THE SYSTEM MEMORY MAP

An electronic system that does not have a microprocessor can be completely described by its schematic, or wiring diagram. This schematic shows all the components of the circuit (ICs, resistors, capacitors, connectors, indicator lamps, etc.), their values or model numbers (i.e., 74LS20, 1 kΩ), and the wired interconnections between all these components. Using this schematic, a technician or engineer can follow the signal flow through the circuit and find out what is working, or what is not working and why not.

Systems with processors also require a schematic, but that is not enough. There is another critical diagram needed to allow the circuit to be studied, called a *memory map*. This map shows all the addresses that the processor can address and what type of memory component, function, or data is located at that address.

A typical memory map is shown in Fig. 11-13 on page 318. This system has a timer IC and a serial I/O support IC, along with its own processor memory. This processor has an 8-bit address bus, and so can address from 0000 0000 to 1111 1111. The map shows the following:

- Locations 0000 0000 to 0001 1111 are ROM, made out of two smaller ROMs. The first location, 0000 0000, contains the starting address of the program, in this case, 0000 0010. The microprocessor IC is designed internally so that when

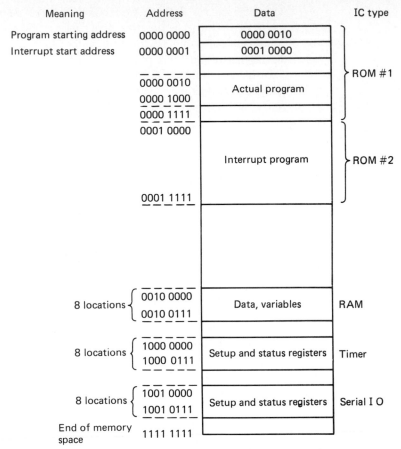

Meaning	Address	Data	IC type
Program starting address	0000 0000	0000 0010	ROM #1
Interrupt start address	0000 0001	0001 0000	
	0000 0010	Actual program	
	0000 1000		
	0000 1111		
	0001 0000	Interrupt program	ROM #2
	0001 1111		
8 locations	0010 0000 / 0010 0111	Data, variables	RAM
8 locations	1000 0000 / 1000 0111	Setup and status registers	Timer
8 locations	1001 0000 / 1001 0111	Setup and status registers	Serial I O
End of memory space	1111 1111		

Fig. 11-13 The memory map of a typical system. The map relates the program addresses with the function and physical locations of these addresses. (Note unused locations).

power is first applied, it reads the contents of memory address 0000 0000, and then begins to execute the instruction at the address stored in that location. This provides a means of making the system start properly.

- Location 0000 0001 contains the interrupt vector, in this case 0001 0000. When an interrupt occurs, the processor begins executing the special interrupt-handling routine located at this address.
- The ROM contents from 0000 0010 to 0000 1000 contain the actual steps of the program. These steps are executed by the processor and cause the processor to implement the desired application.
- Locations 0010 0000 to 0010 0111 are RAM, physically built from a single IC with eight locations. It is here that the processor stores numbers that are being processed, compared, or the results of calculations, as required by the program.
- Locations 1000 0000 through 1000 0111 are for the timer support IC. It is these addresses that the processor must use to set up or read the timer as required by the program.

- Locations 1001 0000 through 1001 0111 are for the serial I/O support IC, which the processor uses to receive data from the user keyboard and send data to the user terminal screen. The rest of the locations are not used in this design.

The system decoder IC is responsible for monitoring the processor address lines and generating CE signals to the correct IC (ROM, RAM, timer, serial I/O) based on the address and memory map (Fig. 11-14). The system designer wires or

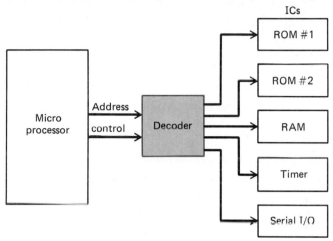

Fig. 11-14 The decoder points to the various ICs, in accordance with the memory map, by looking at the address lines.

programs the decoder to do this, depending on whether the decoder is built from logic gates or if programmable gates such as programmable logic arrays are used. Then, as the processor generates an address, the appropriate IC is enabled automatically.

How the Memory Map Is Used

The memory map is a guide for running tests on systems. As an analogy, if you are looking for someone in a strange town, you need two things: a street map, and their house address. Similarly, if you are trying to test a timer IC, you need to know where it is physically connected, so you can put test probes on the correct points, but you also need to know the memory map address of the timer (assuming memory-mapped I/O) in order to access the timer registers, and test their operation. The memory map tells you where to find that IC.

For example, if the system seems to be taking keyboard inputs properly but appears to be scrambling numbers, it may be that single RAM IC was bad and changing stored bits. The test technician would use a RAM test program, look at the memory map to determine what processor addresses each RAM contains, and then run the test over these addresses. If the system had 4 RAMs from 0000 0000 to 0011 1111, the ICs would cover addresses as follows:

IC Number 1: 0000 0000 to 0000 1111
IC Number 2: 0001 0000 to 0001 1111
IC Number 3: 0010 0000 to 0010 1111
IC Number 4: 0011 0000 to 0011 1111

The memory map would identify the IC by number and show for what addresses it is decoded.

Maps for Programmed I/O

If the system uses programmed I/O, each support IC has an address just as for memory-mapped situation. This address is in the I/O space of the processor, which is accessed when the processor does a programmed I/O cycle. The address lines identify a unique location, while the processor control bus indicates that processor I/O cycle is occurring. Just like the memory map, the I/O map shows these addresses to aid in troubleshooting.

QUESTIONS FOR SECTION 11-7

1. What is a memory map? How does it differ from a schematic?

2. Why is a memory map needed when working with a system?

3. How is a memory map used when troubleshooting a system?

4. How is the memory map information converted in the circuit to actually activate the appropriate ICs? What circuit component does this?

5. What method is used to make sure the processor begins its program in the right place on power-up?

PROBLEMS FOR SECTION 11-7

1. A 4-bit microprocessor is used in a coffee vending machine. The first eight memory locations are ROM, the next two are RAM, and the last two are a parallel I/O support IC. Sketch the memory map showing what ICs are at what addresses. What addresses are unused?

2. For the above example, the program starts at the third ROM location. The first address is the start-up address location indication, and the second is the interrupt vector. The interrupt routine starts at the fifth address. Add this into the memory map.

11-8 A TYPICAL MICROPROCESSOR

Microprocessor ICs are available from many manufacturers. These ICs have varying numbers of data lines, address lines, and capabilities, in order to meet the different cost and performance needs of the system in which the microprocessor is installed. Different microprocessors have similar, but different instruction sets and more or fewer instructions of one type or another (arithmetic, bit manipulation, I/O, etc.).

One popular microprocessor is the MC6802 manufactured by Motorola (Fig. 11-15). This 40-pin IC is considered an 8-bit microprocessor because it handles data as 8-bit units and has an 8-bit data bus. The address bus of the IC is 16 bits wide, which allows $2^{16} = 64K$ individual addresses to be addressed by the IC. The control bus has seven lines, which allow the microprocessor to indicate to the rest

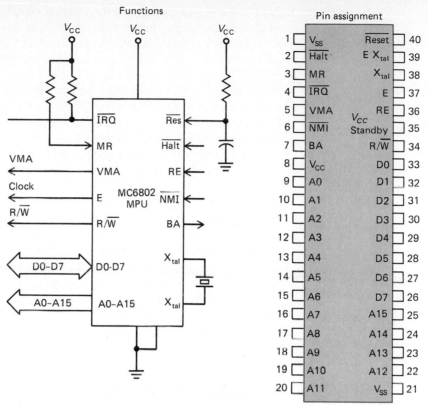

Fig. 11-15 The MC6802 microprocessor has 8 data lines, 16 address lines, plus control lines, in a 40-pin package. *(Courtesy of Motorola, Inc.)*

of the circuit key operational details and receive some priority signals from the circuit. All the signal lines of this chip are TTL-compatible, although the IC is fabricated using MOS technology. It has the capability of handling several types of interrupts. A +5-V supply is required.

The MC6802 accepts a square-wave clock signal of anywhere from 1.0 to 4.0 MHz. A faster clock means that the chip operates at higher rates but also requires faster, more expensive ICs in the rest of the system to keep up with it. The IC signal lines also include:

Address bus: Sixteen lines which are outputs from the MC6802.
Data bus: Eight 3-state lines used for transferring data to and from the MC6802. The lines are three-state so that the data bus can be turned off at the MC6802 as required for DMA and other special functions.
Control bus: These lines are varied in function, and may not be used in all applications:

- Halt, which is used to stop the microprocessor so that the state of all lines can be examined if required for test purposes. This is sometimes connected to a special test switch in the system, which the test technician uses.

- Read-Write, which indicates to the external circuitry if the memory cycle is a Read cycle (when the line is high) or a Write cycle (when the line is low).
- Valid memory address, which tells the external circuitry if the address on the address bus is a valid address, rather than the result of non-memory access cycle.
- Bus available, to tell the external circuitry that the processor has halted as a result of the halt signal being applied. This is used as part of the DMA sequence.
- Interrupt request, used by any external device to signal that it would like to interrupt the processor and the present program step being executed. The processor will normally see this request and respond to it. However, the processor can be programmed by the programmer to ignore this request under special circumstances.
- Memory ready, a signal to the processor IC which can slow down the processor memory access cycle. It is used when the access time of the processor memory ICs is not as fast as the processor would normally require, which occurs when a slower memory IC is used for various cost or technical reasons. The processor expects to read data within a certain amount of time, but the memory ICs may not have it ready. Special circuitry associated with these slower memory ICs signal the processor not to look at the data bus until an additional clock cycle has passed.
- Enable, a handshake line used in conjunction with the memory ready line to show the special slow memory circuitry that the processor has received and acknowledged the memory ready line.

There is a special interrupt line, called nonmaskable interrupt (NMI), which is similar to the interrupt request line discussed above, but which cannot be ignored (masked) by the processor program.

It is the proper interconnection of these many lines to the rest of the circuitry, and the correct operation of all these lines, that causes the microprocessor and its system to work properly. If any line is not working properly, for any reason— short circuit, power problems, bad IC, or severe noise—the processor, and subsequently the system, will not work as intended.

QUESTIONS FOR SECTION 11-8

1. What are the control lines of the Motorola MC6802?

2. What are the functions of these controls lines?

3. What is the nonmaskable interrupt? How does it differ from the regular interrupt?

4. What does the system designer do when the circuit must use memory ICs that are too slow for the normal processor cycle speed?

PROBLEM FOR SECTION 11-8

1. A MC6802-type processor is using a faster memory IC and a slower memory IC. The fast memory is at memory location 0000 through 0111, and the slow memory is at 1000 through 1111. Sketch a block diagram of the control lines used for this system.

11-9 MULTIPLEXED PROCESSOR BUSES

The address and data buses of the MC6802 are physically separate. They do not have circuitry that they share with each other. In some designs, this can result in so many bus lines that a large part of the circuitry outside the microprocessor is devoted to buffers and support ICs for these buses. This can increase the cost of the product and the actual amount of circuit board space that is required for the final circuit.

To overcome this potential problem, some microprocessors use multiplexed address and data lines. This means that the address and data lines share the same pin of the IC, which reduces the number of buffers and so on, needed in the design. The Intel family of microprocessors is a good example of this. The three main elements of the family have varying degrees of bus multiplexing.

The Intel 8085, like the Motorola MC6802 provides an 8-bit data bus and a 16-bit address bus. The Intel IC, however, shares the data bus and eight of the address bus pins (Fig. 11-16, page 324). The external circuitry is able to sort out address information from data because the microprocessor indicates on a separate control line if the signals on the bus are address or data signals. As a result of this multiplexing, the Intel ICs have extra available pins in the 40-pin package which are instead used for additional special control and handshake lines.

The next step in the Intel family is the Intel 8088 microprocessor (Fig. 11-17, page 325). This IC has a 16-bit address bus, and a 16-bit data bus. The data bus bits are shared with the address bus bits, and the data bits are shared even further: they are put in two groups of eight and multiplexed with each other. Thus, eight of the address lines are also used for 1 byte of data, and then for the other byte of data. Once the two data bytes are inside the microprocessor IC, they are handled as a single 16-bit-wide word. This multiplexing scheme makes maximum use of the fact that data bus buffers must be bidirectional and so require more ICs for buffers. Once again, special control lines indicate to the system if the signal on the IC pin is address, data byte number 1, or data byte number 2.

One disadvantage of the multiplexing method is that the overall efficiency of the microprocessor system is lower than with a nonmultiplexed design. Additional clock cycles are needed to transfer address and data signals—they cannot be done all in parallel since the physical pin of the IC can only represent one signal at a time. To meet this problem, the Intel 8086 is available. This microprocessor is nearly identical to the 8088, except that the 16 data lines are not multiplexed into 2 bytes but are instead shared directly with the 16 address bits (Fig. 11-18, page 326). This requires more external ICs but fewer clock cycles, and so the Intel 8086 is faster than the 8088. Otherwise, they are identical and even run the same program instructions without any change. The difference is strictly in the hardware and is not seen by the system user, except that 8088 systems are slower than 8086 systems for a given task. In some applications, where the processor is doing many calculations on data already in the IC, the difference caused by the multiplexed data bus is small, around 10 percent. For applications where there is a lot of data

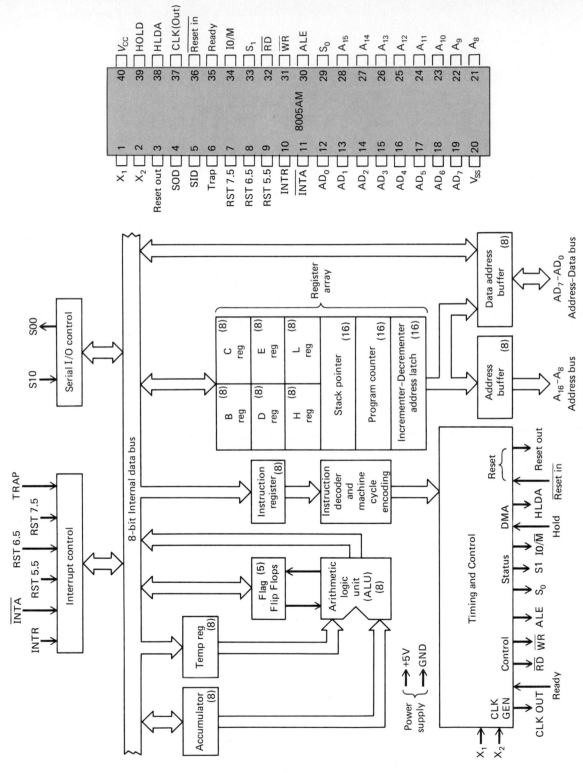

Fig. 11-16 The Intel 8085. Note pins 12 through 19 are both 8 data lines and 8 address lines. Pins 21 through 28 are address lines only. *(Courtesy of Intel Corp.)*

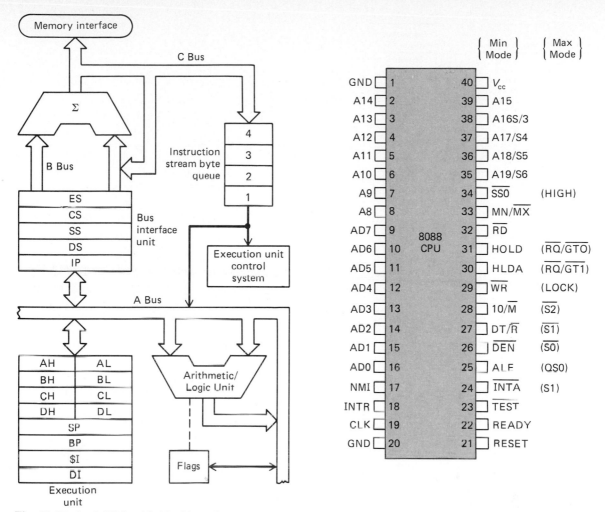

Fig. 11-17 Intel 8088, with 16 address lines and 8 external data lines. Internally, data is handled as 16-bit words, but the data is presented as 2 bytes multiplexed with address lines 0–7. *(Courtesy of Intel Corp.)*

transferring to and from memory, the observed difference may be as high as 50 percent.

From the test and debug perspective, the effect of multiplexing is to complicate the connection of test probes and make it more difficult to interpret the meaning of observed signals. The signals seen on an oscilloscope are constantly changing and may represent completely different functions (address or data values at different times). The control lines indicating what cycle the processor is in— address, data (byte 1 or 2 in the case of the 8088) must also be observed simultaneously with the actual address and data lines. Fortunately, special test equipment is available for most microprocessors to make the physical connection and interpretation job much easier.

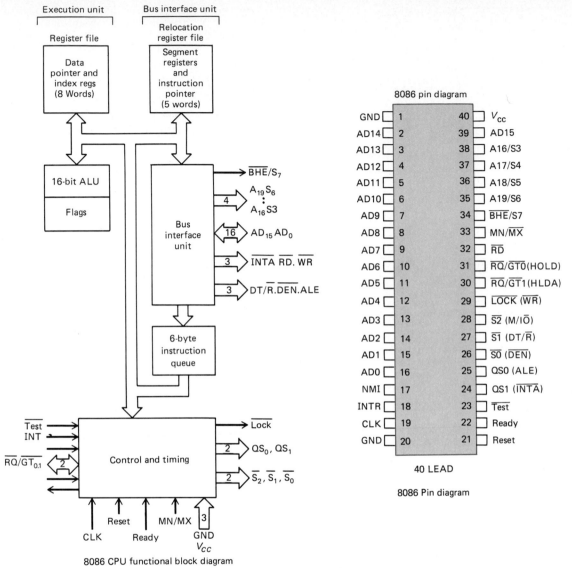

Fig. 11-18 Intel 8086 is identical internally to the 8088. Externally, the 16 data bits are available multiplexed with the 16 address bits, pins 16 through 2 and 39. *(Courtesy of Intel Corp.)*

QUESTIONS FOR SECTION 11-9

1. Why is microprocessor signal multiplexing needed?

2. How does the processor indicate what function (address or data line activity) is occurring?

3. How does the Intel 8088 do its multiplexing?

4. What is the effect on total program execution speed between the 8088 and 8086?

5. What is the drawback of multiplexing from a test viewpoint?

PROBLEMS FOR SECTION 11-9

1. A microprocessor has multiplexed address and data lines. As a result, every Read or Write data access takes twice as long compared to nonmultiplexed. Using the Add instruction, show how this increases the number of clock cycles needed for completing the instruction. What is the difference in execution time of the instruction for each case?

2. If the Intel 8086 did not have multiplexed address and data lines, how many pins would the IC require?

11-10 SINGLE-CHIP MICROCOMPUTER APPLICATIONS

A typical microprocessor-based system requires many more ICs than just the microprocessor itself, since the microprocessor provides only the CPU functions. A complete functioning system also needs memory (both RAM and ROM), bus buffers, a timer-counter, I/O (serial and/or parallel), and decoders for the many ICs. This is true even if the actual amount of computing or processing to be performed is small. The disadvantages of using many ICs for a limited function are the power consumed, cost, physical space required, and reliability. The situation is most obvious in the thousands of applications where the power of the microprocessor is not used for a complete computer (keyboard, CRT screen, disks, printer), but for an intelligent controller which is dedicated to doing one job very well and giving superior performance in that one application, such as motor control, temperature control, or data collection and reporting.

An example shows this more clearly. A very common function is to control temperature so that it stays near the desired value. A standard thermostat does this. The user sets the desired temperature (called the *set point*) and then the thermostat constantly measures the actual temperature. When the actual value is above the set point, the heater (boiler, furnace) is shut off. When the actual value is below the set point, the heater is turned on. While this scheme is very simple, it has a major drawback: The actual temperature will swing widely around the set point. This is because the heater is turned either fully on or off, with no in-between state (Fig. 11-19, page 328).

The result is that either full heat is pumped out, or no heat, and the actual temperature usually goes up and down. When the heater is on, the actual temperature usually goes beyond the set point before the thermostat can shut it off. This overshoot is not only uncomfortable, but it wastes fuel. In many industrial applications, the actual temperature must be controlled to a very close value, typically within 1°, if the final product being manufactured is to come out properly. The on-off control of the simple thermostat would not be enough.

A more complicated alternative is to use a temperature-controlling device (con-

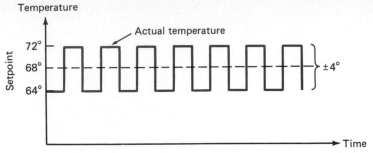

Fig. 11-19 On-off control results in overshoot and large errors, although the average temperature agrees with the set point.

troller) which can turn the heater not just on or off, but also partially on. This is called *proportional control,* where the amount of heat provided is proportional to the difference between the set point and the actual measured temperature. The controller constantly compares the two values. If the difference is large, more heat is called for; if the difference is small, less heat is supplied. The amount of heat provided is thus reduced as the actual measured temperature approaches the desired set point. As a result, the actual temperature does not overshoot the set point, but approaches it smoothly and then stays within a fairly close value of it (Fig. 11-20). The situation is similar to making a car accelerate from 30 to 55 mi/h.

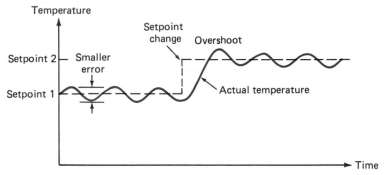

Fig. 11-20 Proportional control has a smaller error and does not overshoot much. Note the smooth transition to a new set point.

First, the driver steps on the gas for maximum acceleration. As the car gets closer to the 55 mi/h speed, the driver lets up gradually on the gas pedal so that the car approaches the 55 mi/h speed without overshooting to 65 and then settling back to 55. A smart controller can also take into account the different heating patterns for different days of the week, or keep the total fuel consumption below some value.

QUESTIONS FOR SECTION 11-10

1. Why can a microprocessor IC not function by itself?

2. What else is needed in a real system application?

3. What is the difference between a controller and a computer?

4. What is the drawback of a simple on-off control?

5. What components are needed for a microprocessor-based controller?

6. What is the relative processing load on the microprocessor and other components in a controller versus that in a general-purpose computer system?

7. What is a microcomputer? Why is it called this?

11-11   A TYPICAL SINGLE-CHIP MICROCOMPUTER

Some low-cost applications require the capabilities of a microprocessor but not large amounts of memory, I/O, or support ICs. This led to the development of the single-chip microcomputer. This IC contains the functions of a microprocessor, and also memory, counter-timers, and I/O, built onto a single die. As a result, the cost for a complete system is very low. The complete IC is called a microcomputer because so many of the individual functions of a computer have been combined into a single IC. Some microcomputers have moderate-speed, 8-bit ADCs included on the IC.

Single-chip microcomputers are being used in thousands of applications, many of which could not afford a microprocessor system otherwise. They are providing equivalent features and new features at a cost much lower than the circuitry they replace. These applications include controllers for motors in robot arms (where the acceleration must be controlled carefully so the arm driven by the motor travels to the final position as fast as possible without overshooting the target); burglar alarm systems (where many points must be checked and a few relatively simple decisions made); and systems used for collecting data from cash registers in a store on a daily basis for later transmission to a main office computer.

The Intel Corporation makes a family of single-chip microcomputers which comes in several varieties, so that the user can select the particular model best suited to the application in terms of cost, time to get the first unit on the market, and other factors. This group, referred to as the 8051 family (Fig. 11-21, page 330), shares the following characteristics:

- Operation from a single +5 V supply.
- 128 × 8 RAM.
- Four 8-bit ports, providing 32 I/O lines.
- Two 16-bit counter-timers.
- Serial communications channel.
- A wide range of instructions in the instruction set.
- A single crystal provides all that is needed for the clock oscillator.

The family has three members, with the only difference in the nature of the memory available:

- The 8031 has no memory built into the IC except the 128 × 8 RAM. The user provides all the memory external to the microcomputer.

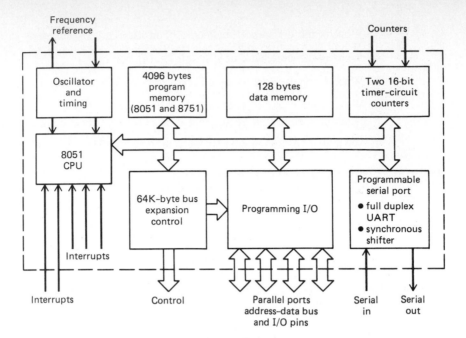

Block diagram

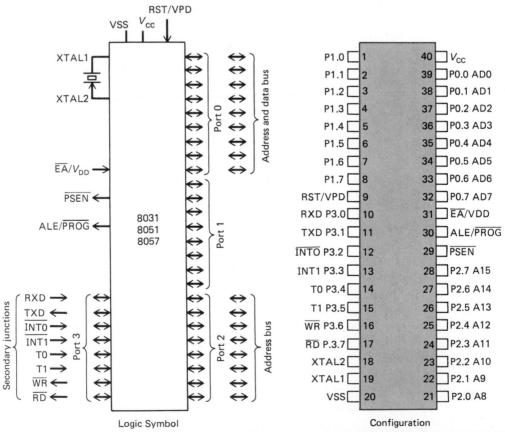

Logic Symbol

Configuration

Fig. 11-21 The Intel 8051 family. Note the various memory and support functions in the IC, along with the CPU. The 8031/8051/8751 differ only in memory configuration. *(Courtesy of Intel Corp.)*

- The 8051 has 4K × 8 of ROM programmed by Intel to user specifications.
- The 8751 has 4K × 8 of EPROM built into the IC.

The last member of the family is relatively expensive compared to the 8051, but the long production start-up time required for a mask-programmed ROM is avoided. The 8051 is economical when the production quantity is very large. The 8031 has the most memory flexibility, since the memory can be any combination of RAM and ROM, but this means that at least one (and often more) external memory IC will be needed.

The electronic pieces needed for a proportional controller as discussed in section 11-10 are:

- A microprocessor to read the actual temperature and calculate the difference between it and the set point. The microprocessor then generates a control output proportional to this difference.
- An ADC to convert the signal from some temperature sensor to digital format for the processor.
- A DAC to take the digital output from the system and provide an analog output for the heating system control.
- A numeric keypad and I/O support ICs, so that the user can enter the set-point temperature and any other important system information.
- A ROM IC, to contain the operational program for the microprocessor.
- A RAM IC to hold the values that are changing or can be changed, such as the set point, the actual value from the ADC, and the results of calculations.
- A clock IC to provide the clock signal to the microprocessor.

The system using these ICs would be relatively costly and large. And yet, the amount of calculation and I/O activity of the system is very low. The temperature is read about once per second, compared, and a few calculations are performed. The I/O need is very small, since the user is entering new values infrequently. The amount of data to be stored in RAM is also low, typically less than a few hundred bytes. Even the program in ROM is small, about 1K to 4K bytes of instruction. The circuit built out of individual ICs would be overbuilt for the application.

The entire controller could be built from an 8051: user keyboard interfaced to an 8-bit port and a DAC for the controlling output also interfaced to another 8-bit port (Fig. 11-22, page 352). The ADC function could be built at low cost by using one of the counter-timers to implement a low-speed ADC using the integrating technique that some ADC ICs use. The result would be a very low cost controller with more than enough performance capability for the application. The serial port could be used to let the temperature controller accept set points not only from the user keyboard but also from another computer system, and to let the controller report the actual temperature to another computer for recordkeeping and energy-monitoring studies.

The final product built around the single-chip microcomputer is small, inexpensive, and reliable. Different capabilities could be provided in different models, and the only difference in the models might be the actual program used with the 8051. The actual physical circuit boards and components could be the same—the software would provide the difference.

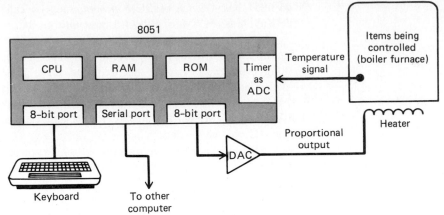

Fig. 11-22 The 8051 microcomputer as a temperature controller for a furnace or boiler. The user has a keyboard for entering the set point only. The serial port allows communication to other computers.

Troubleshooting Single-Chip Microcomputers

The very low number of components in a single-chip microcomputer-based system means that there are few components to fail, and the failure is usually identified quickly. The commonsense approach to troubleshooting usually works:

- All power lines and clock signal must be good.
- Does the system totally refuse to operate or does it work but provide incorrect responses, such as maintaining a wrong temperature?
- All control lines into the IC should be at the proper state so that the IC can actually run.
- Incorrect operation (not totally stopped) could be caused by a bad memory or incorrect inputs reaching the IC from user keyboard, or an ADC (if one is used in the system).
- Most products provide a test routine which the service technician can access via special key sequences. These routines run limited tests on memory and I/O.

QUESTIONS FOR SECTION 11-11

1. What is the difference between the Intel 8031, 8051, and 8751? How would the choice be made between these?

2. How can a microcomputer do analog-to-digital conversions if an ADC is undesirable or too expensive?

3. What could the serial port built into the microcomputer be used for in a typical controller application?

4. Give three examples of how a microcomputer-based controller can provide new cost and performance advantages over the previous method of doing the application.

5. What is the production advantage of a microcomputer-based system? Are there any disadvantages?

6. When testing microcomputer-based products, what is checked? What is the difference in testing procedure when checking these circuits versus microprocessor-based product?

PROBLEMS FOR SECTION 11-11

1. Sketch a speed controller for a car using the 8051. What is measured? What is controlled? What is the user set point?

2. Sketch a microcomputer used in an alarm system with eight doors and windows, a user panel, and an interface to a central computer at the main station.

3. A microcomputer temperature controller operates as detailed in Sec. 11-10. Identify three different failure symptoms and suggest their causes.

SUMMARY

The microprocessor, or computer on a chip, provides many of the functions of a full-scale computer in a single IC. Some additional ICs are needed to make a complete system: memory, I/O, counter-timers, and other support ICs.

The microprocessor cannot function unless it is told exactly what to do. The program, provided by the system user or the product developer, does this. While the processor understands only the 1's and 0's of the digital world, the program may be written in machine language, assembly language, or even a higher-level language such as BASIC or Pascal. Microprocessors are available with a wide range of capabilities and in 4 through 32 data bit sizes. The address bus size determines the maximum number of memory locations for program and data. Different systems use different connections and operations for the input and output that is essential in any system. Memory-mapped, programmed, and DMA are the most common types of I/O.

Troubleshooting a microprocessor-based system often requires an understanding of the correct operation, special instruments, and the memory map, along with the usual schematic. The memory map shows the address of each key element of the system, so that decoders, memory, and support devices can be exercised and checked. In general, troubleshooting and debugging these systems are not easy and are complicated by the many steps that the processor IC executes in order to implement even the smallest instruction. Multiplexed signal lines complicate the situation.

The need for low-cost and low to moderate performance applications often require the use of the single-chip microcomputer, which integrates many system functions onto the same IC as the microprocessor. The single chip is especially useful in dedicated applications where the overall function of the product is fixed and not changed by the product user. Such systems are highly reliable and have only a few points to be checked when problems occur.

REVIEW QUESTIONS

1. What are some features of a computer, whether large or small?

2. What is the difference between a larger computer and a microprocessor?

3. What are two of the many reasons that microprocessors are being designed into new or existing products?

4. How is a microprocessor-based system fundamentally different than one made of hard-wired logic in terms of what it can do? In terms of finding operational problems?

5. Why is hard-wired logic still important?

6. What system pieces must work in unison for the microprocessor-based system to work properly? What does each piece do?

7. What causes the program counter to step to the next value? To change to a very different value? What is the importance of this section?

8. Give three examples of instructions from a typical instruction set.

9. Explain how instructions which refer to addresses of data rather than the data make processors more powerful.

10. Why are interrupts important? What happens when the processor gets one?

11. Does the system user see the instruction set? Why is it important to understand the instruction set?

12. How many processor clock cycles are involved in a simple instruction like Add? What do most of these cycles do?

13. What are the differences between machine code, assembly language, and high-level languages? Why are all three still used?

14. What is processor I/O? How is it different from system I/O?

15. What is the similarity between memory-mapped I/O and programmed I/O? The difference?

16. Why is DMA very different from memory-mapped or programmed I/O?

17. Why is the memory map as important as the schematic when working with processor-based systems? What does it show?

18. How does the system decoder make use of the memory map information?

19. How does the microprocessor program of instructions get started at the right address?

20. Explain how faster and slower memory and support ICs can be used in the same system.

21. What is microprocessor multiplexing? Why is it done? How does it differ from ADC multiplexing?

22. What is the advantage of multiplexing signal lines? The disadvantages?

23. Why are single-chip microcomputer ICs used? What individual ICs can they replace? Where can't they replace microprocessors and related ICs?

24. Give three examples of products that could benefit from a single-chip microcontroller.

25. What can a service technician check in single-chip microcomputer-based products?

REVIEW PROBLEMS

1. A microprocessor system is designed so that location 0000 contains the program starting address and location 0001 has the interrupt vector. The program is at address 1000 through 1110, and the interrupt routine occupies addresses 1100 through 1110 of the program space.
   a. What data is stored in locations 0000 and 0001?
   b. The program is about to execute an instruction at 1010 when an interrupt occurs. What is the next value of the PC? What is the value after that?
   c. What is the last program counter value of the interrupt routine? What is the program counter value after that?

2. Draw a memory map of the system described above. Use binary numbers and also hex numbers for addresses.

3. A system has the following components:

   A processor with 16 address bits and 8 data bits
   Start-up address located at 0000 0000 0000 0000
   256 bytes of ROM at 0000 0000 1000 0000, made from two 128-byte ROMs
   512 bytes of RAM at 0000 0010 0000 0000, using two 256-byte RAMs
   A counter-timer with a 4-byte register at the end of memory

Draw the memory map. Use addresses, and show all IC starting and ending addresses.

4. A processor is doing two Add instructions. The first involves adding the next two numbers in memory. The second involves adding numbers at the locations specified at the next two addresses. The result of both Adds is that the answer is left in the microprocessor. For each case, how many memory cycles are involved? How many are Read cycles and how many are Write cycles?

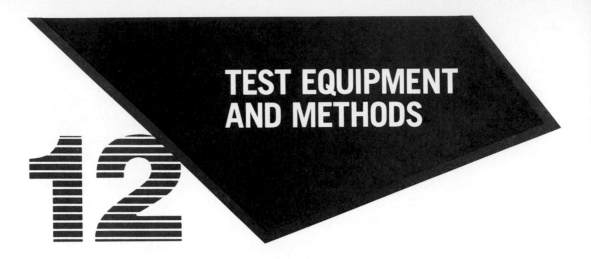

TEST EQUIPMENT AND METHODS

Electronic equipment is normally reliable, but when repair is needed, there are different kinds of instruments available. Finding out the cause of a problem and localizing it from the system is often called troubleshooting. In the field, easy-to-use instruments are available to perform the necessary troubleshooting. Once the trouble is localized, the system can then be repaired. Debugging a product under development needs more flexible but complicated instruments, since the nature of the problems are not fully understood and the system design is not finalized.

Test equipment used includes the digital multimeter, logic probe and pulser, oscilloscope, logic analyzer and personality module, and signature analyzer. Microprocessor systems use different types of test programs for checking out part or most of the system on start-up, or as requested by the test technician. More flexible but harder-to-use test programs are used for products under development.

Hard permanent faults are found using established procedures and plans. Intermittent faults, which come and go, do not have a clear pattern. These faults are located by trying to make the fault more regular, or identifying an underlying pattern of the fault.

12-1 THE TEST SITUATION

Electronic equipment, if properly designed and built, is normally very reliable. Electronic systems contain very few moving parts, so the wear and tear associated with the motion of mechanical pieces does not occur. A typical computer system has hundreds or thousands of ICs. All of these must work properly for the system to work, and they usually do.

The high reliability of digital ICs—meaning the tens or hundreds of thousands of hours that a typical IC will work without a failure—means that it is practical and safe to build a system with several thousand ICs and expect it to guide an airplane, compute bank accounts, or control medical equipment. This high reliability can be seen by a short calculation: If the probability (or chance) of a single

IC failing was even a very low 0.01 percent, then a typical system with 1000 ICs would fail in a few hours. The actual excellent performance of such a system is a testament to the reliability of modern ICs and their circuits.

All this does not mean that electronic systems do not fail—they do. The role of the service engineer is to find what part failed and fix it. The repair may involve replacing the actual defective component, or the circuit board with the component. The service engineer may also be responsible for trying to find the cause of the failure, especially if it is well before a normal failure should occur. The operating environment of the circuit may be causing these premature failures: high temperatures beyond the specified range for the ICs; vibration, which can cause ICs to work loose from their sockets or cause any other mechanical interconnections to fail; or severe voltage spikes on the circuit power-supply lines caused by nearby motors, relays, or ac line problems, which might damage the relatively delicate IC.

The service engineer has to troubleshoot the failed equipment away from his or her office, since it may not be practical to "send it back to the shop." This type of troubleshooting requires that the service engineer carry all the necessary test equipment and spare parts. It also usually requires that the repair be quick, to minimize the inconvenience of downtime to the customer. Troubleshooting and repair in the field means that the service engineer must be well prepared, since a trip back to the office would be time-consuming and costly.

In order to make in-the-field troubleshooting practical, the service engineer often has clearly outlined steps to follow in a service manual. In many cases, specialized test equipment is also used. This equipment is designed to run specific tests on the failed system and give clear indications of what may be wrong. This test equipment usually provides special connections (such as loopback for serial I/O, see Chap. 8) and also the software needed to interact with the test equipment (in the case of microprocessor-based equipment). Most digital systems designed today contain some level of built-in test equipment (BITE) provided by the factory and accessible to the service engineer. This can include program routines to test memory or the CRT screen, for example.

Along with BITE, the most common strategy used by equipment manufacturers is to have the service engineer identify what has failed (and possibly why it has failed) and replace the entire circuit board containing the failed part. This "board-swapping" method is used instead of component replacement for several reasons:

- It is quick.
- It is more practical than unsoldering and replacing a single component on a crowded circuit board, often under less than adequate conditions at the customer's site.
- It would be impractical to carry all of the dozens or hundreds of ICs and other components that a typical system uses. It is more practical to carry the dozen or so circuit boards.
- The replacement boards are factory-built and tested, and so are guaranteed to be working if in the original factory bag. On the other hand, defective individual components sometimes accidentally get mixed in with good ones at the service center, where components are borrowed and returned.

There is another type of work that is performed by electronic technicians that has some similarities to troubleshooting but also many differences. When a new system is under development in the engineering department of a company, first units called *prototypes* must be built. These are then tested to see if the system performs as expected. Often, it does not. There may be an error in the design, or a mistake in how the prototype was wired. The wrong component or a defective one might be installed. Sometimes the prototype works intermittently, and this may be done to a subtle but important point that was overlooked in the design. For example, the access time for a memory IC may be just barely adequate because the designer made a mistake in the calculations of the memory IC speed required. The circuit may work, depending on the exact temperature or power-supply voltage, both of which have a small but definite effect on IC speed.

The electronics technician who builds and tests the prototypes performs a system debug, to get the malfunctions (bugs) out. This technician has some advantages over the field service engineer performing troubleshooting:

- Since the work is done at the engineering development lab, it is usually not a problem to have lots of test equipment available if needed.
- The work is not done while an irritated customer waits; so the pressure is less.
- More spare parts are available.

However, there are some definite drawbacks, too:

- Since the design is new, there can be many problems in the same prototype. These are sometimes subtle and tough to find.
- The test procedure, special test equipment, and BITE equipment probably do not exist yet.
- The operation of the new circuit is often clear only to the designer, and even he or she may have areas that are unclear.
- There may be pressure to get the prototype working so the overall project will be on schedule.

In summary, the tools, techniques, and operating environment for repair of existing equipment are sometimes different from those for new circuits just designed. This chapter will discuss what equipment is used for both and why.

QUESTIONS FOR SECTION 12-1

1. What are three causes of system failure?

2. What are troubleshooting and repair? What are two characteristics of these?

3. What does the service engineer need to perform troubleshooting?

4. What is the repair strategy for equipment in the field? What are three reasons for it?

5. What is debugging? How does it differ from troubleshooting?

6. What are the good points of debugging versus troubleshooting? The bad points?

Some test equipment is so basic that it is used in all aspects of both troubleshooting and debugging. The digital multimeter (DMM) is one such instrument (Fig. 12-1). [It is sometimes called a digital voltmeter (DVM), and while this is not completely correct, it is a common description.] The DMM is used for measuring voltage, current, or resistance in a circuit.

Fig. 12-1 A typical DMM used for checking a circuit board. The user selects voltage, current, or resistance readings. *(Photo Courtesy of John Fluke Mfg. Co.)*

The DMM shows the user the voltage, current, or resistance value with a digital readout, usually up to three or four digits (1.392, 0.475). By using a switch or moving the DMM wires (called test leads) the user selects which of the three physical parameters is to be measured. When measuring voltage, the test leads are connected to the two points of the circuit (one is usually circuit ground, or 0) (Fig. 12-2a). The DMM readout shows the voltage value. For measuring the flow of current, the two DMM leads must be inserted on the path of a wire in the circuit (Fig. 12-2b). For resistance measurements, the two leads go across the component whose value is to be measured or between two points in the circuit as directed by the troubleshooting plan. (Fig. 12-2c). (*Note*: Power must be off when measuring resistance! It is meaningless and dangerous to measure it with power on.)

The virtues of the DMM are:

- It is small, easy to carry, and physically rugged.
- It is simple to use—there are few buttons to push.

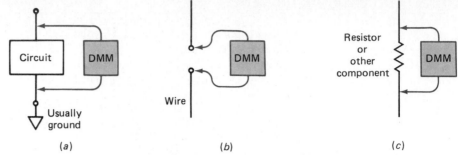

Fig. 12-2 Connection of the DMM for measuring *(a)* Voltage, *(b)* Current, and *(c)* Resistance.

- It runs on internal batteries. This means that no ac outlet is required (a real convenience). Also, it reduces the possibility of electric shock by high voltage flowing from the system being tested through the DMM and back to the power-line.

The DMM is used in the preliminary checks of a circuit. The first thing always tested is the power supply to the circuit—is the voltage value right? If it is not, the circuit may work intermittently or not at all. Most digital ICs allow a power-supply tolerance of ±5 or 10 percent around the ideal value. The digital readout on the DMM quickly shows if this specification is being met.

The resistance function of the DMM allows the technician to check that there are no breaks (called "opens") in a wire or circuit board by verifying that the resistance is low (usually less than 10 Ω). It also helps follow the path of wires or find short circuits, where two wires are touching when they should not be. The resistance function also verifies the condition of mechanical devices such as switches, which have low resistance when on and a very high resistance (typically over 100,000 Ω) when off.

Despite all these features, the DMM is limited in what it can do. Real digital systems have signal lines that are rapidly going between the two logic-level voltages. When connected to these lines, the DMM shows meaningless numbers because it is a relatively slow instrument that makes only a few readings per second. It also shows the value of only one physical point in a system at any time, which makes it difficult to compare values of different points at the same instant.

The DMM is a vital instrument for what it is designed to do, which is to show the value of a voltage, current, or resistance that is relatively steady. Its ease of use, small size, and freedom from a power cord make it an essential first instrument.

The DMM provides a numerical readout, but does not actually show the electrical signal itself. The oscilloscope ("scope") actually draws a picture of the electrical signal on a CRT screen as the signal occurs (Fig. 12-3). The horizontal axis of the screen represents time and the vertical axis is voltage value (Fig. 12-4). The circuitry of the oscilloscope sweeps the CRT beam from left to right with time and moves the beam up and down proportional to the voltage on the oscilloscope test leads (probes). The result is a picture of the signal that is constantly updated as the sweep repeats.

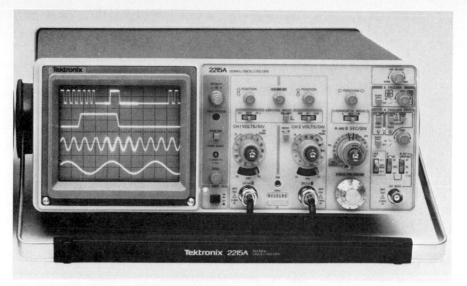

Fig. 12-3 A typical 2-channel oscilloscope. Note the large number of panel switches and knobs the user must set. *(Photo Courtesy of Tektronix, Inc.)*

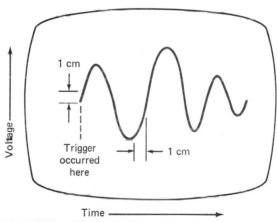

Fig. 12-4 A waveform displayed on the scope screen. The horizontal axis is time; the vertical axis is voltage.

Controls on the oscilloscope allow the user to select the rate at which the CRT beam sweeps. Typically, the range is quite wide, from microseconds per centimeter for very fast signals to seconds per centimeter for very slow ones. There are also adjustments for the number of volts per centimeter on the vertical axis, from about 10 mV/cm to 10 V/cm.

The oscilloscope is an extremely powerful instrument. With it, the exact nature of a signal in a circuit can be studied. Any deviations from the ideal logic 0 and logic 1 values will be seen, as will any disturbance or noise. Oscilloscopes are designed with two input channels to show two signals at the same time. This is used for comparing signals that must have some sort of relationship to each other, such as CE and Read-Write on a memory IC.

The scanning of the CRT is synchronized to the input signal by special triggering circuitry in the scope. This ensures that the display is steady and starts at the proper instant.

In a typical application, a scope might be used to debug a memory circuit in a microprocessor system. The symptom might be that most memory locations read and write without error, but one block of memory locations operates with occasional errors. After checking the power voltage to the memory ICs, the technician would put the scope probes on the signal lines (address, data, control) at the memory ICs to see if the logic 1 and 0 levels were OK, and if they were clean (without noise or distortion). It is important to connect the ground wire of the oscilloscope probe at the IC being checked to ensure that the scope is showing the voltage as it is received at the IC.

To this point, the technician has checked only single-signal lines and not the relationship (timing) between signals. The two-channel capability of the oscilloscope can be used for this. One channel is connected to the CE of a memory IC, and the other is used to investigate other signals from the microprocessor. The address and data lines should be at the correct 1 or 0 value by the time the CE goes active (low, for most ICs). The Read-Write line should also be at its correct value. For ICs, the CE acts as a signal indicating "everything is ready, you can turn on now." If signals arrive after CE goes active, the IC will be confused and incorrect operation will result (Fig. 12-5). Once the technician sees a signal that arrives late,

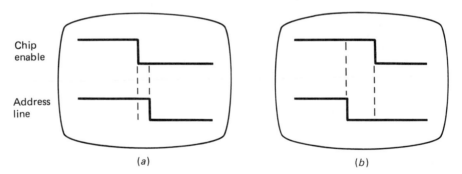

NOTE: Dashed lines are for illustrative purpose only. They do not appear on scope.

Fig. 12-5 The two channels can be used to examine relative timing. *(a)* The address line goes active after CE. *(b)* The address line goes active before CE. (Note: The dashed lines are for illustration only. They do not appear on the scope screen.)

he or she could follow it back to its source with the scope probe. A typical cause might be a signal buffer that was failing and introducing excessive propagation delay.

Most oscilloscopes have two input channels, and some are available with four or even eight. Various triggering modes are also built into the instrument. Some models have provisions for specialized plug-in modules better suited for certain signal types, such as ones that change very quickly.

The most common disadvantages of the oscilloscope are a result of its flexibility. The many controls and switch settings of an oscilloscope result in long setup times and user mistakes. It is sometimes difficult to get a steady, proper display on the screen. Finally, the interpretation of what is on the screen is not obvious, and misinterpretation (which results in incorrect or misleading conclusions) is a frequent occurrence.

The versatility and power of the oscilloscope make it a superb test instrument. But like many very good "all-around" performers, there are some specific test applications which are handled more quickly and efficiently by specific instruments designed for use with digital ICs. The simplest of these are the logic probe and pulses.

QUESTIONS FOR SECTION 12-2

1. What is a DMM? What does it measure?

2. What are the virtues of a DMM?

3. Where is a DMM used? In what part of the test sequence?

4. A power supply is supposed to be 5 V ±5 percent. Is 4.95 V in the good range? What about 5.4 V?

5. Where is the resistance function used besides measuring the actual resistance value of a component?

6. Where do the DMM leads go for voltage measurement? For current measurement? For resistance measurement?

7. What are two limitations of the DMM?

8. What does the oscilloscope show? What does the screen of the scope represent?

9. What are some of the adjustments the user of the scope must make?

10. Why does a scope have two or more input channels?

11. What does triggering do on a scope?

12. How is a scope used in a test situation? How would two channels be used together?

13. What are some of the drawbacks of a scope?

12-3 LOGIC PROBES AND PULSERS

Once the system power supply has been checked, the test technician will usually follow a test plan which involves checking selected points in the system for two things:

- The logic state of IC inputs and outputs.
- To see if there is a change in logic level. This can be a change from a 0 to 1 (or 1 to 0) or a momentary pulse, where the level goes to the opposite state and back to the original state.

The first requirement could be met by a DMM, and often is done that way. However, the DMM digital readout presents too many digits to a technician who wants to rapidly check many points for a simple high or low indication. The second requirement could not be properly filled by the DMM, which reacts too slowly for short pulses to be seen.

The logic probe (Fig. 12-6) is a very simple test instrument that looks like an oversized pen. It has an indicator light that is off (dark) for logic 0 signals, and on

Fig. 12-6 A typical logic probe. Two wires (not shown) connect to the power and ground of the system under test. *(Photo Courtesy of Global Specialties, Inc.)*

(bright) for logic 1. Circuitry within the probe senses pulses even as short as 20 or 50 ns, and stretches this to blink the indicator light for about 0.5 s. The probe usually gets its own power from the system under test.

When using the probe, the tip is touched to the lead of an IC, component, or circuit board track or wire. A quick glance at the light shows the logic level, and a blink shows that a pulse has occurred. The probe is used to compare gate inputs and outputs, for example, to see if they agree with the boolean operation the gate is supposed to perform. The probe could also be used to see if pulses are reaching a counter IC, or if a CE is getting to a support IC.

Many logic probes also offer a "babysitting" feature. Suppose a signal line is supposed to be low, but the symptoms suggest that it may occasionally be pulsing high. Seeing this with the probe would normally require watching the indicator for minutes or hours, which is clearly impractical. In the babysit mode, the indicator does not blink when a pulse passes, but latches and stays on until reset manually by the user. This means the probe can be connected to the circuit, and the technician can go away. If any pulse occurs, the indicator will unfailingly show this fact even hours later.

The test instrument which complements the logic probe is the logic pulser. The pulser is shaped like the probe and also draws its power from the system under test. The function of the pulser is to generate logic 0 or 1 pulses. These can be used in many parts of a circuit: as inputs to counters, to clear or preset flip-flops, and to reset sections of the circuitry.

The logic pulser has a single button which the user pushes to cause a pulse. The pulser puts out a momentary logic 0 pulse, and a momentary logic 1 pulse. The user does not have to decide which is needed, since a logic 0 pulse on a point that is already at 0 has no effect; only the logic 1 pulse is seen by the IC (and vice versa). The pulser automatically drives a point to its opposite state.

In many situations, a pulser and probe are used together. If the technician suspects that a signal buffer is faulty, he or she could use the pulser at the buffer input and the probe at the output. Each time the pulser button is pushed, the probe shows a pulse has occurred. If it does not, the buffer may be faulty. Some advanced pulsers let the user get more than a single pulse for each push of the button—pulse groups of 1, 10, or 100 can be selected. This is very handy for checking the operation of counters, timers, or decoders: the pulser generates the pulses to be counted, and the probe is used to see if the counter has counted up to that number properly.

A note about the names of signals: In digital systems, all signals are either at logic 1 or logic 0, of course. However, it is also necessary to indicate to someone testing the system whether a 0 or 1 means that the signal line is active. Does a 0 on an IC CS actually select that IC, or does a 1? Is Read indicated by a 0 and Write by a 1 for a memory IC, or vice versa? There are two commonly used methods of showing this vital information on the schematic diagram of the system.

The first method is to put a bar over the signal line definition to show that it is logic 0 when active. For example, $\overline{CS}$ means the IC is selected when the line is a logic 0. Read-Write means the signal is a Read when at logic 1 and a Write when it is a logic 0. The second method is to put the letters H and L after the signal line definition: CS L, or Read H–Write L. In either case, the test technician can tell from the signal definition what the proper logic state of the observed point should be.

QUESTIONS FOR SECTION 12-3

1. What is a logic probe? For what is it better than a DMM? Why?

2. How is a logic probe used?

3. What is babysitting? Why is it needed?

4. What is a logic pulser? Describe two places it is used. Why is it used?

5. Why does a pulser put out both a logic 1 and logic 0? Why does this work properly?

6. How are the logic probe and pulser used together?

7. How are logic signals named? What two ways designate that a signal is active low? Why is it important to know?

12-4   LOGIC ANALYZERS

The power of the oscilloscope and the simplicity of the logic probe, pulser, and DMM provide effective tools for troubleshooting and debug. In many situations, however, they provide too much information that is not needed and not enough of the type that is required to solve the problem. Additional electronic instrumentation is therefore available that is better for working with larger digital circuits and microprocessor-based systems.

The logic analyzer is an instrument designed for overcoming the weaknesses in simpler logic probes and harder to use, multichannel oscilloscopes. The need for the logic analyzer comes for these reasons:

- A typical digital system, especially if it contains a microprocessor, may have 16, 32, or even 64 points that have to be observed simultaneously. A regular oscilloscope simply does not have the number of channels.

- Interpreting a screen that is showing many channels is difficult. It requires concentration and mistakes are often made, especially when there is pressure on the person using the instrument.
- In many cases, the details of the shape of the signal are not really important. What is important is seeing when the signal is high or low, and the relative position in time of the many signals with each other.
- Finally, a conventional oscilloscope can show only what is happening from a defined instant in time onward. In other words, the user defines the desired trigger conditions for the sweep of the CRT, and then sees events that occurred after the trigger. However, in many test situations there is a need to look back in time, to see what events led up to the problem. A logic analyzer can do this. The logic analyzer looks like an oscilloscope (Fig. 12-7). It has a CRT screen, front

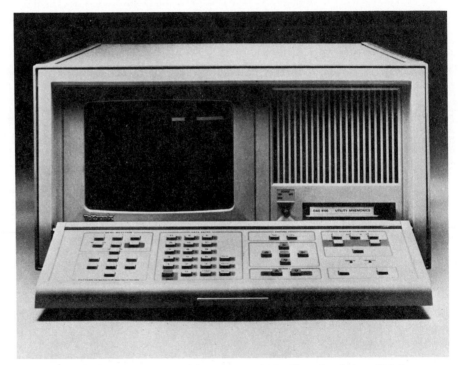

Fig. 12-7 A logic analyzer can show the timing of many key signals in a digital computer system (shown here), or the binary values and instructions for microprocessor lines. *(Photo Courtesy of Tektronix, Inc.)*

panel switches and controls, and connectors for the test probes. But the operation is quite different. The logic analyzer is designed for digital signals that can only have two values, and this is what it shows on the screen. It does not show the exact shape of the signal as it goes from a high to a low. Instead, it shows clean, sharp transitions (Fig. 12-8). This makes it much easier to see the logic state of the many signals at the same time. In fact, the logic analyzer does not have a complicated control for the input signal range, since the only ranges needed are for TTL levels and for CMOS level (if they are not TTL-compatible).

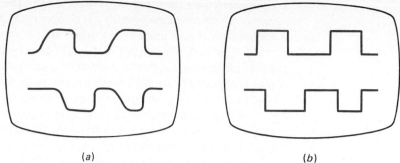

(a)   (b)

Fig. 12-8 Digital signal waveforms on (a) Scope and (b) Logic analyzer. The scope shows more detail, but is harder to interpret.

The logic analyzer also makes use of the clock signal which is present in every microprocessor-based system and all synchronous digital systems. This clock can be used to synchronize the operation of the logic analyzer with the system being tested. The result is that only the desired parts of the signals are shown on the screen.

The kinds of signals that are brought into a logic analyzer can be address lines, data lines, control lines, interrupts, inputs and outputs of gates and flip-flops, and any other key point in the circuit. The analyzer can be used to show the address lines of a microprocessor in order to see which steps the program executes. It can be used to show both the address and the data in order to see not only the program steps but also the program data that is written or read at each step. There are three features that really magnify the power of the logic analyzer:

- It has the ability to use many inputs as qualifiers. These are signals that do not need to be displayed but will control which signals are displayed.
- It can store the states of the signal lines and show them going backward in time to see how the system behaved up to the point of malfunction.
- It can show the information to the user as a series of groups of 1's and 0's rather than as waveform shapes. For some aspects of debugging, this is more convenient and practical. Hexadecimal format can also be selected, which is often used to represent digital values instead of binary (Fig. 12-9).

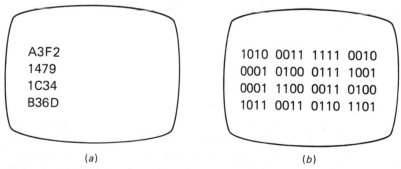

(a)   (b)

Fig. 12-9 A logic analyzer screen showing logic states in (a) Hex format and (b) Binary format.

Look at each of these features in more detail. The qualifiers are used to select precisely which microprocessor cycles or operations are of interest. For example, if the problem centers around the memory ICs, it may be necessary to observe all memory Read and Write cycles. The qualifiers would be connected to the microprocessor line which indicates a memory access, and then only memory cycles would be captured by the logic analyzer. Going a step further, the problem may be related to data being written to memory properly, but not being read back error-free. In this case, another qualifier would be used to show only memory cycles that are Reads, or Writes, and the logic analyzer would show the signal states only for the memory reads. Logic analyzers usually allow the qualifier lines to be ANDed and ORed as desired by the user, so a wide variety of conditions can be set up. By having these qualifiers, the screen of the analyzer is not cluttered with irrelevant circuit logic states or those that are not of interest for the particular test being run.

The qualifiers also define the trigger conditions that the test requires. Perhaps it is necessary to only show the state of the data lines after the processor reaches a certain address in its program. The address bits could be connected to the qualifiers, and the data lines could be the displayed lines. The logic analyzer would then only display after the trigger condition of the proper address had occurred. The use of qualifiers means that the person using a logic analyzer can be selective and display only the needed information.

The logic analyzer can show a timing diagram that has the two logic levels of the circuit. Another way to show this information to a user is to simply show 1's and 0's on the screen. The logic analyzer does this by displaying a 1 where the logic level is high and a 0 where it is low. The states of many lines can thus be easily viewed as groups of 1's and 0's. This method of presenting the data is called a *state diagram*. It shows for each clock cycle the logical state of the various points to which the analyzer probes are connected. These are often the same points that the timing diagram showed, but the visual format may be easier to follow. The state diagram and the timing diagram are basically two ways of describing the same events occurring in the system.

Once the lines into the logic analyzer are considered as 1's and 0's, then their values can be stored in a memory in the analyzer. This leads to the final major feature of the logic analyzer: The ability to show what happened in the past. The trigger and qualifiers can be used to start a display of the digital states of points in the system called *pretriggering*—but they also can be used to stop it, called *posttriggering* (Fig. 12-10). This is how it works: The analyzer is continuously acquiring the values of the inputs and storing them in memory. As the memory fills, the analyzer overwrites the oldest stored values, and the display of the analyzer shows what has happened up to the present. When the triggering event occurs, the analyzer stops taking any more values into memory, so the display shows the events preceding the trigger. This is an extremely powerful feature, and one that cannot be duplicated by a conventional oscilloscope (which has no memory and must display a wide range of signal values, not just 1's and 0's).

The usefulness of the posttriggering is that in many problem situations, once the actual symptom has occurred, the rest of the activity is of no interest. What the technician needs is to find out how the circuit and microprocessor program got to this point, not what happened afterward. In most cases, system performance after

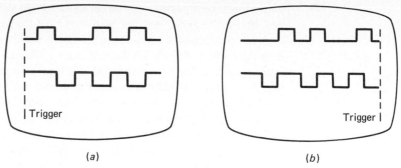

Fig. 12-10 The logic analyzer can use the trigger to start the display or end it. *(a)* Pretrigger and *(b)* Posttrigger.

the fault occurs provides no information because the system may act in an unpredictable, strange manner from that point on. Consider a microprocessor-based system which should be addressing a serial I/O support IC as part of its program. The apparent problem is that occasionally meaningless, incorrect characters are sent out of this serial IC to the peripheral that is attached. The cause of the problem could be many things: a program step that is being read wrong, a decoder in the system that is incorrectly decoding the address lines and enabling the I/O IC at the wrong time, or the actual data going to the IC being garbled. The logic analyzer with the pre- and posttrigger capability would be used. The qualifiers that formed the trigger would be the address in the memory map of the I/O IC, along with the system clock and the memory control lines. The plan would be to have the analyzer capture the data bus values each time the I/O IC is addressed by the processor program. When the I/O IC is enabled, the analyzer freezes the inputs displayed and then it is available for examination. The state of the various input lines to the analyzer will show how the system came to the point where it generated this address. It could have been that the processor misread an instruction because of a bad ROM memory bit, or that the address the processor retrieved from RAM was wrong, or that the decoder itself is defective and generating the CE even though the address lines going into it are not the I/O IC address, or that the data being sent to the I/O IC is wrong because one of the data bus lines has a problem. By watching the sequence of events leading to the observed malfunction, the technician can understand how the problem occurred and what went wrong. The number of signal lines and clock cycles that can be stored and displayed depends on the amount of memory in the logic analyzer. If the analyzer has 32 input lines, then 32 bits are required to store the values at any single clock cycle. To store the signal states at 128 clock cycles would require a memory that is 32 bits wide by 128 words deep. Logic analyzer specifications often call this out. Typical figures are 32×128, 64×64, $16 \times 1K$, and so on. Of course, it may not be practical to display all these 1's and 0's on the CRT at one time, so the analyzer usually has a user-controlled scroll feature so that the subgroups of the stored values can be moved into view.

As the needs and complexities of digital systems have grown, new features have been added to logic analyzers. These include combination pre- and posttrigger, where the trigger position can be set so that logic values are recorded both before and after the trigger event; the ability to have the logic analyzer trigger on more complex conditions (such as the tenth time a certain event occurs); and the ability to trigger when one specified event occurs after another specified event. These more advanced features of the logic analyzer require more thought in setup and planning but can result in some very powerful debugging tools for difficult problems that occur mostly in systems that are under development. These features are generally not needed where troubleshooting equipment exists in the field.

The ability to combine pre- and posttrigger storage might be used to study problems associated with an interrupt. For example, a system might work fine until an interrupt from a timer support IC occurs. The system and program are supposed to see the interrupt, get the interrupt vector, and go to a special interrupt routine which reads the counter value, then resumes the regular program that was running when the interrupt occurred. In this case, the trigger could be the interrupt itself, and the logic analyzer would be connected to store and display the address bus. What should the address bus reveal?

- The program will be executing the program steps.
- The interrupt will occur, and the program will suddenly go to the address where the interrupt vector is stored.
- The program will then go to the interrupt vector and begin executing from that address.
- At the end of the interrupt-handling routine, the program will return to the address it was executing before the interrupt occurred and resume from that point.

In this case, the logic analyzer trigger would be set to capture signals both before and after the interrupt (Fig. 12-10). By studying the address values thus captured, the technician could see what was happening just before the interrupt, if the processor went and got the interrupt vector from the location where it is stored, if the processor then went and began execution from the vector address, and if the processor went back to the regular program after the interrupt steps were completed. The analyzer would show the events preceding and following the trigger event, which in this case is the interrupt. Depending on where the system failed to operate properly, a different cause might be found. If the interrupt occurred but was never recognized by the processor, perhaps the program previously turned off the interrupt via a program step (this is called disabling the interrupt, and it is done when the programmer decides that certain program sequences are too important to be interrupted for any reason). If the processor sees the interrupt, goes to the address where the interrupt vector is stored, and then goes to a wrong location for the interrupt vector, the memory where the vector is stored may be bad. Note that the debugging assumes that the person using the analyzer has a memory map to use as a guide to what should be happening.

QUESTIONS FOR SECTION 12-4

1. Give three reasons for using a logic analyzer.

2. What is the limitation of a trigger signal on a scope? Does this apply to a logic analyzer?

3. What is the difference between the appearance of a signal on a scope versus a logic analyzer? Why is the logic analyzer easier to interpret?

4. What is the Clock input for on a logic analyzer?

5. Which signals from the system can go to a logic analyzer? Why would they be investigated?

6. What are qualifiers? How are they used? Why are they so important?

7. What are three additional features of the logic analyzer?

8. What are the two ways to display signal information? How are they the same? How are they different?

9. How does a logic analyzer show the events before the trigger? Give an example of why this is important. Can a regular oscilloscope do this? Why or why not?

10. What determines the number of signal lines stored, and for how many clock cycles?

11. How many input signals can be stored for a 16×1024 memory in a logic analyzer? How many clock cycles?

12. What are some additional features of the logic analyzer?

13. Why is a memory map needed when using the analyzer?

12-5 PERSONALITY MODULES

The power of the logic analyzer has several drawbacks which can complicate its use. First, there can be many wires to physically attach to the circuit under test. This is time-consuming, error-prone, and aggravating when the wires have to be moved to other points for one test and then again moved for another test. Second, interpreting groups of 1's and 0's (or hex) is always complicated and it is very easy to incorrectly identify an address or the meaning of the word displayed.

The manufacturers of logic analyzers recognized this problem and soon developed a solution. Personality modules are available for most of the more common microprocessors. The modules act as the interface between the system under test and the logic analyzer. The personality module has a clip which attaches to the microprocessor (Fig. 12-11, page 352). This eliminates nearly all of the individual test wires needed, except for a few special ones such as those on certain gates or external points. In many cases, the signals on the microprocessor are enough for test purposes.

More important, the personality module has its own intelligence. It was designed for a specific microprocessor, and so it knows the meaning of the various digital patterns on the data lines and control lines. It recognizes the op codes of the instructions, the states of the various control lines and what they mean, and it can translate these into more meaningful form on the display of the logic analyzer. The

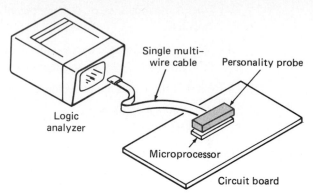

Fig. 12-11 A logic analyzer with a personality probe connected to the microprocessor.

personality module can be directed by the user of the analyzer, via the front panel, to do specific functions. Rather than move wires around, the technician can enter commands which say, in effect, "display all addresses being read," or "display the data as these memory locations are either written to or read from," or "display all data after the processor has read from this specific address." The logic analyzer can really observe and interpret all the signals, under the direction of the analyzer user. The most important capability of an analyzer with a personality module, though, is the ability to translate the binary pattern of the many op codes into the instructions they represent. Recall from the discussion of assembly language programming in Chap. 11 that in assembly language the processor instructions are given in short phrases indicating the operation to be performed and the location of the data. The assembler takes these and converts them into the binary pattern that the microprocessor expects to see. What the logic analyzer with a personality module can do is the opposite: it can "disassemble" the binary codes and put the original program statements on the screen. The user now does not have to try to figure out what the patterns of 1's and 0's going between the microprocessor and the rest of the system mean. The disassembled binary values show the instruction, the addresses specified by the instruction, and the data associated with these in the same format as was used by the person who wrote the instructions when the system was being developed. The translation by the analyzer user of the patterns of 1's and 0's is no longer needed, since the patterns are now in a much more understandable form. This reduces error, minimizes frustration, and increases the efficiency of anyone using the analyzer. The power of the logic analyzer and personality module is so great that many companies will not design products using a specific model of microprocessor unless an analyzer with a personality module for that particular microprocessor type is available.

In a typical application, the technician might be looking for all activity associated with the system CRT controller support IC. Using the memory map, the address of that IC is entered into the analyzer (and its personality module) along with instructions to show the data bus whenever that address appears. Suppose the CRT IC is at 047A (hex) through 047C. The screen of the analyzer would then show things like:

MOV 1000,047A   When data from location 1000 is moved to the CRT IC.

AND 047C,0100   When the status register of the IC (at 047C) is ANDed
                 with a mask to check if a bit in the register is 1.

This ability to disassemble the binary code into the instructions that the system is seeing is especially needed when working with projects that are under development and so are in the debugging stage. Very often, there is no hardware fault, but there is a mistake in the program where the person who wrote the program may have put down the wrong instruction or may have made a logical mistake (thinking that the program should do one thing when really it should do another). This happens quite often, for example, when the program needs to set up and read back status from a support IC. The programmer might misread the data sheets for the IC and incorrectly understand the way the support IC works. Then, the first runs of the program will produce erratic results. The logic analyzer with a personality module can be used to watch the instructions as they execute, confirm what is happening in the system, and help understand what should be happening.

QUESTIONS FOR SECTION 12-5

1. What is a drawback of the logic analyzer?

2. How does the personality module overcome this? What does the personality module do?

3. What does it mean to say that the logic analyzer and personality module can disassemble instructions? Why is this important?

12-6   SIGNATURE ANALYSIS AND ANALYZERS

The logic analyzer is the ideal tool for finding problems in digital systems that are under development in a company's product laboratories. It combines the power of the oscilloscope with the additional features needed to analyze every aspect of the operation of the system, especially when the analyzer has a personality module. However, it is not a convenient tool for troubleshooting products in the field because it requires an understanding of sequence of program instructions used to interact with the various ICs in the system. Usually, this information is provided by the design engineer and the programmer working with the test technician. It is very unrealistic to expect service technicians to understand this level of detail in any single product, and even more impractical for the service technician who is responsible for the repair of many different products.

A better solution to the problem of troubleshooting in the field, on a wide variety of products, was introduced by the Hewlett-Packard Company in 1977. It has since been adapted as a field troubleshooting technique by many other companies. It is called *signature analysis* and uses a simple instrument called a *signature analyzer* instead of an oscilloscope or logic analyzer. The concept of signature analysis is based on various mathematical ideas. Fortunately, using signature analysis does not require an understanding of these any more than using a car requires an understanding of how the engine works.

The problem that signature analysis solves is that it often is very difficult to look at a signal on an oscilloscope, or as a timing or state diagram on a logic analyzer, and tell if it is the right pattern or not. There are many points in a circuit to be checked, and the correct pattern is different at each of them. The digital system is constantly generating long, nonrepeating patterns of 1's and 0's in the course of its normal operation. How does the service documentation show these? How can a service engineer compare these quickly and accurately with what the oscilloscope or logic analyzer shows? Signature analysis is one solution.

The idea behind signature analysis is to provide the technician or engineer with a digital readout that summarizes all the activity on the signal line that the signature analyzer is monitoring. The summary value is then compared to the correct value that is marked in the service manual. If it agrees, then all the signals up to that point in the system are good. If it differs, then there is a problem associated with that point. The troubleshooting guide shows which points should be checked, and in what order. They can be address, data, control lines, or any point in the system.

A typical test procedure using signature analysis requires few steps to get started. The signature analyzer (Fig. 12-12) has a very simple front panel. There are three test leads to be connected: one each to a start signal, a stop signal, and the clock signal. The actual test probe is separate and is moved to different points in the circuit. Once the analyzer is connected, the system being tested is put into a

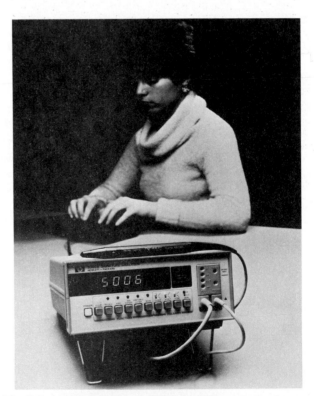

Fig. 12-12 The lightweight signature analyzer is simple to connect, use, and interpret. *(Photo Courtesy of Hewlett-Packard.)*

special test mode of operation, usually by moving some switches on its circuit board to a test position, as called out in the manual. The test mode causes certain specific actions to be taken by the system microprocessor, and these actions require an absolute minimum number of components of the system to be working properly.

Next, the probe of the signature analyzer is touched to a few specified points. If the signature agrees with the manual, then the core of the system (the microprocessor, some RAM, and some ROM) is working, and additional parts of the system can be checked. If the first level does not work, then the problem is with the few components required for minimum operation.

Documentation for testing with signature analysis is simple. The test procedure shows the points to be checked, the order in which they are checked, and what the correct signature should be. It also shows the probable causes of problems if the signature does not agree.

The signature is shown on the analyzer as a four-digit hexadecimal number. Usually, the display uses 7-segment characters and so the set of displayed values is 0, 1, 2, 3, 4, 5, 6, 7, 8, A, C, F, H, P, and U, which are unambiguous on the display. Signatures therefore could look like:

1C3U
021A
3619

Reading and comparing a signature is much faster and more accurate than trying to compare the shape of a signal displayed on an oscilloscope or reading long lists of addresses and data values on the logic analyzer. With the signature, the user makes a relatively simple yes or no type of decision. Note that a wrong signature, even if it is just a little different from the right one, means that the point being probed has a problem associated with it. It does not imply that the point is "almost" OK. If the correct signature is 3A19, and the observed signature is 3A18, the point is just as bad as if the observed signature was 0772.

The advantages of signature analysis in field troubleshooting are the speed and convenience it creates, and the relatively simple documentation required. However, it is practical only for products that are out of the design and debug phase, since any change in the digital circuitry or the associated program of the product may cause the correct signatures to change. Signature analysis must be planned in the design of the product, since some special switch settings are often needed to start the test mode where only the core of the system is checked. For these reasons, signature analysis is not used in the product development lab to debug a design that is still evolving.

The Technical Basis of Signature Analysis

The signature analyzer is a very simple instrument built up of the digital building blocks covered in this book. The heart of the signature analyzer is a shift register of 16 bits. Some of these bits are fed back into the shift register via XOR gates, as shown in the Fig. 12-13. These XORs give the signature its special characteristics. The start and stop lines are used to control when the pulses from the point being probed will be allowed into the shift register, and the clock is used to cause the register to shift its contents and accept the next input bit from the test point. The 16 bits of the shift register go to decoders which convert the groups of 4 bits into hex

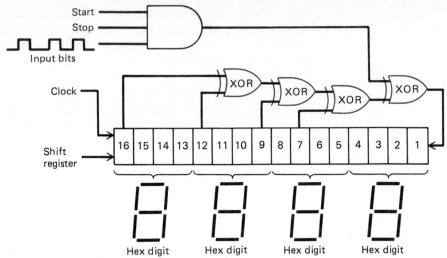

Fig. 12-13 Block diagram of a signature analyzer. A 16-bit shift register and 4 XOR gates accept the input data stream, shifting once per clock bit and entering a new bit from the system under test.

format and then to 7-segment-display format. At the start, the shift register is cleared to all 0's. The display value is shown only after the input bits have all been entered.

The ability of the signature analyzer to present a 4-digit summary of the long, continuous stream of input bits is shown by following what happens in actual operation. If the bit stream consists of a single 1, and then fifteen 0's, the contents of the shift register will look as in Table 12-1 after each clock pulse.

TABLE 12-1
SIGNATURE ANALYZER: EXAMPLE OF SHIFT REGISTER CONTENTS

Bit Number	Input Bit	Shift Register Contents			
1	1	0000	0000	0000	0000
2	0	0000	0000	0000	0001
3	0	0000	0000	0000	0010
4	0	0000	0000	0000	0100
5	0	0000	0000	0000	1000
6	0	0000	0000	0001	0000
7	0	0000	0000	0010	0000
8	0	0000	0000	0100	0000
9	0	0000	0000	1000	0001
10	0	0000	0001	0000	0010
11	0	0000	0010	0000	0101
12	0	0000	0100	0000	1010
13	0	0000	1000	0001	0100
14	0	0001	0000	0010	1001
15	0	0010	0000	0101	0010
16	0	0100	0000	1010	0101
	Signature is:	4	0	A	5

The XOR functions feed back some of the shift register values to the input, and this is what creates the special signature. If the input data stream consisted of two 1's and then fourteen 0's, the shift register contents would look like Table 12-2.

TABLE 12-2
SIGNATURE ANALYZER: EXAMPLE OF SHIFT REGISTER CONTENTS

Bit Number	Input Bit	Shift Register Contents			
1	1	0000	0000	0000	0000
2	1	0000	0000	0000	0001
3	0	0000	0000	0000	0011
4	0	0000	0000	0000	0110
5	0	0000	0000	0000	1100
6	0	0000	0000	0001	1000
7	0	0000	0000	0011	0000
8	0	0000	0000	0110	0000
9	0	0000	0000	1100	0001
10	0	0000	0001	1000	0011
11	0	0000	0011	0000	0111
12	0	0000	0110	0000	0111
13	0	0000	1100	0001	1110
14	0	0001	1000	0011	1101
15	0	0011	0000	0111	1011
16	0	0110	0000	1111	0111
	Signature is:	6	0	U	7

The two nearly similar bit patterns produce completely different signatures. The chances of two different bit streams (the correct one and an incorrect one) producing the same signature are 1 in 65,535. The level of confidence that someone can have when using signature analysis to find problems is very high.

QUESTIONS FOR SECTION 12-6

1. What is the problem with troubleshooting with test equipment such as the scope and logic analyzer?

2. What does the signature analyzer provide the user?

3. What leads does the signature analyzer have besides the actual test probe?

4. What is the first step when using signature analyzer to check out a system?

5. What does a slightly wrong signature mean?

6. Why is the signature analyzer not used in the development lab?

7. What does the shift register in a signature analyzer have that is unusual? What produces the unique signature?

8. How many different signatures are possible? Why?

9. What would be the signature if the 16 input bits are all 0's? (*Note:* Think, don't calculate.)

Microprocessor-based systems usually use the power and flexibility of the system itself to provide various levels of test software. This software performs specific operations that a test engineer has determined would be useful in checking out the operation of the system.

Test software cannot run unless the microprocessor and its memory (both RAM and ROM) are working. In fact, most test routines are kept as short as possible so that the actual amount of memory that must be functioning properly is small. Once this core, or "kernel" is running, the test software can test various parts of the system:

- Memory, by writing and reading bit patterns.
- I/O ports, by using a loop back and writing and reading messages.
- Timers, by setting up a time interval and then checking for an interrupt from the timer within a certain period.
- CRT screens, by putting various patterns on the screen and having the operator check for these.
- Disks used for data storage and the disk support IC, by writing and reading data to and from the disk.

The test software can do a better and more thorough job if the circuitry of the system is designed with test provisions in mind. One example would be a special output bit on the microprocessor that lets the processor generate an output signal that would come back into the system as an interrupt. This would allow the system to check its ability to receive and process interrupts.

There are some things that even a system with a working kernel cannot test without a great deal of additional circuitry (which means additional cost and circuit board space). A switch on the front panel of a system can really only be tested by pushing that switch. If the processor has an output bit which allows it to control a relay which is connected across the switch, so as to simulate the switch being pushed, then all the circuitry from the switch on into the system can be checked. (Of course, the relay may fail and prevent the switch from working when it should.) The other test possibility is to have some external mechanical device that the processor test program could control and which could physically press the switch when necessary for test purposes. This is fairly complicated and expensive. The usual approach is to have the processor software test whatever can be accessed with little or no additional circuitry, except for a few gates or I/O points.

Just as with instrumentation, there are different levels of test software to be used in different situations:

- Start-up self-test software, which is automatically run when the system is powered up. This is designed to give the system user some level of confidence that the system is working properly. If a problem is found, an error message indicating the general type of problem is shown to the user on a CRT or digital display.
- Field test routines go into more detail than the start-up self-test. The field test software is designed to be used by the service technician when doing service work at the office or plant of the equipment customer. The object of this test is

to identify the problem down to the easiest replaceable unit, usually a circuit board. This board can then be exchanged with a spare, and the test run again. Field test routines take longer because they run longer tests on more parts of the system in an attempt to pinpoint the source of the problem accurately. The field test routines may run various memory test patterns and provide special routines for use with the signature analyzer. Field test software is usually started by a special key sequence that the technician knows, or pushing a nonobvious button on the equipment. It usually requires very little in the way of technician assistance after it is started, because of the awkward locations in which the equipment may be installed.

- Engineering lab test routines are the most detailed, most powerful, and have lots of flexibility. Since the product is still in development, it is hard to have test routines that are complete and can accurately point out problems, since the nature of the problems to expect is not yet known. What lab test routines do is provide a set of tools that are the approximate test program equivalent of the voltmeter, logic probe, logic pulser, and oscilloscope. These routines let the technician write any desired data to any memory location, read data from any memory location, set and clear bits in system registers, and run memory tests over different memory addresses. This lets the technician see what pattern of problems exists, which is often necessary when the system problem is not clear-cut. The lab test routines are usually in a special group of instructions called a *monitor*. The monitor is a fairly short program that interfaces the system to a keyboard and CRT, and provides for some very simple but powerful instructions to the system such as:

```
WRITE:          MEMORY LOCATION _____  DESIRED DATA _____
READ:           MEMORY LOCATION _____  DATA READ ____
SET BIT:        MEMORY LOCATION _____  BIT NUMBER _____
READ BIT:       MEMORY LOCATION _____  BIT NUMBER _____
MEMORY TEST:    FROM MEMORY ADDRESS _____
                TO MEMORY ADDRESS _____
                TEST TYPE _____
```

A memory map is required, of course. The monitor also may contain I/O instructions to access programmed I/O devices (as compared to memory mapped I/O). Once the product design has been established, and there is a good understanding of what can fail and how the failure symptoms appear, the test software for the field and user is developed. The experience gained from testing with the monitor is used as a starting point.

QUESTIONS FOR SECTION 12-7

1. Why is a software test program kept as short as possible? What is tested first by the program?

2. What parts of the system need to be specially designed to allow software program tests? What parts are impractical to have a program test?

3. What are the different types of test programs? What are the goals of each?

4. Why do field test programs require minimal input from the technician?

5. Why must development lab test programs be flexible?

6. What is a monitor? What does it allow the technician to do? Why is a memory map needed?

12-8 HARD AND INTERMITTENT FAULTS

Normally, a problem with equipment is caused by the outright failure of some component. This is called a *hard fault*. There are other problems that electronic systems can have which are intermittent, that is, they appear to come and go. The procedure for dealing with these two types of faults has some differences.

Hard faults are consistent, repeatable, and the easiest to find. The system either does not work or fails each time at the same point and under the same set of operating conditions. The cause is usually a bad component, bad connection, or a bad bit in a memory location. The test equipment discussed in this chapter is used in a clear-cut, logical manner to isolate the problem.

Intermittent faults are not easy to set up and repeat. The symptom seems to be there one moment and then not be there later, even though the conditions are the same. In general, such faults are hard to locate. The causes of intermittent problems can be:

- Operation too near the temperature limit for some components, causing them to momentarily malfunction.
- Drift in component specifications caused by changes in temperature. If the design of the circuit is marginal, a few degrees of temperature change may cause the performance of the IC timing, voltage, or current to shift a small amount. This may be enough to put the operation out of the required range.
- Loose connections, whether of actual connectors and wires, or the fine wires within the IC, can make and break as the equipment heats up, or vibrates even a little bit.
- Electrical noise present in the system from other wires, or from the system power supply. This noise may occasionally cause a 0 to look like a 1, or vice versa, if the magnitude of the noise is greater than the difference between logic 0 and logic 1 for the ICs.

The most common approach to dealing with intermittent problems is to try to make the intermittent problem become a hard failure. Some techniques used include heating and cooling the suspected area of the circuit (special aerosols are used), shaking the equipment and wiggling all connections, or reducing the power-supply voltage to a minimum value to ''smoke out'' any marginally performing components.

The other approach is to try to determine if the fault really is intermittent, or if the conditions under which the fault really occurs are not properly understood. Just because a fault appears intermittent does not mean that the real cause is intermittent. For example, a memory bit may be bad and stuck at 0. When the processor stores data and then retrieves it from that location, the system will appear to work properly if the processor did write a 0, since that is what it gets back. However, when it writes a 1 it gets back bad data, and so appears to act incorrectly. The failure symptom appears intermittent because the test technician normally is not looking at each bit and does not know what the processor wrote to that location or the effect on the program of misreading a single bit of data.

For many intermittent failures, the logic analyzer and personality module are used with several qualifiers and a posttrigger to try to see how the system got to the point of failure, and exactly what the address and data values were. The analyzer can watch the operation for hours until the failure occurs, if necessary. By studying the analyzer, the user can see if there is any repeatable pattern to the failure that just is not obvious at first glance.

QUESTIONS FOR SECTION 12-8

1. What are hard fault characteristics?

2. What are intermittent fault characteristics?

3. Why are intermittent faults difficult to find? What are some of their causes? Why?

4. How are intermittent faults dealt with?

5. Are all intermittent faults caused by intermittent failures? How can a hard failure look like an intermittent fault?

SUMMARY

Digital and microprocessor-based systems are fixed or debugged using some of the same test instruments used for nondigital systems, such as the voltmeter and oscilloscope. However, most work on these systems requires some specialized equipment. The requirements of troubleshooting these systems involve monitoring many signal lines simultaneously, comparing signal logic states, and watching for unexpected conditions. In many cases, the logic signals represent processor instructions and data and must be interpreted back to the program software. There is less of a need to actually observe the shape of the waveform on the signal lines.

Another feature of these circuits is that they often require special programs to test parts of the system. This software allows the test technician to address the components, send them data, get them to perform their functions, and observe if the performance is correct.

Repairing equipment that is in the field is very different from debugging systems that are undergoing development in an engineering lab. The types of problems, the equipment used, and the technical resources available are not the same. The troubleshooting and debugging situation is also made more difficult by intermittent problems. Special techniques must be used when dealing with intermittent problems.

REVIEW QUESTIONS

1. Explain three features of troubleshooting. Compare them to debugging.

2. Why is circuit board replacement the most common repair policy?

3. What is the first thing to be checked in a system?

4. Why is the DMM so important? What are the limitations?

5. Why is voltage measured across two points? What is the second point in most cases?

6. Why is current measured through one point?

7. What is the important rule about measuring resistance in a circuit?

8. How is the resistance function used to find open wires, short circuits, and trace wires and circuit board tracks?

9. Why is the oscilloscope needed? What do the two axes of the scope represent?

10. What would be a limitation of a single channel scope?

11. Why is the trigger so critical on a scope? What does it do?

12. What does a scope show that is usually not of interest in many digital debugging situations?

13. What is the advantage of a logic probe when checking the logic state of many points? How does a logic probe also let a point be monitored without constant attention?

14. What does a logic pulser do? Give an example where a logic probe and pulser would be used together on a gate.

15. What does the ''active'' state of a signal mean? How is it marked on the schematic?

16. What is the importance of the posttrigger that a logic analyzer offers? Give an example.

17. Why is a logic analyzer often used in place of a scope?

18. Why are qualifiers so important when using a logic analyzer? Give two examples.

19. What are two of the ways a logic analyzer can display data?

20. What additional advantages does the internal memory of a logic analyzer provide?

21. What are two ways that a personality module enhances the power of the logic analyzer?

22. How does the personality module connect to the system?

23. What is the meaning of *disassemble*? What is disassembled? What does the result look like?

24. Give two reasons why the signature analyzer is used in field test but not product development.

25. What is the concept behind signature analysis? How does it condense many data bits into a few displayed digits?

26. What can be inferred from a slightly wrong signature? From a completely wrong one?

27. Does a signature analyzer catch every error possible? Which kind does it find?

28. Why is software testing required for microprocessor-based products?

29. What parts of a system can be tested by the software? Which cannot? How is the testing done?

30. What are self-test, field test, and engineering development test software? How do they differ?

31. What kinds of tests can usually be called up via engineering test software?

32. What is the difference between hard and intermittent faults?

33. Why are intermittent faults hard to pinpoint? What is usually done to find them?

IC DATA SHEET—4000 SERIES

The figures that follow on the next four pages show the data sheets for the simplest ICs of the 4000 series of CMOS. The data sheet covers the 4001, 4002, 4011, 4012, 4023, and 4025 types. This manufacturer (Motorola) uses the prefix MC1 for the series, and the suffix UB to show that these are unbuffered versions. Thus, the actual part number would be MC14001UB, etc. Simple ICs in a family usually are put into logical groups and go on the same data sheet, since so many of the specifications are the same for all members of the family. The main difference is that for each member, the pins of the IC package represent different logical functions.

 **MOTOROLA**

UB-SUFFIX SERIES CMOS GATES

The UB Series logic gates are constructed with P and N channel enhancement mode devices in a single monolithic structure (Complementary MOS). Their primary use is where low power dissipation and/or high noise immunity is desired. The UB set of CMOS gates are inverting non-buffered functions.

- Quiescent Current = 0.5 nA typ/pkg @ 5 Vdc
- Noise Immunity = 45% of V_{DD} typ
- Supply Voltage Range = 3.0 Vdc to 18 Vdc
- Linear and Oscillator Applications
- Capable of Driving Two Low-power TTL Loads,
  One Low-power Schottky TTL Load or Two HTL Loads
  Over the Rated Temperature Range.
- Double Diode Protection on All Inputs
- Pin-for-Pin Replacements for Corresponding CD4000
  Series UB Suffix Devices
- Formerly Listed without UB Suffix

MC14001UB
Quad 2-Input NOR Gate

MC14002UB
Dual 4-Input NOR Gate

MC14011UB
Quad 2-Input NAND Gate

MC14012UB
Dual 4-Input NAND Gate

MC14023UB
Triple 3-Input NAND Gate

MC14025UB
Triple 3-Input NOR Gate

CMOS SSI

(LOW-POWER COMPLEMENTARY MOS)

UB-SERIES GATES

LOGIC DIAGRAMS

MC14001UB
Quad 2 Input NOR Gate

MC14002UB
Dual 4-Input NOR Gate

MC14011UB
Quad 2 Input NAND Gate

MC14012UB
Dual 4-Input NAND Gate

MC14023UB
Triple 3-Input NAND Gate

MC14025UB
Triple 3-Input NOR Gate

V_{DD} = Pin 14
V_{SS} = Pin 7
for All Devices

L SUFFIX
CERAMIC PACKAGE
CASE 632

P SUFFIX
PLASTIC PACKAGE
CASE 646

ORDERING INFORMATION

MC14XXXUB ——— Suffix Denotes

    L Ceramic Package
    P Plastic Package
    A Extended Operating
       Temperature Range
    C Limited Operating
       Temperature Range

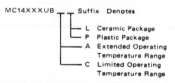

This device contains circuitry to protect the inputs against damage due to high static voltages or electric fields; however, it is advised that normal precautions be taken to avoid application of any voltage higher than maximum rated voltages to this high impedance circuit. For proper operation it is recommended that V_{in} and V_{out} be constrained to the range $V_{SS} \leq (V_{in} \text{ or } V_{out}) \leq V_{DD}$.

Unused inputs must always be tied to an appropriate logic voltage level (e.g., either V_{SS} or V_{DD}).

CMOS UB-SERIES GATES

MAXIMUM RATINGS (Voltages referenced to V_{SS})

Rating	Symbol	Value	Unit
DC Supply Voltage	V_{DD}	-0.5 to +18	Vdc
Input Voltage, All Inputs	V_{in}	-0.5 to V_{DD} +0.5	Vdc
DC Current Drain per Pin	I	10	mAdc
Operating Temperature Range — AL Device	T_A	-55 to +125	°C
CL/CP Device		-40 to +85	
Storage Temperature Range	T_{stg}	-65 to +150	°C

ELECTRICAL CHARACTERISTICS

Characteristic	Symbol	V_{DD} Vdc	T_{low}* Min	T_{low}* Max	25°C Min	25°C Typ	25°C Max	T_{high}* Min	T_{high}* Max	Unit
Output Voltage "0" Level	V_{OL}	5.0	–	0.05	–	0	0.05	–	0.05	Vdc
$V_{in} = V_{DD}$ or 0		10	–	0.05	–	0	0.05	–	0.05	
		15	–	0.05	–	0	0.05	–	0.05	
"1" Level	V_{OH}	5.0	4.95	–	4.95	5.0	–	4.95	–	Vdc
$V_{in} = 0$ or V_{DD}		10	9.95	–	9.95	10	–	9.95	–	
		15	14.95	–	14.95	15	–	14.95	–	
Input Voltage# "0" Level	V_{IL}									Vdc
(V_O = 4.5 Vdc)		5.0	–	1.0	–	2.25	1.0	–	1.0	
(V_O = 9.0 Vdc)		10	–	2.0	–	4.50	2.0	–	2.0	
(V_O = 13.5 Vdc)		15	–	2.5	–	6.75	2.5	–	2.5	
"1" Level	V_{IH}									Vdc
(V_O = 0.5 Vdc)		5.0	4.0	–	4.0	2.75	–	4.0	–	
(V_O = 1.0 Vdc)		10	8.0	–	8.0	5.50	–	8.0	–	
(V_O = 1.5 Vdc)		15	12.5	–	12.5	8.25	–	12.5	–	
Output Drive Current (AL Device)	I_{OH}									mAdc
(V_{OH} = 2.5 Vdc) Source		5.0	-1.2	–	-1.0	-1.7	–	-0.7	–	
(V_{OH} = 4.6 Vdc)		5.0	-0.25	–	-0.2	-0.36	–	-0.14	–	
(V_{OH} = 9.5 Vdc)		10	-0.62	–	-0.5	-0.9	–	-0.35	–	
(V_{OH} = 13.5 Vdc)		15	-1.8	–	-1.5	-3.5	–	-1.1	–	
(V_{OL} = 0.4 Vdc) Sink	I_{OL}	5.0	0.64	–	0.51	0.88	–	0.36	–	mAdc
(V_{OL} = 0.5 Vdc)		10	1.6	–	1.3	2.25	–	0.9	–	
(V_{OL} = 1.5 Vdc)		15	4.2	–	3.4	8.8	–	2.4	–	
Output Drive Current (CL/CP Device)	I_{OH}									mAdc
(V_{OH} = 2.5 Vdc) Source		5.0	-1.0	–	-0.8	-1.7	–	-0.6	–	
(V_{OH} = 4.6 Vdc)		5.0	-0.2	–	-0.16	-0.36	–	-0.12	–	
(V_{OH} = 9.5 Vdc)		10	-0.5	–	-0.4	-0.9	–	-0.3	–	
(V_{OH} = 13.5 Vdc)		15	-1.4	–	-1.2	-3.5	–	-1.0	–	
(V_{OL} = 0.4 Vdc) Sink	I_{OL}	5.0	0.52	–	0.44	0.88	–	0.36	–	mAdc
(V_{OL} = 0.5 Vdc)		10	1.3	–	1.1	2.25	–	0.9	–	
(V_{OL} = 1.5 Vdc)		15	3.6	–	3.0	8.8	–	2.4	–	
Input Current (AL Device)	I_{in}	15	–	±0.1	–	±0.00001	±0.1	–	±1.0	µAdc
Input Current (CL/CP Device)	I_{in}	15	–	±0.3	–	±0.00001	±0.3	–	±1.0	µAdc
Input Capacitance (V_{in} = 0)	C_{in}	–	–	–	–	5.0	7.5	–	–	pF
Quiescent Current (AL Device)	I_{DD}	5.0	–	0.25	–	0.0005	0.25	–	7.5	µAdc
(Per Package)		10	–	0.50	–	0.0010	0.50	–	15.0	
		15	–	1.00	–	0.0015	1.00	–	30.0	
Quiescent Current (CL/CP Device)	I_{DD}	5.0	–	1.0	–	0.0005	1.0	–	7.5	µAdc
(Per Package)		10	–	2.0	–	0.0010	2.0	–	15.0	
		15	–	4.0	–	0.0015	4.0	–	30.0	
Total Supply Current**† (Dynamic plus Quiescent, Per Gate, C_L = 50 pF)	I_T	5.0	$I_T = (0.3 \, \mu A/kHz) f + I_{DD}/N$							µAdc
		10	$I_T = (0.6 \, \mu A/kHz) f + I_{DD}/N$							
		15	$I_T = (0.8 \, \mu A/kHz) f + I_{DD}/N$							

*T_{low} = -55°C for AL Device, -40°C for CL/CP Device.
T_{high} = +125°C for AL Device, +85°C for CL/CP Device.

#Noise immunity specified for worst-case input combination.
Noise Margin for both "1" and "0" level =
  0.5 Vdc min @ V_{DD} = 5.0 Vdc
  1.0 Vdc min @ V_{DD} = 10 Vdc
  1.0 Vdc min @ V_{DD} = 15 Vdc

†To calculate total supply current at loads other than 50 pF:

$$I_T(C_L) = I_T(50 \, pF) + N \times 10^{-3}(C_L - 50)V_{DD}f$$

where: I_T is in µA (per package), C_L in pF, V_{DD} in Vdc, f in kHz is input frequency and N is number of gates per package.
**The formulas given are for the typical characteristics only at 25°C.

CMOS UB-SERIES GATES

SWITCHING CHARACTERISTICS* (C_L = 50 pF, T_A = 25°C)

Characteristic	Symbol	V_{DD} Vdc	Min	Typ	Max	Unit
Output Rise Time	t_{TLH}					ns
$\quad t_{TLH}$ = (3.0 ns/pF) C_L + 30 ns		5.0	–	180	360	
$\quad t_{TLH}$ = (1.5 ns/pF) C_L + 15 ns		10	–	90	180	
$\quad t_{TLH}$ = (1.1 ns/pF) C_L + 10 ns		15	–	65	130	
Output Fall Time	t_{THL}					ns
$\quad t_{THL}$ = (1.5 ns/pF) C_L + 25 ns		5.0	–	100	200	
$\quad t_{THL}$ = (0.75 ns/pF) C_L + 12.5 ns		10	–	50	100	
$\quad t_{THL}$ = (0.55 ns/pF) C_L + 9.5 ns		15	–	40	80	
Propagation Delay Time	t_{PLH}, t_{PHL}					ns
$\quad t_{PLH}, t_{PHL}$ = (1.7 ns/pF) C_L + 30 ns		5.0	–	90	180	
$\quad t_{PLH}, t_{PHL}$ = (0.66 ns/pF) C_L + 22 ns		10	–	50	100	
$\quad t_{PLH}, t_{PHL}$ = (0.50 ns/pF) C_L + 15 ns		15	–	40	80	

FIGURE 1 – SWITCHING TIME TEST CIRCUIT AND WAVEFORMS

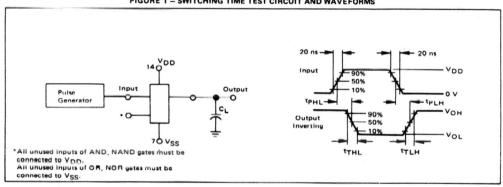

*All unused inputs of AND, NAND gates must be connected to V_{DD}.
All unused inputs of OR, NOR gates must be connected to V_{SS}.

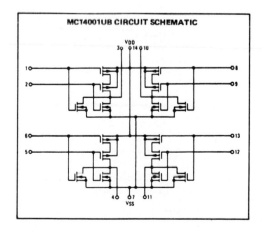

MC14001UB CIRCUIT SCHEMATIC

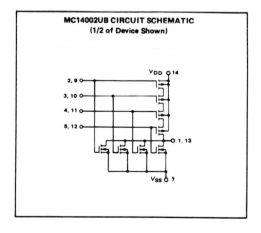

MC14002UB CIRCUIT SCHEMATIC
(1/2 of Device Shown)

CMOS UB-SERIES GATES

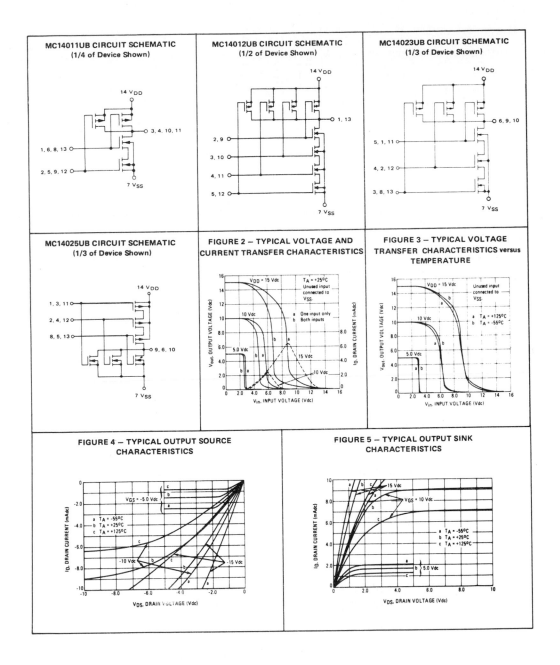

MC14011UB CIRCUIT SCHEMATIC
(1/4 of Device Shown)

MC14012UB CIRCUIT SCHEMATIC
(1/2 of Device Shown)

MC14023UB CIRCUIT SCHEMATIC
(1/3 of Device Shown)

MC14025UB CIRCUIT SCHEMATIC
(1/3 of Device Shown)

FIGURE 2 — TYPICAL VOLTAGE AND CURRENT TRANSFER CHARACTERISTICS

FIGURE 3 — TYPICAL VOLTAGE TRANSFER CHARACTERISTICS versus TEMPERATURE

FIGURE 4 — TYPICAL OUTPUT SOURCE CHARACTERISTICS

FIGURE 5 — TYPICAL OUTPUT SINK CHARACTERISTICS

APPENDIX B

IC DATA SHEET—8253 SERIES

The figures on page 370 through 380 show the data sheets for the Intel 8253 Programmable Interval Timer, discussed in the text. This is a very popular IC which has many modes of operation. Its flexibility and versatility fill many needs in practical microprocessor systems. The IC requires that several specific bit patterns be sent to it to set it up properly, and to report data back to the microprocessor in a specific way. The data sheets describe the messages that are used. They also discuss the special clock, gate, and output lines of the IC and how they are used. The data sheets must also show which physical pins of the IC package are used for which signals of the IC function.

8253/8253-5
PROGRAMMABLE INTERVAL TIMER

- **MCS-85™ Compatible 8253-5**

- **3 Independent 16-Bit Counters**

- **DC to 2 MHz**

- **Programmable Counter Modes**

- **Count Binary or BCD**

- **Single +5V Supply**

- **24-Pin Dual In-Line Package**

The Intel® 8253 is a programmable counter/timer chip designed for use as an Intel microcomputer peripheral. It uses nMOS technology with a single +5V supply and is packaged in a 24-pin plastic DIP.

It is organized as 3 independent 16-bit counters, each with a count rate of up to 2 MHz. All modes of operation are software programmable.

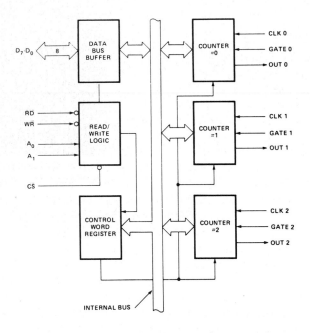

Figure 1. Block Diagram

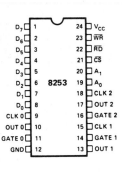

Figure 2. Pin Configuration

FUNCTIONAL DESCRIPTION

General

The 8253 is a programmable interval timer/counter specifically designed for use with the Intel™ Microcomputer systems. Its function is that of a general purpose, multi-timing element that can be treated as an array of I/O ports in the system software.

The 8253 solves one of the most common problems in any microcomputer system, the generation of accurate time delays under software control. Instead of setting up timing loops in systems software, the programmer configures the 8253 to match his requirements, initializes one of the counters of the 8253 with the desired quantity, then upon command the 8253 will count out the delay and interrupt the CPU when it has completed its tasks. It is easy to see that the software overhead is minimal and that multiple delays can easily be maintained by assignment of priority levels.

Other counter/timer functions that are non-delay in nature but also common to most microcomputers can be implemented with the 8253.

- Programmable Rate Generator
- Event Counter
- Binary Rate Multiplier
- Real Time Clock
- Digital One-Shot
- Complex Motor Controller

Data Bus Buffer

This 3-state, bi-directional, 8-bit buffer is used to interface the 8253 to the system data bus. Data Is transmitted or received by the buffer upon execution of INput or OUTput CPU instructions. The Data Bus Buffer has three basic functions.

1. Programming the MODES of the 8253.
2. Loading the count registers.
3. Reading the count values.

Read/Write Logic

The Read/Write Logic accepts inputs from the system bus and in turn generates control signals for overall device operation. It is enabled or disabled by CS so that no operation can occur to change the function unless the device has been selected by the system logic.

$\overline{RD}$ (Read)

A "low" on this input informs the 8253 that the CPU is inputting data in the form of a counters value.

$\overline{WR}$ (Write)

A "low" on this input informs the 8253 that the CPU is outputting data in the form of mode information or loading counters.

A0, A1

These inputs are normally connected to the address bus. Their function is to select one of the three counters to be operated on and to address the control word register for mode selection.

$\overline{CS}$ (Chip Select)

A "low" on this input enables the 8253. No reading or writing will occur unless the device is selected. The $\overline{CS}$ input has no effect upon the actual operation of the counters.

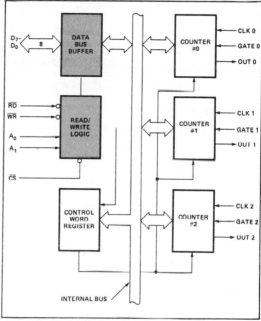

Figure 3. Block Diagram Showing Data Bus Buffer and Read/Write Logic Functions

$\overline{CS}$	$\overline{RD}$	$\overline{WR}$	A_1	A_0	
0	1	0	0	0	Load Counter No. 0
0	1	0	0	1	Load Counter No. 1
0	1	0	1	0	Load Counter No. 2
0	1	0	1	1	Write Mode Word
0	0	1	0	0	Read Counter No. 0
0	0	1	0	1	Read Counter No. 1
0	0	1	1	0	Read Counter No. 2
0	0	1	1	1	No-Operation 3-State
1	X	X	X	X	Disable 3-State
0	1	1	X	X	No-Operation 3-State

Control Word Register

The Control Word Register is selected when A0, A1 are 11. It then accepts information from the data bus buffer and stores it in a register. The information stored in this register controls the operational MODE of each counter, selection of binary or BCD counting and the loading of each count register.

The Control Word Register can only be written into; no read operation of its contents is available.

Counter #0, Counter #1, Counter #2

These three functional blocks are identical in operation so only a single Counter will be described. Each Counter consists of a single, 16-bit, pre-settable, DOWN counter. The counter can operate in either binary or BCD and its input, gate and output are configured by the selection of MODES stored in the Control Word Register.

The counters are fully independent and each can have separate Mode configuration and counting operation, binary or BCD. Also, there are special features in the control word that handle the loading of the count value so that software overhead can be minimized for these functions.

The reading of the contents of each counter is available to the programmer with simple READ operations for event counting applications and special commands and logic are included in the 8253 so that the contents of each counter can be read "on the fly" without having to inhibit the clock input.

8253 SYSTEM INTERFACE

The 8253 is a component of the Intel™ Microcomputer Systems and interfaces in the same manner as all other peripherals of the family. It is treated by the systems software as an array of peripheral I/O ports; three are counters and the fourth is a control register for MODE programming.

Basically, the select inputs A0, A1 connect to the A0, A1 address bus signals of the CPU. The $\overline{CS}$ can be derived directly from the address bus using a linear select method. Or it can be connected to the output of a decoder, such as an Intel® 8205 for larger systems.

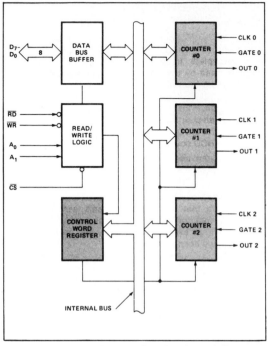

Figure 4. Block Diagram Showing Control Word Register and Counter Functions

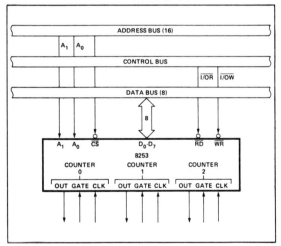

Figure 5. 8253 System Interface

OPERATIONAL DESCRIPTION

General

The complete functional definition of the 8253 is programmed by the systems software. A set of control words must be sent out by the CPU to initialize each counter of the 8253 with the desired MODE and quantity information. Prior to initialization, the MODE, count, and output of all counters is undefined. These control words program the MODE, Loading sequence and selection of binary or BCD counting.

Once programmed, the 8253 is ready to perform whatever timing tasks it is assigned to accomplish.

The actual counting operation of each counter is completely independent and additional logic is provided on-chip so that the usual problems associated with efficient monitoring and management of external, asynchronous events or rates to the microcomputer system have been eliminated.

Programming the 8253

All of the MODES for each counter are programmed by the systems software by simple I/O operations.

Each counter of the 8253 is individually programmed by writing a control word into the Control Word Register. (A0, A1 = 11)

Control Word Format

D_7	D_6	D_5	D_4	D_3	D_2	D_1	D_0
SC1	SC0	RL1	RL0	M2	M1	M0	BCD

Definition of Control

SC — Select Counter:

SC1	SC0	
0	0	Select Counter 0
0	1	Select Counter 1
1	0	Select Counter 2
1	1	Illegal

RL — Read/Load:

RL1	RL0	
0	0	Counter Latching operation (see READ/WRITE Procedure Section)
1	0	Read/Load most significant byte only.
0	1	Read/Load least significant byte only.
1	1	Read/Load least significant byte first, then most significant byte.

M — MODE:

M2	M1	M0	
0	0	0	Mode 0
0	0	1	Mode 1
X	1	0	Mode 2
X	1	1	Mode 3
1	0	0	Mode 4
1	0	1	Mode 5

BCD:

0	Binary Counter 16-bits
1	Binary Coded Decimal (BCD) Counter (4 Decades)

Counter Loading

The count register is not loaded until the count value is written (one or two bytes, depending on the mode selected by the RL bits), followed by a rising edge and a falling edge of the clock. Any read of the counter prior to that falling clock edge may yield invalid data.

MODE Definition

MODE 0: Interrupt on Terminal Count. The output will be initially low after the mode set operation. After the count is loaded into the selected count register, the output will remain low and the counter will count. When terminal count is reached the output will go high and remain high until the selected count register is reloaded with the mode or a new count is loaded. The counter continues to decrement after terminal count has been reached.

Rewriting a counter register during counting results in the following:

(1) Write 1st byte stops the current counting.
(2) Write 2nd byte starts the new count.

MODE 1: Programmable One-Shot. The output will go low on the count following the rising edge of the gate input.

The output will go high on the terminal count. If a new count value is loaded while the output is low it will not affect the duration of the one-shot pulse until the succeeding trigger. The current count can be read at any time without affecting the one-shot pulse.

The one-shot is retriggerable, hence the output will remain low for the full count after any rising edge of the gate input.

MODE 2: Rate Generator. Divide by N counter. The output will be low for one period of the input clock. The period from one output pulse to the next equals the number of input counts in the count register. If the count register is reloaded between output pulses the present period will not be affected, but the subsequent period will reflect the new value.

The gate input, when low, will force the output high. When the gate input goes high, the counter will start from the initial count. Thus, the gate input can be used to synchronize the counter.

When this mode is set, the output will remain high until after the count register is loaded. The output then can also be synchronized by software.

MODE 3: Square Wave Rate Generator. Similar to MODE 2 except that the output will remain high until one half the count has been completed (for even numbers) and go low for the other half of the count. This is accomplished by decrementing the counter by two on the falling edge of each clock pulse. When the counter reaches terminal count, the state of the output is changed and the counter is reloaded with the full count and the whole process is repeated.

If the count is odd and the output is high, the first clock pulse (after the count is loaded) decrements the count by 1. Subsequent clock pulses decrement the clock by 2. After timeout, the output goes low and the full count is reloaded. The first clock pulse (following the reload) decrements the counter by 3. Subsequent clock pulses decrement the count by 2 until timeout. Then the whole process is repeated. In this way, if the count is odd, the output will be high for $(N + 1)/2$ counts and low for $(N - 1)/2$ counts.

MODE 4: Software Triggered Strobe. After the mode is set, the output will be high. When the count is loaded, the counter will begin counting. On terminal count, the output will go low for one input clock period, then will go high again.

If the count register is reloaded between output pulses, counting will continue from the new value. The count will be inhibited while the gate input is low. Reloading the counter register will restart counting beginning with the new number.

MODE 5: Hardware Triggered Strobe. The counter will start counting after the rising edge of the trigger input and will go low for one clock period when the terminal count is reached. The counter is retriggerable. The output will not go low until the full count after the rising edge of any trigger.

Modes \ Signal Status	Low Or Going Low	Rising	High
0	Disables counting	——	Enables counting
1	——	1) Initiates counting 2) Resets output after next clock	——
2	1) Disables counting 2) Sets output immediately high	1) Reloads counter 2) Initiates counting	Enables counting
3	1) Disables counting 2) Sets output immediately high	Initiates counting	Enables counting
4	Disables counting	——	Enables counting
5	——	Initiates counting	——

Figure 6. Gate Pin Operations Summary

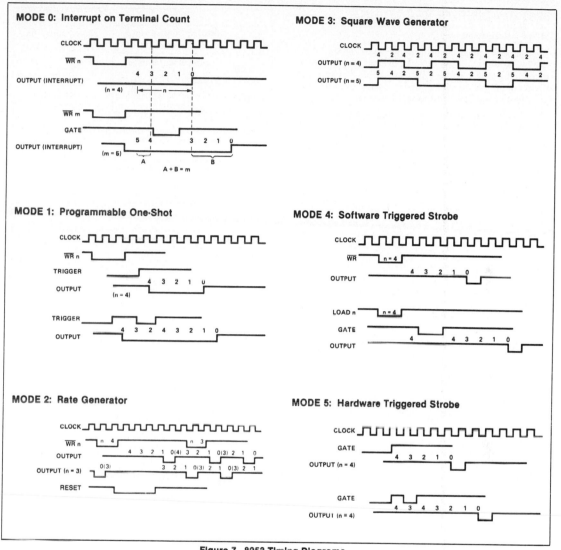

Figure 7. 8253 Timing Diagrams

8253 READ/WRITE PROCEDURE

Write Operations

The systems software must program each counter of the 8253 with the mode and quantity desired. The programmer must write out to the 8253 a MODE control word and the programmed number of count register bytes (1 or 2) prior to actually using the selected counter.

The actual order of the programming is quite flexible. Writing out of the MODE control word can be in any sequence of counter selection, e.g., counter #0 does not have to be first or counter #2 last. Each counter's MODE control word register has a separate address so that its loading is completely sequence independent. (SC0, SC1)

The loading of the Count Register with the actual count value, however, must be done in exactly the sequence programmed in the MODE control word (RL0, RL1). This loading of the counter's count register is still sequence independent like the MODE control word loading, but when a selected count register is to be loaded it <u>must</u> be loaded with the number of bytes programmed in the MODE control word (RL0, RL1). The one or two bytes to be loaded in the count register do not have to follow the associated MODE control word. They can be programmed at any time following the MODE control word loading as long as the correct number of bytes is loaded in order.

All counters are down counters. Thus, the value loaded into the count register will actually be decremented. Loading all zeroes into a count register will result in the maximum count (2^{16} for Binary or 10^4 for BCD). In MODE 0 the new count will not restart until the load has been completed. It will accept one of two bytes depending on how the MODE control words (RL0, RL1) are programmed. Then proceed with the restart operation.

	MODE Control Word Counter n	
LSB	Count Register byte Counter n	
MSB	Count Register byte Counter n	

Note: Format shown is a simple example of loading the 8253 and does not imply that it is the only format that can be used.

Figure 8. Programming Format

			A1	A0
No. 1		MODE Control Word Counter 0	1	1
No. 2		MODE Control Word Counter 1	1	1
No. 3		MODE Control Word Counter 2	1	1
No. 4	LSB	Count Register Byte Counter 1	0	1
No. 5	MSB	Count Register Byte Counter 1	0	1
No. 6	LSB	Count Register Byte Counter 2	1	0
No. 7	MSB	Count Register Byte Counter 2	1	0
No. 8	LSB	Count Register Byte Counter 0	0	0
No. 9	MSB	Count Register Byte Counter 0	0	0

Note: The exclusive addresses of each counter's count register make the task of programming the 8253 a very simple matter, and maximum effective use of the device will result if this feature is fully utilized.

Figure 9. Alternate Programming Formats

Read Operations

In most counter applications it becomes necessary to read the value of the count in progress and make a computational decision based on this quantity. Event counters are probably the most common application that uses this function. The 8253 contains logic that will allow the programmer to easily read the contents of any of the three counters without disturbing the actual count in progress.

There are two methods that the programmer can use to read the value of the counters. The first method involves the use of simple I/O read operations of the selected counter. By controlling the A0, A1 inputs to the 8253 the programmer can select the counter to be read (remember that no read operation of the mode register is allowed A0, A1-11). The only requirement with this method is that in order to assure a stable count reading the actual operation of the selected counter must be inhibited either by controlling the Gate input or by external logic that inhibits the clock input. The contents of the counter selected will be available as follows:

first I/O Read contains the least significant byte (LSB).

second I/O Read contains the most significant byte (MSB).

Due to the internal logic of the 8253 it is absolutely necessary to complete the entire reading procedure. If two bytes are programmed to be read then two bytes must be read before any loading WR command can be sent to the same counter.

Read Operation Chart

A1	A0	RD	
0	0	0	Read Counter No. 0
0	1	0	Read Counter No. 1
1	0	0	Read Counter No. 2
1	1	0	Illegal

Reading While Counting

In order for the programmer to read the contents of any counter without effecting or disturbing the counting operation the 8253 has special internal logic that can be accessed using simple WR commands to the MODE register. Basically, when the programmer wishes to read the contents of a selected counter "on the fly" he loads the MODE register with a special code which latches the present count value into a storage register so that its contents contain an accurate, stable quantity. The programmer then issues a normal read command to the selected counter and the contents of the latched register is available.

MODE Register for Latching Count

A0, A1 = 11

D7	D6	D5	D4	D3	D2	D1	D0
SC1	SC0	0	0	X	X	X	X

SC1,SC0 — specify counter to be latched.

D5,D4 — 00 designates counter latching operation.

X — don't care.

The same limitation applies to this mode of reading the counter as the previous method. That is, it is mandatory to complete the entire read operation as programmed. This command has no effect on the counter's mode.

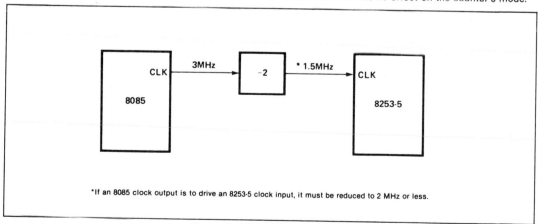

*If an 8085 clock output is to drive an 8253-5 clock input, it must be reduced to 2 MHz or less.

Figure 10. MCS-85™ Clock Interface*

ABSOLUTE MAXIMUM RATINGS*

Ambient Temperature Under Bias .... 0° C to 70° C
Storage Temperature .... −65° C to +150° C
Voltage On Any Pin
  With Respect to Ground .... −0.5 V to +7 V
Power Dissipation .... 1 Watt

*NOTICE: Stresses above those listed under "Absolute Maximum Ratings" may cause permanent damage to the device. This is a stress rating only and functional operation of the device at these or any other conditions above those indicated in the operational sections of this specification is not implied. Exposure to absolute maximum rating conditions for extended periods may affect device reliability.

D.C. CHARACTERISTICS   (T_A = 0°C to 70°C, V_{CC} = 5V ±10%)

Symbol	Parameter	Min.	Max.	Unit	Test Conditions
V_{IL}	Input Low Voltage	−0.5	0.8	V	
V_{IH}	Input High Voltage	2.2	V_{CC}+.5V	V	
V_{OL}	Output Low Voltage		0.45	V	Note 1
V_{OH}	Output High Voltage	2.4		V	Note 2
I_{IL}	Input Load Current		±10	μA	V_{IN} = V_{CC} to 0V
I_{OFL}	Output Float Leakage		±10	μA	V_{OUT} = V_{CC} to 0V
I_{CC}	V_{CC} Supply Current		140	mA	

CAPACITANCE   (T_A = 25°C, V_{CC} = GND = 0V)

Symbol	Parameter	Min.	Typ.	Max.	Unit	Test Conditions
C_{IN}	Input Capacitance			10	pF	fc = 1 MHz
$C_{I/O}$	I/O Capacitance			20	pF	Unmeasured pins returned to V_{SS}

A.C. CHARACTERISTICS   (T_A = 0°C to 70°C, V_{CC} = 5.0V ±5%, GND = 0V)

Bus Parameters (Note 3)

READ CYCLE

Symbol	Parameter	8253 Min.	8253 Max.	8253-5 Min.	8253-5 Max.	Unit
t_{AR}	Address Stable Before $\overline{READ}$	50		30		ns
t_{RA}	Address Hold Time for $\overline{READ}$	5		5		ns
t_{RR}	$\overline{READ}$ Pulse Width	400		300		ns
t_{RD}	Data Delay From $\overline{READ}$[4]		300		200	ns
t_{DF}	$\overline{READ}$ to Data Floating	25	125	25	100	ns
t_{RV}	Recovery Time Between READ and Any Other Control Signal	1		1		μs

A.C. CHARACTERISTICS (Continued)

WRITE CYCLE

Symbol	Parameter	8253		8253-5		Unit
		Min.	Max.	Min.	Max.	
t_{AW}	Address Stable Before $\overline{\text{WRITE}}$	50		30		ns
t_{WA}	Address Hold Time for $\overline{\text{WRITE}}$	30		30		ns
t_{WW}	$\overline{\text{WRITE}}$ Pulse Width	400		300		ns
t_{DW}	Data Set Up Time for $\overline{\text{WRITE}}$	300		250		ns
t_{WD}	Data Hold Time for $\overline{\text{WRITE}}$	40		30		ns
t_{RV}	Recovery Time Between $\overline{\text{WRITE}}$ and Any Other Control Signal	1		1		μs

CLOCK AND GATE TIMING

Symbol	Parameter	8253		8253-5		Unit
		Min.	Max.	Min.	Max.	
t_{CLK}	Clock Period	380	dc	380	dc	ns
t_{PWH}	High Pulse Width	230		230		ns
t_{PWL}	Low Pulse Width	150		150		ns
t_{GW}	Gate Width High	150		150		ns
t_{GL}	Gate Width Low	100		100		ns
t_{GS}	Gate Set Up Time to CLK↑	100		100		ns
t_{GH}	Gate Hold Time After CLK↑	50		50		ns
t_{OD}	Output Delay From CLK↓[4]		400		400	ns
t_{ODG}	Output Delay From Gate↓[4]		300		300	ns

NOTES:
1. I_{OL} = 2.2 mA.
2. I_{OH} = −400 μA.
3. AC timings measured at V_{OH} 2.2, V_{OL} = 0.8.
4. C_L = 150pF.

A.C. TESTING INPUT, OUTPUT WAVEFORM

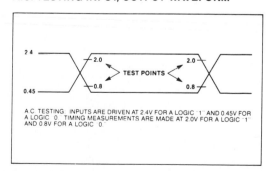

A.C. TESTING: INPUTS ARE DRIVEN AT 2.4V FOR A LOGIC "1" AND 0.45V FOR A LOGIC "0". TIMING MEASUREMENTS ARE MADE AT 2.0V FOR A LOGIC "1" AND 0.8V FOR A LOGIC "0".

A.C. TESTING LOAD CIRCUIT

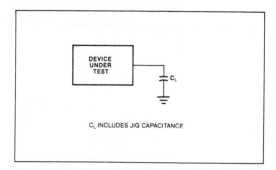

C_L INCLUDES JIG CAPACITANCE

WAVEFORMS

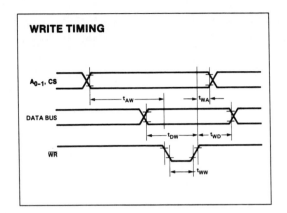

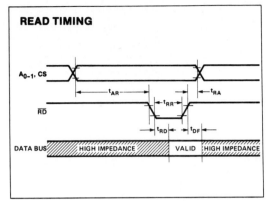

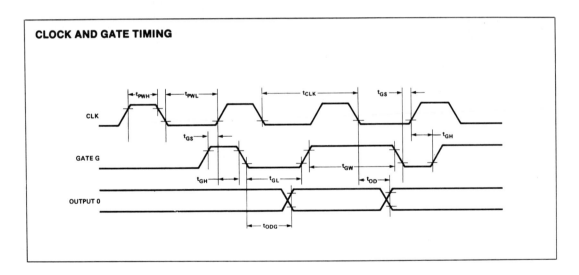

APPENDIX C

LOGIC SYMBOLS

The logic symbols used in this book are used by many companies and are widely understood. They follow well known standards MIL-STD-806C and ASA Y32.14-1962 and work well for basic, simple gates. Unfortunately, they do not give enough of the right kind of information to the user for the more complicated ICs such as flip-flops, registers, decoders, memories, counters, and support ICs.

A new set of standard symbols and notation was introduced to provide more information in the same space on a schematic drawing. These newer symbols are not in widespread use, but it is expected that they will be used by more companies and their designers in the next few years.

The major difference in the new symbols is that they use special notations to indicate the functions of the IC and its signal lines. The older symbols rely more on symbol shape. Fig. C-1 compares examples of the existing symbols with the newer ones.

An edge triggered D-type flip-flop such as the 7474 (which has two independent flip-flops in the same IC package) is shown in the new format in Fig. C-2. The new symbols are especially powerful for the complex ICs because there are notations to show which lines control the IC, how they are grouped, where the enables are, which signals depend on others, and which signals are sensitive to logic level versus the occurrence of a transition (edge) between logic levels.

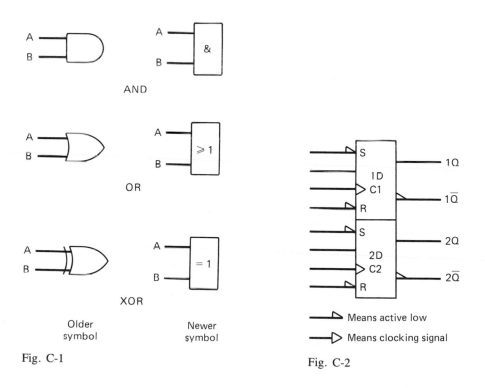

Fig. C-1

Older symbol

Newer symbol

Fig. C-2

⊳ Means active low

▷ Means clocking signal

access time: the time it takes for an IC (such as a memory IC) to respond to a processor read cycle and present the desired data onto the data bus

accumulator: a register in a microprocessor into which data words are placed to be operated upon. The results are also placed there.

active: the logic state (high or low; 0 or 1) at which a signal performs the function it is named for. Example: an active low enable line causes an IC to be enabled when the line is low.

address: a specific storage location in a memory IC determined by the address lines, which point to one of many possible locations within the memory

analog: a signal which is continuous over its total range, and not restricted to a specific set of values

analog to digital converter (ADC): an IC which takes an analog value and converts it to its digital representation

AND: a boolean function which provides a logic 1 output if and only if both input values are 1

arithmetic and logic unit (ALU): an IC which can provide the capability for performing arithmetic operations and boolean operations on input bits

assembly language: a programming language which is converted through a special program into the 1s and 0s of program instructions used by a processor

asynchronous: refers to signals and events that can occur at any time without reference to a system clock

binary: a digital signal which can take only one of two values: high or low, 1 or 0, on or off

binary coded decimal (BCD): a format for representing decimal numbers in the binary system

bit: a single numeric position in a binary number; a binary digit

boolean: a system of logic which develops a result using specific rules operating on the input values

buffer: *(a)* an IC which takes a binary value and boosts its current, voltage, or some other specific signal characteristic, but does not change the logic value (except for inverting buffers, which always invert the signal) *(b)* a collection of registers which allow binary signals to enter and exit independently, to act as an interface between systems

bus: a common signal path used between many ICs and parts of a system

byte: a group of 8 bits

cathode ray tube (CRT): a large glass tube which forms the screen of a video terminal

checksum: a binary pattern that goes with a collection of bits, such as in a memory IC, to help verify that those bits are correct and not changed as a result of noise, IC problems, or other causes

chip enable, chip select: a control line to an IC which turns the IC on so that the IC knows to pay attention to the other signal lines at the proper time

clock: a system pacing function which generates the "ticks" for all the operations in a synchronous system; a function in a system that allows the system to keep track of the time

combinational: a logic circuit where various inputs and outputs are combined directly, without any memory, delay, or timing elements or clock

compiler: a special program which takes a high level programming language such as FORTRAN and converts it to the binary format the processor requires

complement: the inverse (negation) of a binary value

complementary metal-oxide-silicon (CMOS): a technology for producing one type of IC family which features low power use and high noise immunity, among other features

component: an individual piece of a circuit, such as an IC, a resistor, or a transistor

control bus: the bus which the processor uses to signal other components in the system regarding memory cycles, read and write cycles, and other processor activity

counter: the function of providing a digital output equal to the number of pulses received

custom IC: an IC which is designed for a specific application, and so is not general purpose but ideal for the application

data: a bit or group of bits which conveys information

decoder: *(a)* a circuit which converts one numeric format to another *(b)* a circuit or component which uses address lines and other qualifiers to enable a specific IC and no other

demultiplexer: a circuit which takes an input line and distributes it to one of many outputs, as specified by a select code

dice/die: the heart of an IC—the individual piece of silicon with the desired circuit pattern, which then goes into the IC package

digital: a signal which can take on only specific values, or information which is presented as one value of a group of allowed values

digital to analog converter (DAC): a component that converts a signal in digital format to its equivalent in analog format

digital multimeter or voltmeter (DMM or DVM): a fundamental piece of test equipment which digitally shows values of voltage, current, and resistance

discrete: individual transistors, resistors, etc. used to make a circuit, as compared to an IC used for the circuit design

dynamic: *(a)* the situation where the signal lines in a circuit are changing *(b)* a type of memory IC which requires constant refreshing of the contents

electrically erasable PROM (EEPROM): a PROM which can be erased by special voltages applied to its pins, rather than by removing it from the system

enable: a signal which activates another function or component

erasable PROM (EPROM): a PROM which can be erased using an ultraviolet light shining through a window in the IC package

error detection and correction (EDC): the ability of a system to detect errors in groups of bits and correct them

exclusive OR (XOR): a boolean function which makes the output a 1 if and only if one input is 1 and the other is 0

execute: the second part of the instruction cycle where the fetched instruction is actually implemented by the processor

fall time: the time it takes for a binary signal to go from logic 1 to logic 0

fault, hard: a system failure which is repeatable and consistent

fault, intermittent: a system problem which is hard to repeat consistently, but comes and goes for unknown reasons

fetch: the first part of the instruction cycle where the instruction and associated data are read from memory

firmware: a sequence of program steps for a processor which has been permanently burned into a ROM

flip-flop: a digital circuit component which provides one bit of memory, so the state of an input signal can be retained even after the input has changed

ground: the point in a system at 0 volts with respect to the earth

hardware: the actual circuitry of the system, built using ICs and other components

hexadecimal (hex): a number format which groups bits into groups of 4, for easier notation (0, 1, . . . 9, A, B, C, D, E, F)

high: a boolean logic value, usually referring to the logic level at a higher voltage than the other logic state. Also called a 1.

high level language: a language for writing instructions (programs) for a processor which is more English-like than assembly language. Examples: BASIC, FORTRAN

IEEE-488: a standard for allowing communication between various pieces of test equipment

implied-OR: the boolean function that results when 2 open collector gate outputs are wired together, often called "wired-AND"

input/output (I/O): any method or path for getting data into and out of a component or system

instructions: the sequence of steps that a processor is given to implement as its program

integrated circuit: a complete collection of transistors, gates, and other functions built onto a single piece of silicon

integrating ADC: an ADC which counts the number of clock pulses that occur, while the input signal is compared to a ramping voltage

interface: the physical meeting point between any two systems or circuits

interpreter: a special program which converts a high level language into the pattern of 0s and 1s understood by the processor

latch: a circuit element which stores the state of a signal line, usually a flip-flop

least significant bit (LSB): the rightmost (units) bit in a binary number

light emitting diode (LED): a diode which gives off light when voltage is applied; used for indicator lamps and digital displays

linearity error: an error in a DAC or ADC where the converted value is in error by a constant percentage from the correct value

liquid crystal display (LCD): a form of digital display which can be used as the visual interface to a user

logic: circuitry which implements the boolean functions

logic analyzer: a test instrument which is connected to many signal lines simultaneously and displays them in a practical format for study

logic probe: a simple test instrument which shows the state of a digital signal line

logic pulser: a complement for the logic probe, this instrument can force a logic signal to its opposite state

low: a boolean logic value, usually referring to the logic level at a lower voltage than the other logic state. Also called a 0

mask: a group of bits which will combine in a boolean operation with another group of bits, to highlight or suppress some bits, as required by the application

memory: a large collection of latches which can store the logic levels of many bits

memory map: a diagram showing each address of a microprocessor memory and what component and function is at that address

microcomputer: a computer built using a microprocessor as its central IC, and which performs many of the functions of a large computer, except on a reduced scale

microprocessor: an IC which provides many of the functions required for a computer, but requires some additional components to function as a computer

most significant bit (MSB): the leftmost column position in a binary number

multiplexer (MUX): an IC which selects and steers one of several input signals to its output. The signals can be either analog or digital, depending on the mux design

NAND: a boolean function which takes the AND of the inputs and then inverts (negates) it

noise: undesired voltage fluctuations that are superimposed on the desired signal, and can cause system problems or logic errors

NOR: a boolean function which takes the OR of the inputs and then inverts (negates) it

NOT: a boolean function which simply provides an output opposite the input (negates, inverts)

octal: a number format which puts bits into groups of 3, for easier notation

offset error: an error in a DAC or ADC where the converted value is in error by a constant amount from the correct value

open collector: a type of output circuit used within a TTL IC, to allow the IC to drive certain non-TTL loads

OR: a boolean function which provides a 1 output if at least one input is a 1

oscilloscope: a sophisticated test instrument which shows shape of signals by displaying their voltage versus time on a small CRT screen

parallel: a group of bits available simultaneously on multiple signal lines

parity: a checking bit added to a group of bits, to help verify the group has no errors; can be odd or even

personality module: an enhancement to a logic analyzer which makes the analyzer work with a specific microprocessor in a more efficient way

pinout: the physical location of the IC signals on the package of the IC

program: *(a)* the sequence of steps that a processor must execute to perform the application *(b)* to burn a specific pattern of bits into a PROM, EPROM, or EEPROM

program counter: the part of a microprocessor which generates the addresses and so points to the next instruction in memory to be implemented

programmable IC: an IC whose contents can be burned into it, so they do not disappear when power is removed from the IC

Programmable ROM (PROM): a ROM IC whose contents can be burned into it

PROM burner: a device which fixes the required pattern of bits into a programmable IC

propagation delay: the time it takes for a signal to go from the input of a gate function to the output

pullup: a resistor used on the unused input of an IC to ensure that the input is at logic 1

qualifier: special signal(s) that is used to additionally identify a desired set of unique conditions to a decoder or logic analyzer

random access memory (RAM): a collection of memory latches that can be read from and written to as needed by the processor

read only memory: a collection of memory latches that permanently store the logic levels, and cannot be changed

read/write memory: a more technically correct term for RAM

refresh: the need for reading and writing on a regular basis the contents of dynamic RAM memories, to restore the charge

register: a small group of latches, used to store the logic values of several bits

rise time: the time it takes for a logic signal to go from the logic 0 state to the logic 1 state

sample and hold: a circuit used with ADCs to hold the value of the analog signal while the ADC performs the conversion

select code: the binary pattern used with a mux or demux to select the one signal of many that the mux or demux selects or controls

semicustom IC: an IC which is partially completed, the final interconnections of which can be done to specific application requirements

sequence generator: a system or instrument which generates a sequence of 1s and 0s repeatedly

sequential logic: logic which generates outputs based on present and past inputs

serial: a group of bits available one after the other on a single signal line

seven segment digit: a digit to be displayed formed out of seven individual segments

signature analysis and analyzer: the method of taking a serial bit stream, and summarizing it with a shorter number, to check for errors and changes in the original bit stream (and the instrument for doing it).

software: the sequence of instructions (program) which a processor implements

static: *(a)* the logic conditions and signal values on an IC when the signals are not changing *(b)* a type of RAM which can be run as slow as required or even stopped

strobe: a short signal used to signal to an IC that the other signals at the IC are at their proper logic levels and should be accepted at this time

successive approximation ADC: an ADC which develops the digital equivalent of the analog input by comparing the analog value to a digital value in an internal register, starting from the MSB and successively moving over to the LSB

synchronous: a system where all the ICs are paced by a common clock, and activity does not occur between the clock pulses

three state: a type of IC output where the output can either be 1, 0, or a high impedance mode

timer: a function which measures how much time has elapsed between two events

totem pole: the output structure used inside of many TTL gates

transistor transistor logic (TTL): a family of ICs which uses one transistor driving another in the output stage

universal asynchronous receiver/transmitter (UART): an IC which acts as the interface between a synchronous system and an asynchronous communications line

V_{CC}, V_{DD}, V_{PP}, V_{SS}: the designations for the power supply for various logic families

wafer: a large disk of silicon which is made into many ICs

wired-AND: connecting together two or more open collector outputs to provide an AND function. More properly called "implied-OR"

XOR: see exclusive OR

CHAPTER 2

SECTION 2-3

1. (A AND B) OR (C AND D) = Y

A	B	A AND B	C	D	C AND D	Y = OR
0	0	0	0	0	0	0
0	0	0	0	1	0	0
0	0	0	1	0	0	0
0	0	0	1	1	0	0
0	1	0	0	0	0	0
0	1	0	0	1	0	0
0	1	0	1	0	0	0
0	1	0	1	1	1	1
1	0	0	0	0	0	0
		etc ↓			↓	↓
		0			0	0
1	1	1	1	1	1	1

3. Truth table for four-input AND gate: total of 16 input states, 15 produce 0, 1 produces a 1.

A	B	C	D	Y = A AND B AND C AND D
0	0	0	0	0
0	0	0	1	
0	0	1	0	
0	0	1	1	
0	1	0	0	
0	1	0	1	
0	1	1	0	
0	1	1	1	
1	0	0	0	
1	0	0	1	
1	0	1	0	
1	0	1	1	
1	1	0	0	
1	1	0	1	↓
1	1	1	0	0
1	1	1	1	1

5. NOT[NOT(NOT A)] is 0 for A = 1.

A	NOT A	NOT(NOT A)	NOT(NOT(NOT A))
1	0	1	0

7. Truth table for A OR B = C, then C OR D, versus A OR B OR C:

A	B	C = A OR B	D	C OR D
0	0	0	0	0
0	1	1	0	1
1	0	1	0	1
1	1	1	0	1
0	0	0	1	1
0	1	1	1	1
1	0	1	1	1
1	1	1	1	1

Same as (A OR B OR D)

9. Boolean of Fig. 2-11:
NOT(A AND B) AND (C OR D) = Y

SECTION 2-4

1. A OR A = A

A	A	A OR A
0	0	0
1	1	1

3. A AND (NOT A) = 0

A	NOT A	A AND (NOT A)
0	1	0
1	0	0

SECTION 2-5

1. OR with Inhibit:

A	B	Inhibit	Y
0	0	0	0
0	1	0	1
1	0	0	1
1	1	0	1
0	0	1	0
0	1	1	0
1	0	1	0
1	1	1	0

8 input conditions,
3 outputs = 1, 5 outputs = 0

3. NOT(NAND A) = AND A

5. (A NAND B) XOR (C NOR D) = Y

SECTION 2-7

1. Decimal to binary: 27 is 1 1011; 45 is 10 1111, 100 is 110 0100, 247 is 1111 0011.

3. Binary to decimal: 0110 is 6; 1 1001 is 25; 1100 is 12; 1101 is 13.

5. 1 1111 = 31 decimal

SECTION 2-8

1. BCD to decimal: 0110 1001 0011 is 693; 0111 1001 0011 is 793; 0101 1000 0110 is 586.

3. Decimal to BCD: 27 is 0010 0111; 45 is 0100 0101; 100 is 0001 0000 0000; 247 is 0010 0100 0111.

5. Hex to binary: 3 is 0011; 7 is 0111; 9 is 1001; A is 1010; D is 1101; E is 1110; F is 1111.

7. Binary to hex: 0110 is 6; 1111 is F; 0011 is 3; 1001 is 9.

9. Hex to binary: 123 is 0001 0010 0011; A32F is 1010 0011 0010 1111.

11. Binary to hex: 0111 0011 is 73; 1101 0010 is D2.

REVIEW PROBLEMS

1. (A AND B) OR C = Y   A = I go; B = He goes; C = She goes.

3.

A	B	C	Y
0	0	0	0
0	1	0	0
1	0	0	0
1	1	0	1
0	0	1	1
0	1	1	1
1	0	1	1
1	1	0	1

5. (A XOR B) AND (NOT A) = Y

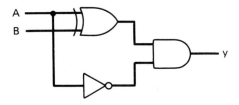

7. NOT((NOT A) AND (NOT B)) = Y

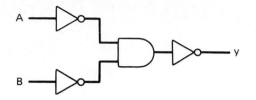

9.

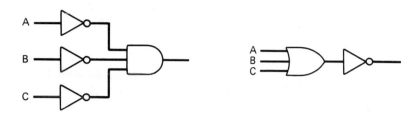

11.

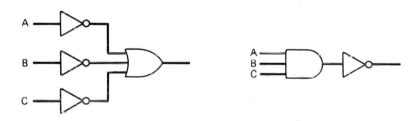

13. Decimal to BCD: 22 is 0010 0010; 44 is 0100 0100; 63 is 0110 0011; 64 is 0110 0100; 65 is 0110 0101.

15. Hex to binary: FA is 1111 1010; 83 is 1000 0011; A3C is 1010 0011 1100; 3456 is 0011 0100 0101 0110.

17.

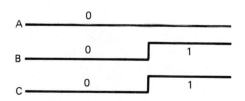

CHAPTER 3

SECTION 3-1

1.

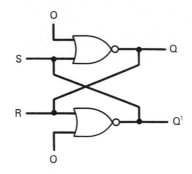

SECTIONS 3-2 and 3-3

1.

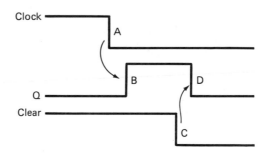

A: External event occurs, causes Q to go high.

B: Q goes high, is seen by computer.

C: Computer completes task, sets Clear = 0.

D: Flip-flop sees Clear = 0, so Q goes low again.

3. The NOT (Invert) Function

SECTION 3-4

1. If $S_n = Q'_n$ and $R_n = Q_n$, then S cannot equal R.

S_n	R_n	Q_{n+1}
0	1	Q'_n
1	0	Q_n

3.

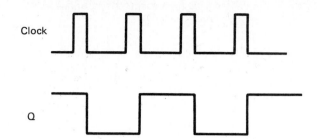

Clock

Q

5. $\dfrac{D_n = Q'_n \quad Q_{n+1}}{\begin{array}{ll} 1 & Q'_n \\ 0 & Q_n \end{array}}$ Same as *T*-type flip-flop

SECTION 3-6

1. 1100 1111
1101 0000
1101 0001
1101 0010
1101 0011
1101 0100
1101 0101
1101 0110
1101 0111
1101 1000
1101 1001
1101 1010
1101 1011
1101 1100
1101 1101
1101 1110
1101 1111
1110 0000

3. Same as Fig. 3-16, with one more "divide by 2" stage.

SECTION 3-7

1. Divide-by-7 counter

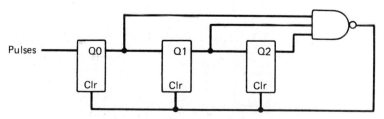

3. 100 pulses in 1 s requires 7 stages ($2^7 = 128$).
100 decimal = 110 0100 binary

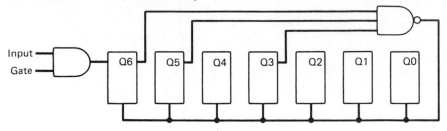

5. (a) will reset when S3 = S4 = 1 or 1 1000 binary = 24 decimal
(b) will reset when S0 = S1 = S2 = 1 or 0 0111 binary = 7 decimal

SECTIONS 3-8 and 3-9

1.

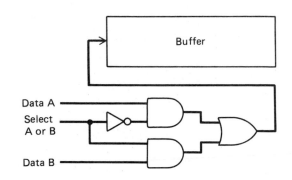

3.

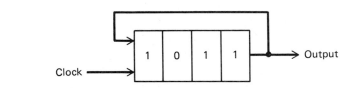

5. $9 \times 8 = 72$

7.

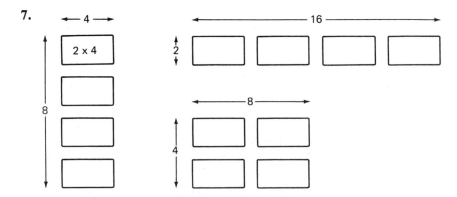

9.

1:	no output
2:	no output
3:	no output
4:	no output
5:	0
6:	10
7:	010
8:	0010
9:	0 0010
10:	10 0010
11:	110 0010
12:	0110 0010

REVIEW PROBLEMS

1.

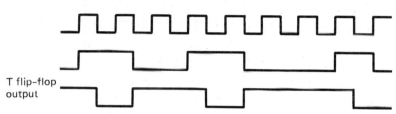

T flip-flop
output

3. 0 1100, 0 1101, 0 1110, 0 1111, 1 0000

5. To 500: 9 stages, which count to $512 - 1 = 511$. To 525: 10 stages, which count to $1024 - 1 = 1023$.

7. A 6-bit counter counts to 31. The gate time would be $31/200 -$ approx 0.15 s.

9.

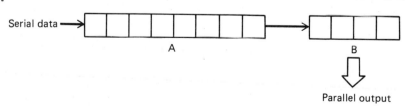

Serial data →

A

B

Parallel output

1. Shift 4 bits from A to B
2. Read B in parallel
3. Repeat steps 1 and 2

11. 0, 1, 1, 0, 0, 0.

CHAPTER 4

SECTION 4-1

1.

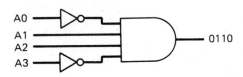

3. It decodes: 11 10X0 i.e. 11 1000 and 11 1010.
It does not decode: 11 1001.
It does decode   11 10X0   The MSB is not desired.

5.

AO
A1
A2
A3
Read/Write

1011 (Read)

1110 (Write)

SECTION 4-2

1. Code 01: data is 0.
Code 10: data is 1.

3. 0110 is 6, data is 0.
1000 is 8, data is 1.

5.

A	B	C	C'	Mux 1 Output	Mux 2 Output	Selected Output
0	0	0	1	0	1	0
0	1	0	1	0	1	1
1	0	0	1	0	1	2
1	1	0	1	0	1	3
0	0	1	0	1	0	4
0	1	1	0	1	0	5
1	0	1	0	1	0	6
1	1	1	0	1	0	7

Select C acts to enable either mux A or mux B, but never both.

SECTION 4-3

1. Essentially same as Section 4-2, Problem 4.

3. 40 (decimal) = 1 0100 (binary)
Select code is 1 0100.

SECTION 4-4

1.

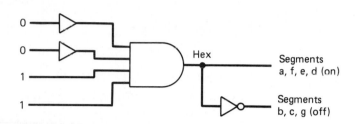

SECTION 4-5

1. Six-digit LED, not muxed: $6 \times 7 = 42$
muxed: $7 + 6 = 13$

REVIEW PROBLEMS

1.

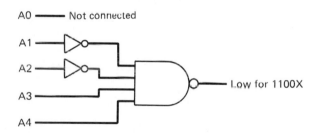

3. Add a qualifier Read to 1 1000 decoder;
add a qualifier Write' to 1 1010 decoder.

5. One-hundredth input: use code 99 = 0110 0011.
Two-hundredth input: use code 199 = 1100 0111.

7. $2^7 = 128$. Yes for 128 outputs; no for 256 outputs.

9.

BCD 0110 is 6.

BCD 0010 is 2.

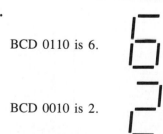

CHAPTER 5

SECTION 5-8

1. (a) add 3 + 3 (b) 6 + (−4) (c) 6 + 6 + 6 + 6 (d) 12 + (−3) + (−3) + (−3) + (−3)

3. 101 + 101 = 1010, 011 + 110 = 1001, 001 + 100 = 101. 001 + 010 = 011.

SECTION 5-10

1. Mask is 1000 0001; AND result is 1000 0000.

3. 1010 OR 0111 is 1111; 0011 OR 1001 is 1011; 0000 OR 0100 is 0100.

5. 1010 AND 0111 is 0010; 0011 AND 1001 is 0001; 0000 AND 0100 is 0000.

7. 100 is larger than 011; 110 is larger than 101; 1011 is larger than 0110; 0100 is larger than 0011.

SECTION 5-11

1. The logic level of A < B is low, since A is not less than B.

REVIEW PROBLEMS

1. 11, 11, 110, 111, 1111, 1101, 1000, 0110, 0010, 1101 0101, 1111 1101, 1 1101 0001

3. 11, 101, 110, 1011, 1 0001, 1101, 1 0000, 1100, 0010, 0010, 0010 0001, 1 0000 1001

5. (call first number A, second B): A greater; A greater; A greater; A greater; A greater; A greater; equal; equal; B greater; A greater; B greater; B greater

7. There are several choices. Use 1111 0000 and OR, see if all 1's is the result; use 0000 1111 and AND, see if result is 0000 1111.

CHAPTER 6

SECTION 6-3

1. 4.8 V + 0.4 V = 5.2 V (OK); but 4.8 V − 0.4 V = 4.4 V (too low).

3. To get a fan-out of 30:

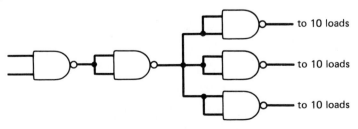

SECTION 6-4

1.

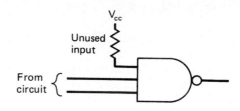

3. The source capability is less than 2 mA; the sink capability is greater than 2 mA. Therefore the load must be connected so that the IC is sinking current:

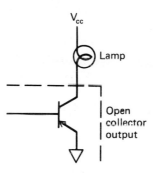

REVIEW PROBLEMS

1. (a) Output is 1, always. **(b)** Output is equal to the complement of the input, so it is an inverter. **(c)** Output is 1, always.

3.

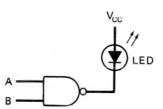

When $A = B = 0$, the output is high, the LED is off.
When $A = B = 1$, the output is low, the LED is on.
When $A = 0$, $B = 1$ the output is high, the LED is off.

5. For $Y = A$ OR B, blow fuses F2, F3, F4, F5, F6, F8.
For $Y = A$ OR (NOT B), blow fuses F2, F3, F4, F5, F6, F7.

7.

Characteristic	Custom IC	Semicustom IC	Programmable IC
Initial Cost	High	Med.	Low
Cost per IC	Low	Med.	Med. to High
Time for 1st IC	Long	Med.	Short
Flexibility	High	Med. to High	Med.
Difficulty to Test	High	High	Med.

CHAPTER 7

REVIEW PROBLEMS

1. 64K bytes has 16 address lines; 8 data lines. A 16K × 4 memory has 12 address lines and 4 data lines.

3. (a)

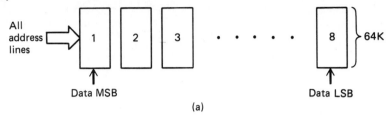

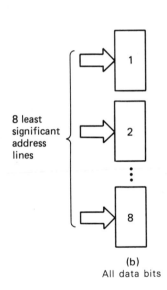

(a)

(b)

(b)
All data bits

5. (a) bad decoder **(b)** bad memory IC **(c)** bad memory IC **(d)** bad data buffer
  (e) bad decoder **(f)** bad decoder

7.

Address	g	f	e	d	c	b	a
0	0	1	1	1	1	1	1
1	0	0	0	0	1	1	0
2	1	0	1	1	0	1	1
3	1	0	0	1	1	1	1
4	1	1	0	1	1	0	0
5	1	1	0	1	1	0	1
6	1	1	1	1	0	0	1
7	0	0	0	0	1	1	1
8	1	1	1	1	1	1	1
9	1	1	0	1	1	1	1

9. (a)

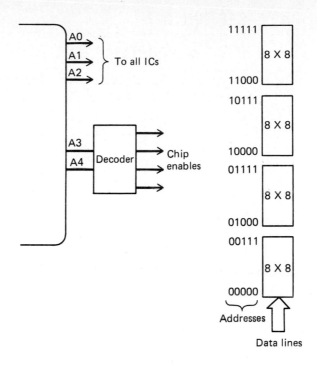

(b)

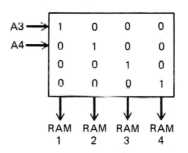

CHAPTER 8

SECTION 8-5

1. 0 1000001 0 1 (Spaces are just shown for clarity; the pulses are really sent as a continuous stream.)

3. 0, 0, 1, 1, 0, 1, 1

5. **(a)** 0101 1110   **(b)** XX00 1110   (X means ''don't care'')

SECTION 8-7

1.

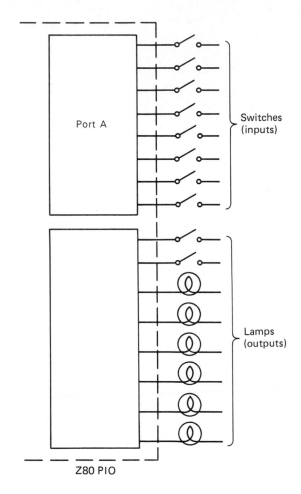

Z80 PIO

3. 1101 0111 (interrupt control) 1111 1100 (mask control)

5. Bad PIO or bad CE decoder; bad PIO or bad circuitry connected external to PIO; bad PIO or bad circuitry connected external to PIO.

REVIEW PROBLEMS

1. **(a)** 0 11000010 1
(b) 0 11000010 1 1
(c) 0 11000010 11   (note—same pattern as b)
(d) 0 11000011 1 11

3. Two possibilities are: 0000 1100; 0110 1111.

CHAPTER 9

SECTION 9-1

1. (a) C/B goes low, Z goes low.
   (b) C/B stays high, Z goes low.

SECTION 9-3

1. 1111

3. Binary: 0000 0000 0110 0100; BCD: 0000 0001 0000 0000

5. (a) 0111 X111
   (b) 0000 0000 at address 1001 (LSByte)
       0001 0000 at address 1001 (MSByte)
       Note: Other answers are possible.

SECTION 9-5

1. 24, one per character

3. (a) 9   **(b)** 24   **(c)** 12

SECTION 9-9

1. 1K bits

3.

Data word	Parity Over Overall	1st byte	2nd byte	Bytes
0100 1100 1111 0010	1	0	0	1
1111 0000 1111 1111	1	0	0	1
0101 1010 0101 1101	0	1	0	0

SECTION 9-10

1.

Data Bits	Check Bits
(a) 0000 0000 0000 0000	0 0 1 1 0 0
(b) 1111 1111 1111 1111	0 0 1 1 0 0
(c) 0110 1000 1011 0011	0 0 1 1 0 1

3. (a)

A	B	A⊕B	Even Parity
0	0	0	0
0	1	1	1
1	0	1	1
1	1	0	0

(b) A	B	A⊕B	C	A⊕B⊕C	Even Parity
0	0	0	0	0	0
0	1	1	0	1	1
1	0	1	0	1	1
1	1	0	0	0	0
0	0	0	1	1	1
0	1	1	1	0	0
1	0	1	1	0	0
1	1	0	1	1	1

REVIEW PROBLEMS

1. A total of 65,535 counts, for 65.535 s.

3. There would be 500 counts in 100 ms.
500 (decimal) is 1 1111 0100
500 (BCD) is 0101 0000 0000

5. (a) $80 \times 7 = 560$
(b) 9
(c) $25 \times 80 = 4000$
(d) $7 \times 9 = 63$
(e) $7 \times 9 \ 25 \times 80 = 252,000$
(f) part e) $\times 30 = 7,560,000$
(g) line counter is 9; character counter is 80; row counter is 25.

7. (a) 16 without EDC   **(b)** 16 + 6 with EDC

9. (a) 1   **(b)** 0   **(c)** 0   **(d)** 0   **(e)** 1

CHAPTER 10

SECTION 10-1

1.   6 bits : $2^6 - 1$ steps = 63 steps
  8 bits : $2^8 - 1$ = 255
12 bits : $2^{12} - 1$ = 4095
14 bits : $2^{14} - 1$ = 16,383

3. 6 bits : resolution of 63 steps

$$\frac{250 \text{ lb}}{63 \text{ steps}} = 3.96 \text{ lb/step}$$

5. 12 bits : resolution of 4095 steps

$$\frac{700°}{4095 \text{ steps}} = 0.171°$$

SECTION 10-2

1. 10 bit = 2^{10} = 1024   step size = $\dfrac{10}{1024}$ = about 0.01 V

0 to 5 volt is effectively 9 bits.
0 to 2.5 volt is effectively 8 bits.
Step size is still 0.01 V.

3.

Binary	Regular Format	Sign Bit Format
0111	7	7
1111	15	−7
0101	5	5
1101	13	−5

SECTION 10-3

1. 1 kΩ LSB, 500 Ω, 250 Ω, 125 Ω, 62.5 Ω, 31.25 Ω

3. The current per bit is $\dfrac{10\text{ V}}{32\text{ k}\Omega}$ = .3125 mA

Output currents: 27 × .3125 =   8.44 mA
                    52 × .3125 = 16.25 mA
                    12 × .3125 =   3.75 mA
                    63 × .3125 = 19.69 mA

SECTION 10-4

1. For half volume of 0111, use 0011.
For twice volume, use 1110.

3. (a) bad scaling circuit, bad reference.
(b) bad LSB latch or internal DAC switch.
(c) bad latches for some bits, or bad internal switches.

5. 0.1% of 10 V = 0.01 V

7. $\dfrac{1}{4096}$ = .00024 = .024%. Use 12 bits.

SECTION 10-6

1. (a) 4 bits = 0 to 15 = 1 V/bit
(b) Internal values: 1, 0, 1, 1
(c) 4 × 20 μs = 80 μs

SECTION 10-7

1. 10 bits = 1024 counts at 100 Hz = 10.24 s
Half scale = 512 counts = 5.12 s

3. Increasing clock frequency yields more counts/s.
Increasing counter size yields more bits for the range, but more counts to be done.
Increasing counter by 1 bit doubles counts; therefore, doubles clock frequency.

SECTION 10-8

1. 3 bits

3. At 100 V/s, signal changes 1 V in 0.1 s.

SECTION 10-9

1. (a) No start signal or no end of conversion signal.
(b) Bad scaling circuit or bad resistor.
(c) The MSB is stuck at 0.
(d) Bad scaling or clock.
(e) Input signal not connected to A/D—probably an open circuit or wire.

REVIEW PROBLEMS

1. $2^{16} - 1$ steps = 65,535
Resolution 0.00153%

3. $50 \div (2^{10} - 1) = 0.0488$ lb

5.

28 (decimal)	= 01 1100 (binary)
50	= 11 0010 (binary)
47	= 10 1111 (binary)
-20	= 11 0100 (binary with sign)
-28	= 11 1100 (binary with sign)
-10	= 10 1010 (binary with sign)

7. (a) Memory is 8×4
(b) 16 Hz
(c) 4-bit DAC
(d) 0000 4 times, then 1111 4 times for square wave
1111 2 times, then 0000 6 times for 25%

9. For 10 bits, 1 bit is about 0.01 V.
$10.000 - 9.999 = 0.001$ This meets the spec.
$10.2 - 10.000 = 0.2$ This is not good.

11. In $\dfrac{1}{10}$ s it can change 100 V/s $\times$ 0.1 s = 10 V.

CHAPTER 11

SECTION 11-4

1. **(a)** 1110 **(b)** 1111 **(c)** 0010 to get the data vector; 1110 to start interrupt

3. 0111 1000

5. None of these addresses appear as data. They do appear as addresses from which data is read.

SECTION 11-7

1.

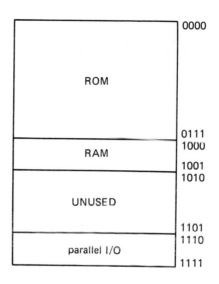

SECTION 11-8

1.

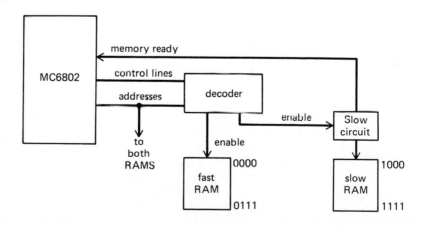

SECTION 11-9

1. Every Read or Write takes two times as many clock cycles.
The ADD instruction has 8 cycles, of which 7 are memory Read or Write.
$7 \times 2 + 1 = 15$ cycles

SECTION 11-11

1.

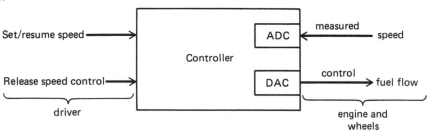

REVIEW PROBLEMS

1. **(a)** 0000 contains 1000; 0001 contains 1100.
   (b) PC values are 0001, then 1100.
   (c) PC values are 1110, then 1010.

3.

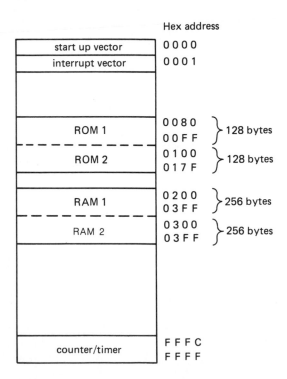

Index